Whistler and Alfvén Mode Cyclotron Masers in Space

The subject of wave–particle interactions occurring in space plasmas has developed strongly, both observationally and theoretically, since the discovery of the Van Allen radiation belts of energetic charged particles trapped in the Earth's magnetosphere over 40 years ago. These wave–particle interactions are recognized today as being a most important research topic in space plasma physics. This is the first book to provide a full and systematic description of the physical theory of whistler and Alfvén cyclotron masers acting in planetary magnetospheres, and in the Sun's outer atmosphere.

The book introduces current research topics by examining significant problems in the subject. It gives sufficient detail on the topic for readers to go on to apply the methods presented to new problems, helping them with their own research.

This book is a valuable reference for researchers and graduate students working in space science, solar–terrestrial physics, plasma physics, and planetary sciences.

Cambridge Atmospheric and Space Sciences Series
Editors: J. T. Houghton. M. J. Rycroft and A. J. Dessler

This series of upper-level texts and research monographs covers the physics and chemistry of different regions of the Earth's atmosphere, from the troposphere and stratosphere, up though the ionosphere and magnetosphere and out to the interplanetary medium.

VICTOR TRAKHTENGERTS (1939–2007) was Head of the Sector of Ionospheric and Magnetospheric Physics at the Institute of Applied Physics Russian Academy of Sciences, and Professor of the State University, Nizhny Novgorod, Russian Federation. He was a member of the Editorial Boards of *Geomagnetism and Aeronomy* and the *Journal of Atmospheric and Solar-Terrestrial Physics*.

MICHAEL RYCROFT is a Visiting Professor at the School of Engineering, Cranfield University, UK, and the International Space University in Strasbourg, France. He was Editor-in-Chief of the *Journal of Atmospheric and Solar–Terrestrial Physics* from 1989 to 1999, and since 2002 has been Managing Editor of the overview journal *Surveys in Geophysics*.

Both authors are well known in the field, each having published more than 200 papers, some of them jointly. Since 1997, both of them have been key members of international teams involved in three collaborative INTAS Projects and two NATO Science Programmes.

Cambridge Atmospheric and Space Sciences Series

EDITORS

Titles in print in this series

M. H. Rees
Physics and chemistry of the upper atmosphere

R. Daley
Atmosphere data analysis

J. K. Hargreaves
The solar–terrestrial environment

J. R. Garratt
The atmosphere boundary layer

S. Sazhin
Whistler-mode waves in a hot plasma

S. P. Gary
Theory of space plasma microinstabilities

I. N. James
Introduction to circulating atmospheres

T. I. Gombosi
Gaskinetic theory

M. Walt
Introduction to geomagnetically trapped radiation

B. A. Kagan
Ocean–atmosphere interaction and climate modelling

D. Hastings and H. Garrett
Spacecraft–environment interactions

J. C. King and J. Turner
Antarctic meteorology and climatology

T. E. Cravens
Physics of solar system plasmas

J. F. Lemaire and K. I. Gringauz
The Earth's plasmasphere

T. I. Gombosi
Physics of space environment

J. Green
Atmospheric dynamics

G. E. Thomas and K. Stamnes
Radiative transfer in the atmosphere and ocean

R. W. Schunk and A. F. Nagy
Ionospheres: Physics, plasma physics, and chemistry

I. G. Enting
Inverse problems in atmospheric constituent transport

R. D. Hunsucker and J. K. Hargreaves
The high-latitude ionosphere and its effects on radio propagation

M. C. Serreze and R. G. Barry
The Arctic climate system

N. Meyer-Vernet
Basics of the solar wind

V. Y. Trakhtengerts and M. J. Rycroft
Whistler and Alfvén mode cyclotron masers in space

WHISTLER AND ALFVÉN MODE CYCLOTRON MASERS IN SPACE

V. Y. TRAKHTENGERTS

and

M. J. RYCROFT

Cranfield University, UK

CAMBRIDGE
UNIVERSITY PRESS

CAMBRIDGE UNIVERSITY PRESS
Cambridge, New York, Melbourne, Madrid, Cape Town, Singapore, São Paulo, Delhi

Cambridge University Press
The Edinburgh Building, Cambridge CB2 8RU, UK

Published in the United States of America by Cambridge University Press, New York

www.cambridge.org
Information on this title: www.cambridge.org/9780521871983

First published 2008

Printed in the United Kingdom at the University Press, Cambridge

A catalogue record for this publication is available from the British Library

ISBN 978-0-521-87198-3 hardback

Contents

Preface

The purpose of this monograph is to formulate a quantitative and self-consistent theoretical approach to wave–particle interactions occurring in space plasmas, and present a logical development of the subject. In the Earth's magnetosphere, Nature has given us a plasma laboratory that is accessible to observations made by radio, magnetic and electric instruments on the ground, and a great variety of instruments aboard rockets and Earth-orbiting satellites. Spacecraft are making similar observations in the more distant solar system.

To understand such observations as fully as possible, with colleagues around the world we have been challenged to produce a rigorous description of the energetic charged particle distribution function interacting with electromagnetic waves across a wide frequency spectrum. The space plasma is, as a rule, a non-equilibrium system with sources and sinks of energy and charged particles. As such, electromagnetic waves are generated via the process of the stimulated emission of radiation. Together with the electrodynamic properties of the space plasma, determined by variations of the magnetic field and plasma density, this constitutes a maser system. It exerts a strong influence on the state of the space plasma.

Cyclotron masers (CMs) are a shining example of such maser systems operating in the Universe. Whether in the Earth's magnetosphere or Jupiter's, in the solar corona or in the laboratory, CMs are exciting systems to marvel at, to wonder about and to investigate in detail. Such is the theme of this book. We analyse waves in a resonant cavity (here termed the eigenmodes), the excitation conditions and different wave generation regimes. In these wave–particle interactions, feedback processes, which are inherently nonlinear, have to be taken into account. Energetic electrons interact in CMs with electromagnetic waves and are precipitated into the atmosphere; electrons can be accelerated by these waves to produce secondary radiation. Similar results hold for the interaction between energetic ions and hydromagnetic (Alfvén) waves.

During this book's preparation, we have good reason to be most grateful for support received from the Institute of Applied Physics of the Russian Academy of Sciences in Nizhny Novgorod, Russia, the Russian Foundation for Basic Research, NATO and INTAS in Brussels, Belgium, the Royal Society of London, United Kingdom, the International Space University in Strasbourg, France, the International Space Science Institute in Bern, Switzerland, and the Institute of Atmospheric Physics, Prague, Czech Republic. We are most grateful to Lyudmila Semenova for her excellent typing of the entire book. Special thanks are also given to Andrei Demekhov, Victor's colleague at Nizhny Novgorod, for his detailed reading of the text, and a considerable amount of work on the illustrations and the references. Every effort has been made to secure the necessary permissions to reproduce copyright material in this work, though in some cases it has proved impossible to trace or receive replies from copyright holders. If any omissions are brought to our notice, we shall include appropriate acknowledgements on reprinting or in subsequent editions. Finally we thank all the staff at Cambridge University Press who have been involved in our book.

For their devotion and patience, we especially thank our wives, Galja and Mary, respectively.

Post Script, added by Michael Rycroft and Andrei Demekhov

We are greatly saddened to report the death of our good colleague and dear friend Victor Yurievich Trakhtengerts on 4 December 2007; he had battled against cancer for more than three years. At that time this book had been completed, and was being finally checked and collated. Poignantly, it was submitted electronically to the Cambridge University Press a few days later. We consider that it is an appropriate testament to Victor's many achievements in research on space plasma physics over his entire career. We hope that his keen physical insight and imagination, and his superb analytical skills, are as evident to the reader as they are to us. We trust that the publication of this book may stimulate others to continue to pursue research in this fascinating field.

Chapter 1

Introduction

It is customary today to call a generator or amplifier of electromagnetic waves a maser (*M*icrowave *A*mplification by the *S*timulated *E*mission of *R*adiation) if its operation is based on the stimulated emission of distributed oscillators. Electrons and ions, rotating around an ambient magnetic field, are the oscillators in so-called cyclotron masers (CMs).

There are two types of CMs in space, which differ considerably from each other. In the Earth's magnetosphere the first type operates on open field lines in the auroral region at heights between 10^3 and 10^4 km, in plasma cavities where the plasma density is so low that the electron plasma frequency is much less than the electron cyclotron frequency. Here auroral kilometric radiation (AKR) is generated by energetic electrons in such a maser system (Fig. 1.1a). The eigenmodes of these auroral CMs are electromagnetic waves with frequencies close to the electron cyclotron frequency; the wave vector $\mathbf{k}$ is almost perpendicular to the geomagnetic field $\mathbf{B}$. These auroral CMs are rather similar to a family of laboratory devices, termed gyrotrons (Fig. 1.1b). The operation of these devices is based on the cyclotron interaction of electrons, moving along a homogeneous magnetic field through an evacuated region inside a geometrical cavity resonator. A specific feature of a laboratory CM is the cyclotron interaction of a well-organized beam of electrons rotating around a homogeneous magnetic field with a monochromatic electromagnetic wave having a spatially fixed field structure. As a rule, such an electron beam has a very narrow spread of particle energies and pitch-angles. The investigation of such devices has a rich history, beginning with the pioneering paper of Twiss (1958) on the cyclotron instability. Very important contributions both to the theoretical development of the subject and to the design and construction of such devices have been made by the Gorky (now Nizhny Novgorod) group of scientists under the

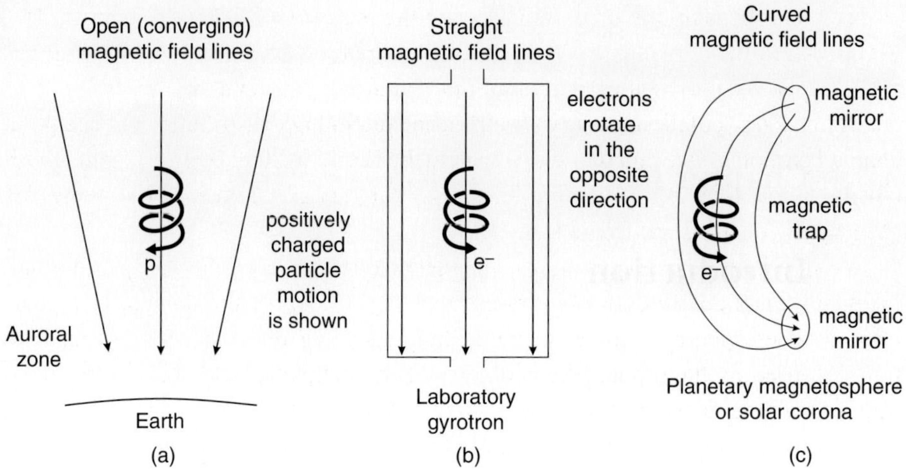

Figure 1.1 The three most typical magnetic field configurations for a cyclotron maser.

leadership of Gaponov-Grekhov. The first papers in this direction referred to the years 1959–1965 (Gaponov 1959, 1960; Petelin, 1961; Gaponov and Yulpatov, 1962; Yulpatov, 1965). The investigations of the cyclotron instability in a relativistic plasma by Zheleznyakov (1960a, b) were important for further applications to laboratory and space plasmas. It is necessary to mention as well the papers by Schneider (1959) and Bekefi *et al.* (1961). The gyrotrons and their modifications now find a wide range of applications in the fields of plasma experiments, controlled thermonuclear fusion, radar transmitters, and plasma and chemical technology. Details of these can be found in the review by Gaponov-Grekhov and Petelin (1980).

The specific feature of a space gyrotron where AKR is generated is the presence of cold electrons whose number density is comparable with, or slightly greater than, the number density of energetic electrons. A very important aspect of space gyrotrons is the inhomogeneity of both the magnetic field strength and the plasma parameters, the detailed investigation of which is an active field of research today.

Different aspects of space gyrotrons have been investigated when applied to AKR (Wu and Lee, 1979), Jupiter's decametric radiation and planetary radio emissions from the other outer planets (Zarka, 1992), and solar microwave bursts (Melrose and Dulk, 1982). A gyrotron-like radiation is also believed important for various stellar radiation bursts (Benz *et al.*, 1998; Trigilio *et al.*, 1998; Bingham *et al.*, 2001, 2004).

The second type of CMs in space, either electron or ion masers, operates along closed magnetic flux tubes (Fig. 1.1c) filled by a dense cold plasma. For such CMs the electron plasma frequency is greater than the electron gyrofrequency. In the Earth's magnetosphere these CMs function within the plasmasphere and in filaments of dense cold plasma outside (and sometimes attached to) the plasmasphere. Both types of CMs occur widely in natural plasmas – they exist in the magnetospheres of planets, in the solar corona and in the plasma envelopes of active stars.

The main focus of this book is on CMs of the second type. They are large scale – yet unseen to the human eye – features of planetary magnetospheres. Via single or multi-hop wave propagation and amplification along geomagnetic flux tubes, they determine the population (energy spectrum and pitch-angle distribution) of energetic charged particles distributed in space around the Earth. In other cosmic plasmas with closed (dipole likc) magnetic field lines, they can crucially influence the properties of energetic charged particles there too.

In many laboratory experiments, magnetic mirrors are produced by additional magnetic fields due to current-carrying coils at the ends of the device. In a planetary magnetosphere or for a magnetic loop reaching into the solar corona, the trap is caused by the dipole-like geometry of the magnetic field. Such a magnetic trap, containing dense plasma, serves as a CM cavity; the high-energy fraction of charged particles is the active substance for the CM. The ends of the magnetic flux tube, immersed in the planetary ionosphere, serve as mirrors for the electromagnetic waves. The eigenmodes in space CMs are whistler-mode or Alfvén waves, which possess one very important property; they are guided by the magnetic field. The situation is similar to that in fibre optics when light is guided by dielectric filaments. Cyclotron resonance occurs when the electric field of a circularly polarized wave propagating through the plasma exactly matches the Doppler-shifted cyclotron motion of a charged particle. It takes place when the Doppler-shifted wave frequency coincides with the particle gyrofrequency, and can occur either for whistler-mode waves and energetic electrons, or for Alfvén (hydromagnetic) waves and energetic ions.

CM operation is based on the cyclotron instability (CI), which is due to the transverse anisotropy of the charged particle distribution function; this exists when the effective temperature transverse to the magnetic field direction, $T_\perp$, exceeds the longitudinal temperature, $T_\parallel$ (the anisotropy factor $\alpha = T_\perp/T_\parallel$ is greater than unity). The pioneering paper by Sagdeev and Shafranov (1960) on this instability played a crucial role in developing the concept of space cyclotron masers. Sagdeev and Shafranov introduced this anisotropy factor $\alpha > 1$ as the universal quantitative measure for the momentum inversion of energetic charged particles, which serves as the necessary condition for space CM operation, just as the population inversion does for conventional lasers. It is important to recognize that this anisotropy, with $\alpha > 1$, is a natural feature of adiabatic traps because of the existence of the loss cone. This is the region in pitch-angle space with values near zero, where the energetic charged particles are lost through the ends of the magnetic trap by collision with atoms or molecules of the upper atmosphere. There are many other causes of the transverse anisotropy of energetic charged particles in space, such as magnetic compression, charged particle transport within planetary magnetospheres, acceleration by global electric fields, stochastic acceleration mechanisms, and so on. Most sources supply energetic particles with a wide spread of energies and pitch-angles. In space, as we shall see, the inhomogeneous magnetic field and broad range of energies and pitch-angles radically change the wave–particle interaction process from the laboratory case with its well-organized beam and monochromatic wave.

The second important step in the theoretical development of the anisotropic cyclotron instability (CI) was formulation of the quasi-linear (QL) theory of the CI by Vedenov *et al.* (1962) and Sagdeev and Galeev (1969). They introduced the idea of diffusion paths in velocity (phase) space, along which the charged particles resonant with cyclotron waves could reach the edge of the loss cone and be lost from a magnetic trap.

As applied to space CMs, operating via the geomagnetic trap, the CM theory was developed in relation to two geophysical phenomena, namely natural ELF (from 0.3 to 3 kHz) and VLF (from 3 to 30 kHz) electromagnetic emissions (Helliwell, 1965; Kimura, 1967) and the Earth's radiation belts (Van Allen *et al.*, 1958). Early estimates (Trakhtengerts, 1963) showed that the intensity of ELF/VLF emissions was very much (more than 50 dB) higher than the intensity of the thermal equilibrium emission from the Van Allen radiation belt electrons. The idea of quantitatively connecting these audio frequency radio emissions with the cyclotron instability (CI) of radiation belt electrons first appeared in 1963 (see below). The subject has been actively developed in the succeeding four decades.

An enormous quantity of valuable experimental data on energetic charged particles and ELF/VLF waves in the Earth's magnetosphere has been accumulated over the last four decades. Calculations have used these data to demonstrate the very important role of wave–particle interactions in determining the state of the radiation belts, and hence their lifetime. We concentrate our attention here on self-consistent models, which can give not only detailed and exhaustive information on the spectral and dynamical characteristics of waves generated in such CMs but also explain the dynamics of trapped and precipitated energetic charged particles.

ELF/VLF radio signals observed on the ground and in space alike include both broadband noise-like electromagnetic emissions, such as hiss and quasi-periodic noise bursts (Fig. 1.2), and narrowband, or discrete, emissions having a

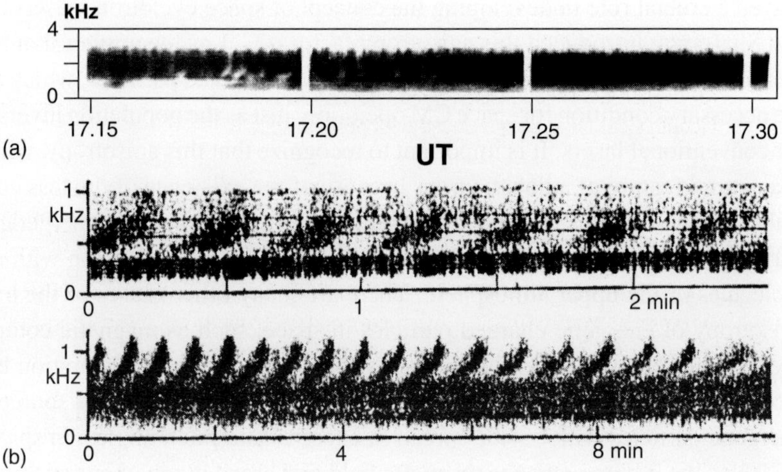

Figure 1.2 Examples of the dynamical spectrum of noise-like ELF/VLF emissions: (a) hiss band (taken from Helliwell, 1965); (b) quasi-periodic emissions (Sato and Fukunishi, 1981). (Copyright American Geophysical Union, reproduced with permission.)

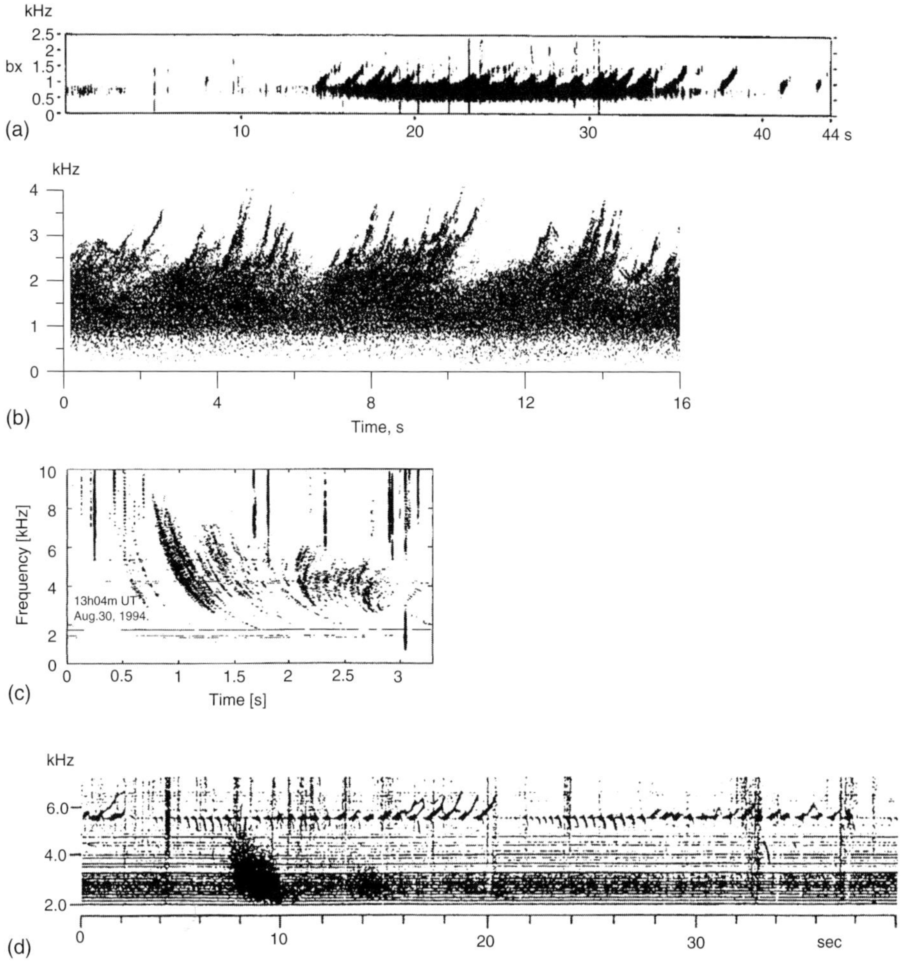

Figure 1.3 Examples of the dynamical spectrum of discrete ELF/VLF signals: (a) chorus observed on the satellite GEOS-1 (Hattori *et al.*, 1991); (b) chorus observed on the ground (Manninen *et al.*, 1996); (c) whistlers from a lightning discharge (Ohta *et al.*, 1997). (Copyright American Geophysical Union, reproduced with permission); (d) triggered VLF emissions (Helliwell and Katsufrakis, 1974). (Copyright American Geophysical Union, reproduced with permission.)

quasi-monochromatic structure. Among these are natural signals, such as successions of rising frequency tones named chorus, and whistlers generated by lightning discharges, as well as man-made signals from VLF transmitters propagating in the whistler-mode, and triggered VLF emissions from both natural and man-made signals (Fig. 1.3).

The CM theory to explain such observations developed in two practically independent directions. The first of these is based on a quasi-linear (QL) theory of gyroresonant wave–particle interactions, which is valid for noise-like emissions when the electromagnetic field can be considered to be the sum of independent wave packets with stochastic phases and a broad frequency spectrum. This radiation interacts with the energetic particles as it does with particles in many energy level systems. In such a situation the mathematical description of CM operation is very similar to the balance approach of quantum generators (Khanin, 1995). In particular, as with quantum generators, the fundamental generation regime in CMs is a relaxation oscillation, when both the wave intensity and charged particle flux are modulated with a period depending on the properties of both the particle source and the wave and particle sinks. The bandwidth of the oscillation, and hence the quality factor, also depend on these source and sink properties. This is the basis of many other wave generation regimes of CMs (e.g. periodic, spike-like and stochastic). It leads to an explanation of the different modulation phenomena in natural electromagnetic emissions and energetic charged particle dynamics.

Trakhtengerts (1963), Brice (1963, 1964), Andronov and Trakhtengerts (1964), and Kennel and Petschek (1966) published the first fundamental papers on the QL theory of magnetospheric CMs. These studies were based on the classical plasma physics of the above mentioned papers by Sagdeev and Shafranov (1960) and Vedenov *et al.* (1962). Important contributions to the development of this theory in the Earth's magnetosphere were made by Trakhtengerts (1966, 1967), Kennel (1969), Gendrin (1968), Tverskoy (1968), Cornwall *et al.* (1970), Coroniti and Kennel (1970), Roux and Solomon (1971), Lyons *et al.* (1972), and Schulz and Lanzerotti (1974). The monograph by Bespalov and Trakhtengerts (1986a) partly summarizes the state of QL theory.

The second direction develops a nonlinear theory of monochromatic wave–particle interactions, generalizing it for the case of an inhomogeneous magnetic field. Here the first paper specifically considering space (magnetospheric) plasmas was that of Dungey (1963). New experimental results from the VLF transmitters in Russia and at Siple Station, Antarctica, stimulating development of the nonlinear theory, were published over a period of more than 25 years from 1967; for a review, see Molchanov (1985) and Helliwell (1993). Helliwell (1967, 1970) first formulated the key idea about second-order cyclotron resonance as a generation mechanism for discrete VLF emissions in the magnetosphere. Important and original contributions to the quantitative development of this nonlinear theory were made by Dysthe (1971), Nunn (1971, 1974), Sudan and Ott (1971), Budko *et al.* (1972), Helliwell and Crystal (1973), Karpman *et al.* (1974a, b), Roux and Pellat (1978), and others; more detailed references are given throughout the book. These theories are somewhat similar to those of laboratory CMs in the case of a homogeneous magnetic field, but have very important novel features when the magnetic field is inhomogeneous.

Both directions – quasi-linear theory for broadband waves and nonlinear theory for monochromatic waves – lead to the explanations of certain phenomena, noise-like emissions in the first case and triggered emissions in the second. Natural electromagnetic phenomena, produced by CM operation and observed on the ground or in space, reveal the very interesting phenomenon that noise-like emissions may sometimes become quasi-monochromatic signals, or vice versa. Thus, in some sense, disorder produces order. Some early observations here are those of Helliwell (1969), Reeve and Rycroft (1971) and Burtis and Helliwell (1976). The modern theory of a CM enables such a change to be understood. This transition can be due to the appearance of a step deformation on the distribution function at a certain velocity, when energetic charged particles interact with a noise-like emission. So, starting from noise-like emissions, we come to the situation when it is necessary to use the theory of monochromatic wave interaction with a specific particle distribution, taking phase effects into account. The very interesting backward wave oscillator (BWO) generation regime can then be realized; wave reflection and positive feedback are organized in a BWO generation regime via a charged particle beam whose specific properties are prepared by the noise-like emission.

In writing this book we have pursued the objective of describing analytically, and from a unified viewpoint, both directions of CM theory and also their interrelation. We start with the analyses of wave eigenmodes and their excitation conditions in CMs. Then the nonlinear equation of motion of a single charged particle moving in a homogeneous magnetic field and in the field of a monochromatic wave is investigated. This approach is generalized, successively, for the case of an inhomogeneous magnetic field, of two waves and of many waves. Collective effects are taken into account using a collisionless kinetic (Boltzmann) equation for the distribution function of charged particles. In such an approach the quasi-linear description is obtained as a generalization of the case of many waves with random phases. The kinetic equation for the particle distribution function and the electromagnetic wave equation with a current of energetic particles are solved self-consistently. Thus, the first nine chapters of this book cover the fundamentals, which gradually become more and more complex. The treatment is developed in a logical manner; so as not to disrupt the flow, key references are given only at the beginning and end of each chapter.

The latter part of this book (Chapters 10–13) discusses several applications of CM theory to different electromagnetic emissions, to the loss of energetic charged particles from the Earth's magnetosphere and to their precipitation into the upper atmosphere. Both noise-like and discrete emissions with frequencies ranging from 0.1 up to $\sim$ 10 Hz interacting with ions, or from 0.3 up to $\sim$ 10 kHz interacting with electrons, are explained. The analysis of particular cases, such as pulsating aurora or the interaction involving detached cold plasma regions, are examples of a rather sophisticated theory and a detailed comparison with experimental data. Applications of CM theory to Jupiter's magnetosphere and to the solar corona seem to be promising. Laboratory experiments and their quantitative interpretation give additional confidence in the CM theory.

In summary, space plasma experiments are a source of information on this fascinating and important physical phenomenon of cyclotron maser operation. The first steps have been taken in the application of this theory to different space plasma situations, and the dynamical phenomena discussed deserve further investigations. We hope that our book may stimulate these, whether they be experimental researches, theoretical studies or computational simulations.

Chapter 2

Basic theory of cyclotron masers (CMs)

The purpose of this chapter is to formulate the self-consistent set of equations (expressed in cgs units) which describe the theoretical behaviour of cyclotron masers operating in a plasma constrained by a dipole magnetic field. Thus the equation of motion of a non-relativistic charged particle is considered; the first adiabatic invariant of motion is the magnetic moment, and the second is the bounce integral. The charged particle distribution function approach is then presented.

The general dispersion relation for electromagnetic waves propagating in a plasma is derived. From this expressions are obtained for the refractive index and polarization parameters of whistler-mode waves and of Alfvén waves. These lead to the wave eigenmodes (natural, or resonant, oscillations) in an operating cyclotron maser. Due to the current of hot particles in gyroresonance with the wave, a monochromatic wave is an evolving wave packet. Its amplitude changes slowly with time; the wave packet propagates at the group velocity.

The theory is derived first for a homogeneous plasma for which the plasma density and the ambient magnetic field do not vary with any spatial coordinate. Secondly it is extended to an inhomogeneous plasma, specifically to plasma and energetic charged particles confined by a dipole magnetic field such as the Earth's.

The necessary references to books and papers will be given at the start of each chapter. Much reference material has been presented in the book by Akasofu and Chapman (1972).

2.1 Attributes of a CM. The Earth's magnetic field in space

To construct CM theory we begin with a description of the basic elements of cyclotron masers – a cavity, an active substance and electromagnetic eigenmodes.

As has been pointed out in Chapter 1, a CM cavity is formed by a magnetic trap, filled with plasma, which includes two components, a cold plasma which determines the wave eigenmodes and a fraction of energetic charged particles which serve as the active material. The magnetic trap is an adiabatic trap; this means that the spatial scale l of the trap is much larger than the gyroradius ρ_{B} of the energetic particles contained in the trap:

$$l \gg \rho_{\mathrm{B}}. \tag{2.1}$$

This condition is readily fulfilled in CMs in space. Also the mean free path l_{c} between binary collisions of the energetic particles is much larger than the magnetic trap length l:

$$l_{\mathrm{c}} \gg l. \tag{2.2}$$

We consider a single magnetic trap, which has just one minimum value of magnetic field B along the flux tube of interest. Such a trap possesses all the principal features of a CM.

The magnetic field in near-Earth space, in the magnetosphere, has a rather complex configuration. Except near the magnetopause, the magnetic field is essentially a dipolar magnetic field, expressed as

$$\vec{B} = \frac{M}{r^3}\left(-2\sin\lambda\,\vec{r}_0 + \cos\lambda\,\vec{\lambda}_0\right) \tag{2.3}$$

or

$$B = \frac{M}{r^3}\left(1 + 3\sin^2\lambda\right)^{1/2} \tag{2.4}$$

where the vector $\vec{B}$ is written in spherical polar coordinates whose origin is at the centre of the planet as shown in Fig. 2.1. Here, M is the magnetic moment, r is the radial distance, and λ is the latitude. The dipole magnetic field has axial symmetry.

In the case of the Earth

$$M = 0.311 R_0^3 \text{ G}, \quad R_0 = 6370 \text{ km} \tag{2.5}$$

where R_0 is the Earth's radius. The equation for a magnetic field line has the form

$$r = LR_0 \cos^2\lambda \tag{2.6}$$

where L, the so-called L-shell, is McIlwain's (1961) parameter, as illustrated in Fig. 2.1.

The important magnetic trap parameter is the magnetic mirror ratio of the maximum value B_{M} (on the planetary surface) to its lowest (minimum) value B_{L} (in the equatorial plane):

$$\sigma = \frac{B_{\mathrm{M}}}{B_{\mathrm{L}}} = L^3\,(4 - 3/L)^{1/2}. \tag{2.7}$$

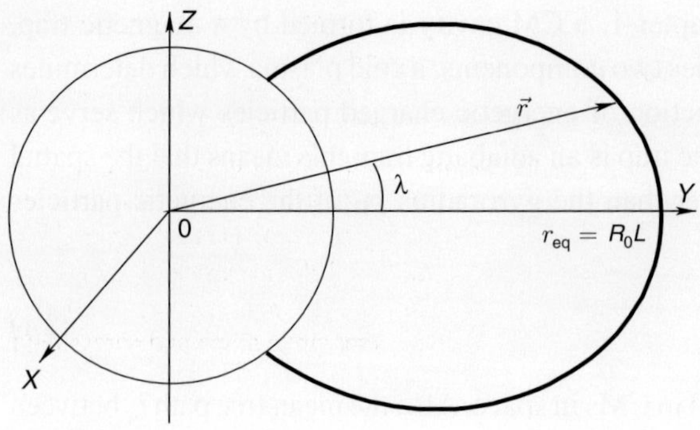

Figure 2.1 The spherical polar coordinate system whose origin is at the centre of the planet; r is the radial distance, λ is the latitude, r_{eq} is the distance of the magnetic field line in the equatorial plane.

A good approximation for the magnetic field variation along a flux tube in the vicinity of the equatorial plane is a parabolic variation:

$$B = B_{\mathrm{L}}(1 + z^2/a^2) \tag{2.8}$$

where the coordinate z is measured from the central cross-section of the magnetic trap. For a dipole magnetic field (2.4), the scale length a is given by

$$a = LR_0/2.12. \tag{2.9}$$

2.2 The motion of charged particles in a CM

The inequalities (2.1) and (2.2) permit us to use a guiding centre approach (Northrop, 1963) to describe charged particle motion in an inhomogeneous magnetic field. In this approximation a particle's cyclotron motion is expressed as the Larmor rotation around a guiding centre, which moves along a magnetic field line and drifts slowly (in longitude) across the magnetic field direction. Figure 2.2 shows this with respect to an origin which, for convenience, is also taken to be the centre of the planet

$$\vec{r} = \vec{R}_{\mathrm{c}} + \vec{\rho}_{B\alpha}. \tag{2.10}$$

Further, we are limited by non-relativistic considerations. In this case the equations of motion have the form:

$$\vec{\rho}_{B\alpha} = \rho_{B\alpha}(\vec{x}_0 \sin\varphi_\alpha, \vec{y}_0 \cos\varphi_\alpha), \quad \vec{v}_{\perp\alpha} = v_\perp(\vec{x}_0 \cos\varphi_\alpha, -\vec{y}_0 \sin\varphi_\alpha) \tag{2.11}$$

$$\frac{d\varphi}{dt} = \dot{\varphi}_\alpha = -\omega_{B\alpha}(\vec{R}_{\mathrm{c}}), \quad \frac{m_\alpha v_\perp^2}{2B} = J_{\perp\alpha} = \mathrm{const.} \tag{2.12}$$

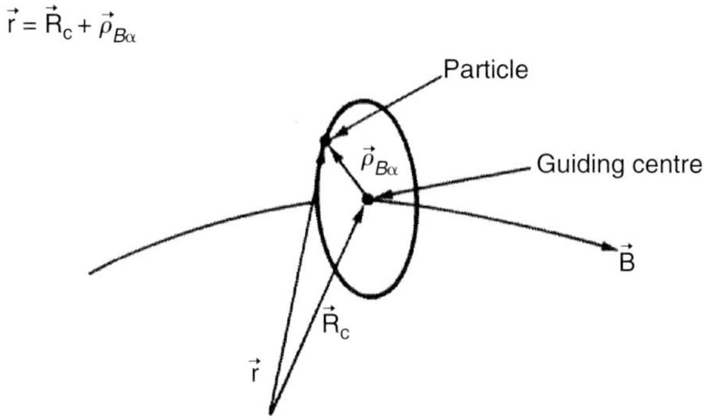

Figure 2.2 Charged-particle motion in a slowly changing magnetic field can be treated by the guiding centre approximation where the motion is broken down into the gyration of a particle around the field in a circle of radius $\rho_{B\alpha}$ and the motion of the centre of that circle along and across field lines.

$$\frac{dv_\|}{dt} \equiv \dot{v}_\| = -\frac{e_\alpha}{m_\alpha}\frac{\partial \Phi}{\partial z} - \frac{J_{\perp\alpha}}{m_\alpha}\frac{\partial B}{\partial z}, \qquad \dot{R}_{c\|} = \dot{z} = v_\|, \qquad \dot{\vec{R}}_{c\perp} = \vec{v}_D \tag{2.13}$$

where $\omega_{B\alpha} = \dfrac{e_\alpha B}{m_\alpha c}$ is the gyrofrequency, e_α and m_α are the charge and mass of a particle, respectively, α denotes the particle type (e for electrons, and i for ions), $\rho_{B\alpha} = v_{\perp\alpha}/|\omega_{B\alpha}|$, $v_\|$ and $\vec{v}_\perp(v_x, v_y)$ are the velocity components along and across the magnetic field, respectively, $\vec{x}_0$ and $\vec{y}_0$ are unit vectors across the magnetic field in the local coordinate system, connected with the guiding centre of the particle, $J_\perp$ is the first adiabatic invariant, and Φ is the electric potential which may be present in the system; $\vec{v}_D$ is the drift velocity across the magnetic field. It follows from (2.13) that, for a plasma particle whose kinetic energy is W, the total energy of the charged particle is conserved

$$w = \frac{m}{2}\left(v_\|^2 + v_\perp^2\right) + e\Phi \equiv W + e\Phi = \text{const.} \tag{2.14}$$

Let us define the pitch-angle θ between $\vec{v}$ and the magnetic field $\vec{B}$ so that

$$v_\| = v\cos\theta, \qquad v_\perp = v\sin\theta. \tag{2.15}$$

If the electric potential Φ is constant along the magnetic field line, from (2.12) and (2.14) we have

$$\mu \equiv \frac{\sin^2\theta(z)}{B(z)} B_L = \sin^2\theta_L = \text{const.} \tag{2.16}$$

where B_L is the minimal value of the magnetic field, which is reached at the equator for the case of a dipole magnetic field, and where θ_L is the pitch angle.

Accordingly, the charged particle is reflected along the magnetic field line at the magnetic reflection point R, where $v_\| = 0$. Hence

$$\sin^2\theta_R = 1 \text{ and } B_R = B(z)/\sin^2\theta(z). \tag{2.17}$$

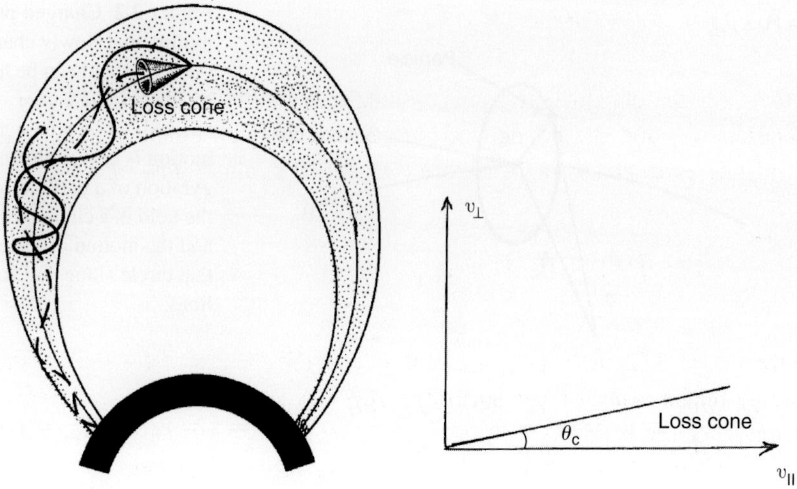

Figure 2.3 Motion of charged particles in the radiation belt. Charged particles with their velocity inside the loss cone (dashed line) are absorbed by the atmosphere. Particles outside the loss cone (solid line) are reflected by the magnetic mirror and trapped by the geomagnetic field.

It is convenient to characterize the charged particle motion relative to the magnetic trap's central cross-section which is crossed by all trapped particles. The most important characteristic of an imperfect magnetic trap is the loss cone. This defines the range of pitch-angles, where trapped particles are absent (see Fig. 2.3).

The edge of the loss cone θ_c can be found from (2.17), if we put $B_R = B_M$ where B_M is the magnetic field value at the foot of the magnetic trap. In the case of planetary magnetospheres this boundary corresponds to ionospheric heights, where the atmospheric density is sufficiently large that energetic particles are lost by collisions with atmospheric atoms or molecules. Thus

$$\sin^2\theta_c = B_L/B_M \text{ and } \sigma = 1/\sin^2\theta_c \tag{2.18}$$

where B_L is the magnetic field value at the central cross-section of the magnetic trap and σ is the magnetic mirror ratio.

The loss cone determines the transverse anisotropy of a particle distribution function in a CM independently of the particle source properties. In planetary magnetospheres it connects magnetospheric and ionospheric processes through the fluxes of precipitated energetic charged particles. These particles have a velocity distribution inside the loss cone. Such particles can appear due to pitch-angle diffusion towards lower values of pitch-angle during the process of wave–particle interactions in a CM.

If the magnetic and electric fields are slowly changing functions of $\vec{r}$ and t, the total charged particle energy (2.14) is not conserved. However, the first and the second adiabatic invariants are conserved, if the following inequality is obeyed:

$$\tau \gg T_{\mathrm{B}} > 2\pi/|\omega_{B\alpha}| \tag{2.19}$$

where τ is the characteristic time of magnetic field change. The second adiabatic invariant $J_{\|}$, and the bounce period between magnetic mirror points in the two hemispheres and back again T_{B}, are determined by the following relations:

$$J_{\|} = \oint v_{\|}\, dz, \quad T_{\mathrm{B}} = \oint \frac{dz}{v_{\|}}. \tag{2.20}$$

Here $v_{\|} = \pm(2/m)^{1/2}(W_{\mathrm{L}} - e_{\alpha}\Delta\Phi - J_{\perp}B)^{1/2}$; W_{L} is the value of the kinetic energy in the central cross-section of the magnetic flux tube, $\Delta\Phi$ is the electric potential difference between z and $z = 0$, and $J_{\perp} = mv^2\mu/2B_{\mathrm{L}}$. For an electron, $e_{\alpha} = -e$; thus e is the magnitude of the charge on the electron.

Now we consider the drift motion of a charged particle across the magnetic field. According to Lyons and Williams (1984) the drift velocity $\vec{v}_{\mathrm{D}} = \dot{\vec{R}}_{\mathrm{c}\perp}$ can be written as

$$\vec{v}_{\mathrm{D}} = \frac{c}{e_{\alpha}B^2}\left[\vec{B} \wedge \left\langle -e_{\alpha}\vec{E} + \frac{W}{B}(1 + \cos^2\theta)\nabla B\right\rangle\right] \tag{2.21}$$

where $\vec{E} = -\nabla\Phi$ is the electric field. In natural magnetic traps the drift motion is only important for times much larger than the bounce period T_{B}. In this case we shall use the averaged drift velocity:

$$\langle\vec{v}_{\mathrm{D}}\rangle = \frac{1}{T_{\mathrm{B}}}\oint \vec{v}_{\mathrm{D}}\, dz/v_{\|}(z). \tag{2.22}$$

In particular, for a dipole magnetic field (2.3)

$$\langle\vec{v}_{\mathrm{DM}}\rangle = \vec{\psi}_0 \frac{6W}{e_{\alpha}B_{\mathrm{L}}\, L\, R_0}(0.35 + 0.15\sin\theta_{\mathrm{L}}) \tag{2.23}$$

and

$$\langle\vec{v}_{\mathrm{DE}}\rangle = c\,\frac{\left[\vec{E}_{\perp\mathrm{L}} \wedge \vec{B}_{\mathrm{L}}\right]}{B_{\mathrm{L}}^2} \tag{2.24}$$

as given by Lyons and Williams (1984); here $\vec{\psi}_0$ is the unit vector, corresponding to the local time angle ψ_{LT} (Fig. 2.4).

Problem

Calculate the energy change of a charged particle across and along the dipole magnetic field using the parabolic approximation (2.7)–(2.8) during particle transport in

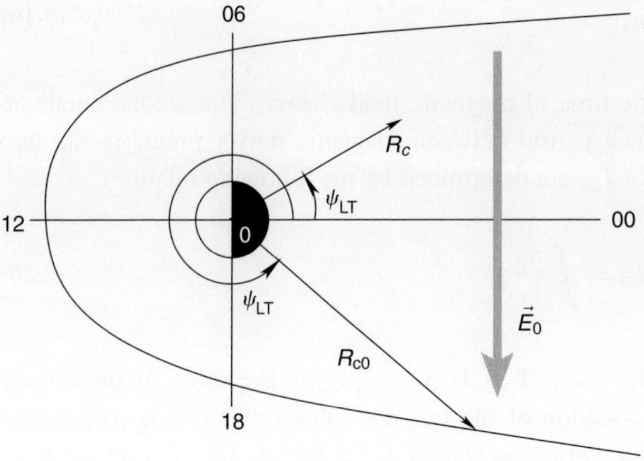

Figure 2.4 The equatorial cross-section of the magnetosphere as seen from the northern magnetic pole; ψ_{LT} is the local time angle. The dawn-to-dusk electric field across the magnetosphere, $\vec{E}_0$, is shown.

a magnetosphere. The electric potential Φ is taken to be constant along a magnetic field line.

Solution

The first and the second adiabatic invariants (2.12), (2.20), and the energy integral (2.14) are used. From (2.14) under $\Phi(z) = \text{const}$

$$v_{\|} = \pm\sqrt{\frac{2}{m}\Big(W - J_{\perp} B(z)\Big)}\,. \tag{2.25}$$

Substituting (2.7) and (2.25) into (2.20), we find:

$$T_{\mathrm{B}} = 2\pi a/v_{\perp\mathrm{L}}, \quad J_{\|} = \pi a v_{\|\mathrm{L}}^2/v_{\perp\mathrm{L}}. \tag{2.26}$$

If the particle is starting from $L = L_0$, for any other value L we have from (2.12) and (2.26):

$$\left(\frac{W_{\perp\mathrm{L}}}{W_{\perp 0}}\right) = \frac{B_{\mathrm{L}}}{B_{\mathrm{L}0}}, \quad \frac{W_{\|\mathrm{L}}}{W_{\|0}} = \left(\frac{W_{\perp\mathrm{L}}}{W_{\perp 0}}\right)^{1/2}\frac{a_0}{a_{\mathrm{L}}} = \left(\frac{B_{\mathrm{L}}}{B_{\mathrm{L}0}}\right)^{1/2}\frac{a_0}{a_{\mathrm{L}}}, \tag{2.27}$$

$$\alpha_1 = \frac{W_{\perp\mathrm{L}}}{W_{\|\mathrm{L}}} = \left(\frac{B_{\mathrm{L}}}{B_{\mathrm{L}0}}\right)^{1/2}\frac{a_{\mathrm{L}}}{a_0}\left(\frac{W_{\perp 0}}{W_{\|0}}\right) \tag{2.28}$$

where the anisotropy factor α_1 of a single particle has been introduced. Using (2.7) and (2.8) we find:

$$\left(\frac{W_{\perp\mathrm{L}}}{W_{\perp 0}}\right) = \left(\frac{L_0}{L}\right)^3, \quad \left(\frac{W_{\|\mathrm{L}}}{W_{\|0}}\right) = \left(\frac{L_0}{L}\right)^{2.5}, \quad \alpha_1 = \left(\frac{L_0}{L}\right)^{1/2}\alpha_{10}. \tag{2.29}$$

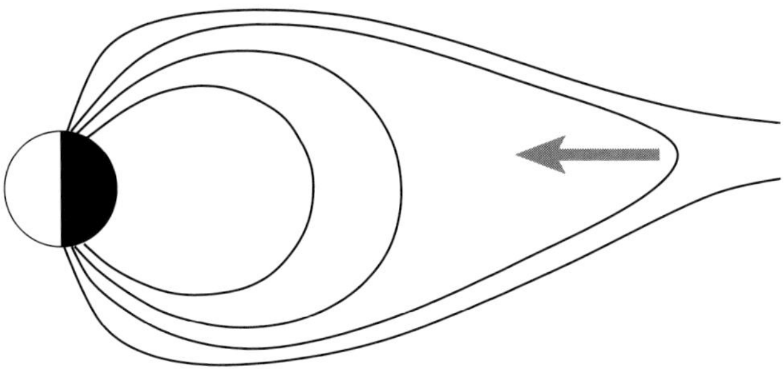

Figure 2.5 Dipolarization of the geomagnetic field during a magnetic substorm: stressed magnetic field lines shorten as they become closer to the dipole configuration.

In the case of magnetic compression, $B_{\rm L}$ and $a_{\rm L}$ grow with time as occurs, for example, during the breakup phase of a magnetospheric substorm, when dipolarization of the magnetic field takes place, as shown in Fig. 2.5. If $B_{\rm L}$ and a are changing slowly compared with T_B (2.27), it is possible to use the formulae (2.28)–(2.29), in which $B_{\rm L}$ and $a_{\rm L}$ are growing in time. The particular growth of $W_\perp$, $W_\|$ and α_1 depends on the details of the compression process.

2.3 The distribution function

To describe collective effects, a distribution function $\mathcal{F}(\vec{r}, \vec{v}, t)$, expressing the density of particles in phase space $(\vec{r}, \vec{v})$, is used. The integral of $\mathcal{F}$ over $\vec{v}$ gives the charged particle number density n:

$$\int \mathcal{F}\, d^3\vec{v} = n. \tag{2.30}$$

Evolution of the distribution function is determined by a kinetic equation, corresponding to the phase volume conservation law:

$$\frac{d\mathcal{F}}{dt} = \frac{\partial \mathcal{F}}{\partial t} + \mathrm{div}_{\vec{r}}(\vec{v}\mathcal{F}) + \mathrm{div}_{\vec{v}}(\dot{\vec{v}}\mathcal{F}) = 0. \tag{2.31}$$

If $\vec{r}$ and $\vec{v}$ are canonical variables, that is if $\vec{r}$ and $\vec{v}$ satisfy the relations:

$$\dot{r}_i = -\frac{\partial \mathcal{H}}{\partial v_i}, \quad \dot{v}_i = \frac{\partial \mathcal{H}}{\partial r_i} \tag{2.32}$$

where $\mathcal{H}$ is the Hamiltonian, Equation (2.31) takes the form:

$$\frac{\partial \mathcal{F}}{\partial t} + \vec{v}\frac{\partial \mathcal{F}}{\partial \vec{r}} + \dot{\vec{v}}\frac{\partial \mathcal{F}}{\partial \vec{v}} = 0. \tag{2.33}$$

In particular, $\vec{r}$ and $\vec{v}$ are canonical variables if we consider wave–particle interactions in a homogeneous magnetic field. In this case, putting z along the magnetic field and using cylindrical coordinates $v_\parallel$, $v_\perp$, φ in velocity space, we can write, since the system is uniform across the magnetic field:

$$\frac{\partial \mathcal{F}}{\partial t} + v_\parallel \frac{\partial \mathcal{F}}{\partial z} + \dot{v}_\perp \frac{\partial \mathcal{F}}{\partial v_\perp} + \dot{\varphi} \frac{\partial \mathcal{F}}{\partial \varphi} + \dot{v}_\parallel \frac{\partial \mathcal{F}}{\partial v_\parallel} = 0 \tag{2.34}$$

where $\dot{v}_\perp$ and $\dot{v}_\parallel$ due only to wave–particle interactions. This is the basic kinetic equation for our further consideration.

For an inhomogeneous magnetic field, the variables in (2.31)–(2.33) are canonical as well (see Belyaev, 1958; Northrop and Teller, 1960; Gurevich and Dimant, 1987). According to Belyaev (1958), the kinetic equation for the distribution function of guiding centres can be written as:

$$\frac{\partial \mathcal{F}}{\partial t} + v_\parallel \frac{\partial \mathcal{F}}{\partial z} + \vec{v}_{\mathrm{D}} \frac{\partial \mathcal{F}}{\partial \vec{R}_{\perp c}} + \dot{v}_\perp \frac{\partial \mathcal{F}}{\partial v_\perp} + \dot{\varphi} \frac{\partial \mathcal{F}}{\partial \varphi} + \dot{v}_\parallel \frac{\partial \mathcal{F}}{\partial v_\parallel} = 0 \tag{2.35}$$

where the terms $\dot{v}_\perp$ and $\dot{v}_\parallel$ include wave–particle interactions and the forces arising from the inhomogeneous magnetic field (see (2.12)–(2.13)). It is convenient to write Equation (2.35) in the variables of the first adiabatic invariant $J_\perp$ and the total energy $w = W + e_\alpha \Phi$, which are both conserved in the absence of waves:

$$\frac{\partial \mathcal{F}}{\partial t} + v_\parallel \frac{\partial \mathcal{F}}{\partial z} + \vec{v}_{\mathrm{D}} \frac{\partial \mathcal{F}}{\partial \vec{R}_{\perp c}} + \dot{J}_\perp \frac{\partial \mathcal{F}}{\partial J_\perp} + \dot{w} \frac{\partial \mathcal{F}}{\partial w} + \dot{\varphi} \frac{\partial \mathcal{F}}{\partial \varphi} = 0. \tag{2.36}$$

In Equation (2.36) $\dot{J}_\perp$ and $\dot{w}$ are due only to wave–particle interactions. The active substance in the CM is, as a rule, prepared by large-scale and slowly changing processes, such as transport within the magnetosphere or magnetic compression. In this case it is useful to consider the bounce-averaged distribution function $\langle \mathcal{F} \rangle$, depending on the motion integrals: $J_\perp$, $J_\parallel$ and $W + e_\alpha \Phi$. The kinetic equation for $\langle \mathcal{F} \rangle$ is very simple in the absence of waves:

$$\frac{\partial \langle \mathcal{F} \rangle}{\partial t} + \langle \vec{v}_{\mathrm{D}} \rangle \frac{\partial \langle \mathcal{F} \rangle}{\partial \vec{R}_{\perp c}} = 0. \tag{2.37}$$

In principle, the solution of the kinetic equations (2.31)–(2.37) can be found from the phase volume conservation law (Liouville's theorem), according to which:

$$\mathcal{F}(\vec{r}, \vec{v}, t) = \mathcal{F}_0\Big(\vec{r}_0(\vec{v}, \vec{r}, t, t_0),\ \vec{v}_0(\vec{v}, \vec{r}, t, t_0)\Big) \tag{2.38}$$

where $\mathcal{F}_0$ is the initial distribution function, and the relations between $\vec{r}_0(\vec{r}, \vec{v}, t)$ and $\vec{v}_0(\vec{r}, \vec{v}, t)$ are found from the phase trajectory equations. For example, in the case of Equation (2.37) such equations are

$$dt = \frac{dR_{\mathrm{c}}}{v_{\mathrm{D}\,R_{\mathrm{c}}}(R_{\mathrm{c}}, \psi_{\mathrm{LT}})} = \frac{R_{\mathrm{c}}\, d\psi_{\mathrm{LT}}}{v_{\mathrm{D}\psi_{\mathrm{LT}}}(R_{\mathrm{c}}, \psi_{\mathrm{LT}})}, \tag{2.39}$$

the coordinate system being shown in Fig. 2.4; ψ_{LT} is the local time angle. From these equations we have the formal solution ($t_0 = 0$):

$$R_c = R_c(R_{c0}, \psi_{\mathrm{LT}_0}, t), \qquad \psi_{\mathrm{LT}} = \psi_{\mathrm{LT}}(R_{c0}, \psi_{\mathrm{LT}_0}, t). \tag{2.40}$$

Thus, if the initial distribution function is given by

$$\mathcal{F}\Big|_{t=0} = \mathcal{F}_0(J_\perp, J_\parallel, \psi_{\mathrm{LT}_0}, L_0) \tag{2.41}$$

the distribution function at any time can be found as

$$\mathcal{F}(J_\perp, J_\parallel, R_c\psi_{\mathrm{LT}}, t) = \mathcal{F}_0\Big(J_\perp, J_\parallel, R_{0c}(R_c, \psi_{\mathrm{LT}}, t), \psi_{\mathrm{LT}_0}(R_c, \psi_{\mathrm{LT}}, t)\Big) \tag{2.42}$$

where $R_{0c}(R_c, \psi_{\mathrm{LT}}, t)$ and $\psi_{\mathrm{LT}_0}(R_c, \psi_{\mathrm{LT}}, t)$ are determined from (2.40). The above method is suitable if we are dealing with one or two quasi-monochromatic waves, interacting with charged particles. In the case of many waves the problem is similar to the many-body problem in mechanics, when statistical methods are employed. Thus, we come to the quasi-linear theory considered in later chapters.

Problem
Find the stationary distribution function $\mathcal{F}$ of electrons (or ions) and the anisotropy factor α inside the magnetosphere, if the boundary distribution function is given by:

$$R_c = R_{0c}, \quad \mathcal{F} = \mathcal{F}_1(J_\perp, J_\parallel)\,\delta(\sin\psi_{\mathrm{LT}} - \sin\psi_{\mathrm{LT}_0}). \tag{2.43}$$

A uniform electric field $\vec{E}_0$ is directed from the dawnside of the magnetosphere to the dusk side (Fig. 2.4); charged particle injection occurs at a local time defined by ψ_{LT_0}.

Solution
According to the above description, the solution can be written as:

$$\mathcal{F}(J_\perp, J_\parallel, R_c, \psi_{\mathrm{LT}}) = \mathcal{F}_0\Big(J_\perp, J_\parallel, \psi_{\mathrm{LT}_0}(\psi_{\mathrm{LT}}, R_c, R_{c0})\Big). \tag{2.44}$$

The function $\psi_{\mathrm{LT}_0}(\psi_{\mathrm{LT}}, R_c, R_{c0})$ can be found from (2.39). However, for the stationary case, it is better to use the energy integral (2.14), in which we must put $\Phi = -E_0 R_c \sin\psi_{\mathrm{LT}}$. Taking into account that $R_c = LR_0$ and $W = J_\perp B_L + W_{\parallel L}$, and using (2.27), we have:

$$\frac{J_\perp}{L^3} + \frac{K}{L^{2.5}} - \frac{e_\alpha E_0 R_0}{B_M} L\sin\psi_{\mathrm{LT}} = \frac{J_\perp}{L_0^3} + \frac{K}{L_0^{2.5}} - \frac{e_\alpha E_0 R_0}{B_M} L_0 \sin\psi_{\mathrm{LT}_0} \tag{2.45}$$

where $K = \dfrac{4.4}{\pi R_0}\left(\dfrac{2m}{B_M}\right)^{1/2} J_\perp^{1/2} J_\parallel$ and B_M is the magnetic field at the foot of the flux tube.

We have demonstrated that many natural phenomena in space plasmas lead to the formation of anisotropic energetic particle distributions with $\alpha > 1$. Below we shall show that such distributions are unstable and lead to the excitation of electromagnetic waves. To consider this process, we must include a wave–particle interaction in the kinetic equation (2.31) and analyze the energy exchange between waves and particles. Before that we review the properties of electromagnetic waves in a plasma such as pervades a CM.

2.4 Electromagnetic waves in a magnetized plasma

Wave–particle interactions are described by a system of self-consistent equations, including a kinetic equation for the particle distribution function and Maxwell's equations for the waves, with the currents and charges being determined by the particle distribution function. In this section we briefly consider some properties of Maxwell's equations which are necessary for further analysis.

There are many excellent books, beginning with Budden (1964), Ginzburg (1970) and up to the recent edition of Stix (1992), Gary (1993), Sturrock (1994), Zheleznyakov (1996), Baumjohann and Treumann (1996) and Gurnett and Bhattacharjee (2005), where both general and some different specific aspects of electromagnetic wave propagation through a plasma and wave–wave and wave–particle interactions are considered rigorously, and giving many important details. We particular recommend to readers the *Handbook of Plasma Physics* (general editors M.N. Rosenbluth and R.Z. Sagdeev, 1984), which considers these issues most fully. More special questions concerning propagation and generation of VLF/ELF emissions are considered below in this book and in the books by Helliwell (1965), Bespalov and Trakhtengerts (1986a), Molchanov (1985), Sazhin (1993) and in *Proceedings of a Workshop on the Very Low Frequency (VLF) Phenomena* edited by Hughes *et al.* (2003). In order to retain continuity of the material which is essential for a clear understanding the basic elements of CM theory as used in later sections and chapters, we shall give here the foundation of electromagnetic theory taking the survey of Shafranov (1963) as the basis for our further considerations.

For our aims it is convenient to present Maxwell's equations in the c.g.s. Gaussian system of units in the following form (see Jackson, 1962):

$$\operatorname{curl}\vec{B} = \frac{4\pi}{c}\vec{j}_{\rm c} + \frac{4\pi}{c}\vec{j}_{\rm h} + \frac{1}{c}\frac{\partial\vec{E}}{\partial t} \tag{2.46}$$

$$\operatorname{curl}\vec{E} = -\frac{1}{c}\frac{\partial\vec{B}}{\partial t} \tag{2.47}$$

$$\operatorname{div}\vec{B} = 0, \quad \operatorname{div}\vec{E} = 4\pi(\rho_{\rm c} + \rho_{\rm h}) \tag{2.48}$$

where $\vec{B}$ and $\vec{E}$ are the wave magnetic and electric fields and $(\vec{j}_{\rm c,h}, \rho_{\rm c,h})$ are the electric currents and charges related, respectively, to the cold plasma (subscript 'c') and to the energetic (hot) charged particles (subscript 'h'):

$$\vec{j}_{\mathrm{c,h}} = \sum_{\alpha} \int e_\alpha \mathcal{F}_\alpha^{\mathrm{c,h}} \, \vec{v} \, d^3\vec{v}, \quad \rho_{\mathrm{c,h}} = \sum_{\alpha} \int e_\alpha \mathcal{F}_\alpha^{\mathrm{c,h}} \, d^3\vec{v} \tag{2.49}$$

where the summation is over all types, α, of charged particles. We have used the c.g.s. system of units and take into account that the magnetic field $\vec{H} = \vec{B}$ in a plasma, leaving $\vec{B}$ as the symbol for the magnetic field.

Equations (2.46)–(2.47) can be presented in the form of one second-order equation:

$$\nabla^2 \vec{E} - \mathrm{grad}\,(\mathrm{div}\,\vec{E}) - \frac{1}{c^2}\frac{\partial^2 \vec{D}}{\partial t^2} = \frac{4\pi}{c^2}\frac{\partial \vec{j}_{\mathrm{h}}}{\partial t} \tag{2.50}$$

where the electric displacement $\vec{D}$ is related to $\vec{E}$ and $\vec{j}_{\mathrm{c}}$ by the relation:

$$\frac{1}{c^2}\frac{\partial \vec{D}}{\partial t} = \frac{4\pi}{c^2}\vec{j}_{\mathrm{c}} + \frac{1}{c^2}\frac{\partial \vec{E}}{\partial t}. \tag{2.51}$$

In space plasmas the wavelength is much less than the spatial scale characterizing the inhomogenity of the system (magnetic field or plasma density). We can therefore use the approximation of a smoothly non-uniform medium, i.e. the geometrical optics approximation, to describe the electromagnetic waves.

For a homogeneous medium a spatial–temporal Fourier transform is applied to the system of Equations (2.46)–(2.49). This corresponds to the expansion of the electromagnetic field over a sum of independent wave harmonics:

$$\left\{ \begin{matrix} \vec{E}(\vec{r}, t) \\ \vec{B}(\vec{r}, t) \end{matrix} \right\} = \int d\omega \, d\vec{k} \left\{ \begin{matrix} \vec{E}(\omega, \vec{k}) \\ \vec{B}(\omega, k) \end{matrix} \right\} \exp(i\omega t - i\vec{k}\,\vec{r}\,). \tag{2.52}$$

$\vec{E}$ and $\vec{D}$ are related by the well-known linear relation:

$$D_\alpha(\omega, \vec{k}) = \varepsilon_{\alpha\beta}(\omega) \, E_\beta(\omega, \vec{k}) \tag{2.53}$$

where $\varepsilon_{\alpha\beta}$ are the components of the permittivity tensor of a cold plasma, and the subscripts α, β run through x, y, z. In the coordinate system with the magnetic field $\vec{B}$ being along the z axis, the permittivity tensor $\varepsilon^0_{\alpha\beta}$ is given by:

$$\varepsilon^0_{\alpha\beta} = \begin{pmatrix} \varepsilon & ig & 0 \\ -ig & \varepsilon & 0 \\ 0 & 0 & \eta \end{pmatrix} \tag{2.54}$$

where

$$\varepsilon = 1 - \sum_{\alpha} \frac{\omega_{p\alpha}^2}{\omega^2 - \omega_{B\alpha}^2}, \quad g = \sum_{\alpha} \frac{\omega_{B\alpha}}{\omega} \frac{\omega_{p\alpha}^2}{\omega^2 - \omega_{B\alpha}^2}, \quad \eta = 1 - \sum_{\alpha} \frac{\omega_{p\alpha}^2}{\omega^2}. \tag{2.55}$$

Here, $\omega_{p\alpha} = \left(4\pi e_\alpha^2 n_\alpha / m_\alpha\right)^{1/2}$ is the plasma frequency for charged particles of type α, and $\omega_{B\alpha} = e_\alpha B / m_\alpha c$ is their gyrofrequency. Then Equation (2.50) can be written for the Fourier component $\vec{E}(\omega, \vec{k})$ in the form:

$$k^2\vec{E} - \vec{k}(\vec{k}\vec{E}\,) - \frac{\omega^2}{c^2}\vec{D} = -\frac{4\pi i\omega}{c^2}\vec{j}_{\mathrm{h}}. \tag{2.56}$$

The wave magnetic field $\vec{B}_\sim$ is:

$$\vec{B}_\sim = \frac{c}{\omega}\left[\vec{k} \wedge \vec{E}_\sim\right]. \tag{2.57}$$

In a Cartesian coordinate system, with the projection of $\vec{E}$ onto the α-axis being E_α, we have from (2.56):

$$\mathcal{D}_{\alpha\beta}(\vec{k}, \omega)E_\beta = -\frac{4\pi i\omega}{c^2} j_{h\alpha} \tag{2.58}$$

where the matrix $\mathcal{D}_{\alpha\beta}$ is given by:

$$\mathcal{D}_{\alpha\beta} = k^2\delta_{\alpha\beta} - \frac{\omega^2}{c^2}\,\varepsilon_{\alpha\beta} - k_\alpha k_\beta \tag{2.59}$$

where $\delta_{\alpha\beta}$ is the unit diagonal tensor.

We use for solution of the system of equations (2.58) the 'eigenmode' method, when the field $\vec{E}$ is represented as a sum of eigenmodes

$$\vec{E}(\omega, \vec{k}\,) = \sum_l E_l(\omega, \vec{k}\,)\vec{a}_l \tag{2.60}$$

where the normalized polarization vector $\vec{a}_l$ of an eigenmode is determined from the homogeneous system (2.58) (with $\vec{j}_{h\alpha} = 0$).

Putting $\vec{k}$ along the z-axis and the magnetic field $\vec{B}$ in the xz-plane, we obtain from (2.58) and (2.59), with $\vec{j}_{\mathrm{h}} = 0$:

$$\left(N_l^2 - \varepsilon_{xx}\right)a_x - \varepsilon_{xy}a_y - \varepsilon_{xz}a_z = 0 \tag{2.61}$$

$$-\varepsilon_{yx}a_x + \left(N_l^2 - \varepsilon_{yy}\right)a_y - \varepsilon_{yz}a_z = 0 \tag{2.62}$$

$$-\varepsilon_{zx}a_x - \varepsilon_{zy}a_y - \varepsilon_{zz}a_z = 0 \tag{2.63}$$

where we define the refractive index N_l as the l-th eigenmode solution for the system (2.61)–(2.63) to obtain the dispersion relation:

$$\frac{k^2c^2}{\omega^2} = N_l^2(\vec{a}, \omega). \tag{2.64}$$

The components ε_{ij} for this chosen coordinate system are equal to:

$$\begin{aligned} &\varepsilon_{xx} = \varepsilon \cos^2 \chi + \eta \sin^2 \chi, && \varepsilon_{xy} = -\varepsilon_{yx} = -ig \cos \chi, \\ &\varepsilon_{yy} = \varepsilon, && \varepsilon_{xz} = \varepsilon_{zx} = (\varepsilon - \eta) \sin \chi \cos \chi, \\ &\varepsilon_{yz} = -\varepsilon_{zy} = -ig \sin \chi, && \varepsilon_{zz} = \varepsilon \sin^2 \chi + \eta \cos^2 \chi \end{aligned} \tag{2.65}$$

where χ is the wave normal angle between $\vec{k}$ and $\vec{B}$. Putting the determinant of the system (2.61)–(2.63) equal to zero, we find:

$$N_{1,2}^2 = \frac{1}{2}(\eta_{xx} + \eta_{yy}) \pm \sqrt{\frac{(\eta_{xx} - \eta_{yy})^2}{4} + \eta_{xy}\eta_{yx}} \tag{2.66}$$

where

$$\begin{aligned} \eta_{xx} &= \varepsilon\eta(\varepsilon \sin^2 \chi + \eta \cos^2 \chi)^{-1}, \\ \eta_{xy} &= -\eta_{yx} = ig\eta \cos \chi\left(\varepsilon \sin^2 \chi + \eta \cos^2 \chi\right)^{-1} \\ \eta_{yy} &= \left[\varepsilon\left(\varepsilon \sin^2 \chi + \eta \cos^2 \chi\right) - g^2 \sin^2 \chi\right]\left(\varepsilon \sin^2 \chi + \eta \cos^2 \chi\right)^{-1}. \end{aligned} \tag{2.67}$$

For electromagnetic eigenmodes with $[\vec{a} \times \vec{k}\,] \neq 0$ it is convenient to define the polarization vector components as the ratio of the x- and z-components of $\vec{a}$ to its y-component:

$$\vec{a} = a_y(i\alpha_x, 1, i\alpha_z). \tag{2.68}$$

From Equation (2.63) we have:

$$i\alpha_z = (-i\alpha_x\varepsilon_{zx} - \varepsilon_{zy})\varepsilon_{zz}^{-1}. \tag{2.69}$$

Eliminating α_z from (2.61) and (2.62) with the help of (2.69), we find:

$$\begin{aligned} &(N_l^2 - \eta_{xx})i\alpha_x - \eta_{xy} = 0, \\ &-\eta_{yx}\, i\alpha_x + N_l^2 - \eta_{yy} = 0. \end{aligned} \tag{2.70}$$

After eliminating N_l^2, we have:

$$\alpha_x^2 + i\,\frac{\eta_{xx} - \eta_{yy}}{\eta_{yx}}\,\alpha_x + \frac{\eta_{xy}}{\eta_{yx}} = 0. \tag{2.71}$$

Thus, the polarization vectors of the two normal waves (eigenmodes) in a cold magnetized plasma are connected by the relations:

$$\alpha_{x1} + \alpha_{x2} = \frac{(\eta_{xx} - \eta_{yy})}{i\eta_{yx}}, \quad \alpha_{x1}\alpha_{x2} = \frac{\eta_{xy}}{\eta_{yx}} = -1. \tag{2.72}$$

Taking (2.67) into account, we obtain the final formulae for N_l^2 and α_i from (2.66) and (2.70):

$$N_{1,2}^2 = \frac{(\varepsilon^2 - g^2 - \varepsilon\eta)\sin^2\chi + 2\varepsilon\eta \pm \sqrt{(\varepsilon^2 - g^2 - \varepsilon\eta)^2 \sin^4\chi + 4g^2\eta^2\cos^2\chi}}{2(\varepsilon\sin^2\chi + \eta\cos^2\chi)} \tag{2.73}$$

$$\alpha_x = g\eta\cos\chi\Big[N_{1,2}^2(\varepsilon\sin^2\chi + \eta\cos^2\chi) - \varepsilon\eta\Big]^{-1} \tag{2.74}$$

$$\alpha_z = (\eta - N_{1,2}^2)g\sin\chi\Big[N_{1,2}^2(\varepsilon\sin^2\chi + \eta\cos^2\chi) - \varepsilon\eta\Big]^{-1}. \tag{2.75}$$

The $\pm$ sign indicates birefringence of the plasma.

Further we shall be interested in quasi-longitudinal wave propagation, when $\vec{k}$ is not far from being parallel to $\vec{B}$, and the following inequality holds:

$$4\eta^2 g^2\cos^2\chi \gg (\varepsilon^2 - g^2 - \varepsilon\eta)^2 \sin^4\chi. \tag{2.76}$$

Then, expression (2.73) for N_l^2 takes the form:

$$N_l^2 \simeq \frac{\eta(\varepsilon \pm g\cos\chi) + \Delta\sin^2\chi}{\varepsilon\sin^2\chi + \eta\cos^2\chi} \tag{2.77}$$

where $\Delta = \frac{1}{2}\left(\varepsilon^2 - g^2 - \varepsilon\eta\right)$.

The formulae (2.74)–(2.75) are simplified as well:

$$\alpha_x = \pm\left(1 \mp \frac{\Delta\sin^2\chi}{g\eta\cos\chi}\right), \qquad \alpha_z = \frac{\pm(\eta - \varepsilon)\cos\chi - g}{(\eta\cos^2\chi + \varepsilon\sin^2\chi)}\sin\chi. \tag{2.78}$$

The expressions (2.77)–(2.78) describe all possible ‘cold’ modes in a CM. They give the refractive index and polarization parameters for electromagnetic waves propagating in a magnetoionic medium.

Now we consider the solution for the inhomogeneous system of equations (2.58) when $j_{\mathrm{h}} \neq 0$. For that we first analyze some general properties of the eigenmodes (the normal waves) of the system (2.61)–(2.63), as follows:

$$\Big\{N_l^2(\delta_{\alpha\beta} - n_\alpha n_\beta) - \varepsilon_{\alpha\beta}\Big\}\, a_{\beta l} = 0, \quad \vec{n} = \vec{k}/k. \tag{2.79}$$

Let us consider the complex conjugate to (2.79):

$$\Big\{N_m^2(\delta_{\alpha\beta} - n_\alpha n_\beta) - \varepsilon_{\beta\alpha}\Big\}\, a_{\beta m}^* = 0. \tag{2.80}$$

Multiplying (2.79) by $a_{\alpha m}^*$ and (2.80) by $a_{\alpha l}$, we have:

$$\begin{aligned} &N_l^2(\delta_{\alpha\beta} - n_\alpha n_\beta)a_{\alpha m}^* a_{\beta l} - \varepsilon_{\alpha\beta} a_{\beta l} a_{\alpha m}^* = 0,\\ &N_m^2(\delta_{\alpha\beta} - n_\alpha n_\beta)a_{\alpha m}^* a_{\beta l} - \varepsilon_{\alpha\beta} a_{\beta l} a_{\alpha m}^* = 0. \end{aligned} \tag{2.81}$$

Here the order of the subscripts has been exchanged in the second equation. Subtracting the second equation in (2.81) from the first, we obtain:

$$\left(N_l^2 - N_m^2\right)(\delta_{\alpha\beta} - n_\alpha n_\beta)a_{\alpha m}^* a_{\beta l} = 0. \tag{2.82}$$

It follows that, for the condition $m \neq l (N_l^2 \neq N_m^2)$, the multiplier $(\delta_{\alpha\beta} - n_\alpha n_\beta)$ $a_{\alpha m}^* a_{\beta l} = 0$. When $l = m$ this multiplier is not zero and can be put equal to unity, the normalization condition for $\vec{a}_l$. Thus, since the polarization vector a satisfies the normalization condition

$$(\delta_{\alpha\beta} - n_\alpha n_\beta)a_{\alpha m}^* a_{\beta l} = \delta_{lm} \tag{2.83}$$

it follows from (2.79) and (2.83) that

$$\varepsilon_{\alpha\beta} a_{\beta l} a_{\alpha m}^* = N_l^2 \delta_{lm}, \quad N_l^2 = \frac{k_l^2 c^2}{\omega_l^2}. \tag{2.84}$$

If we substitute (2.68) into (2.83), we find:

$$a_{yl} a_{ym}^* (1 + \alpha_{xl}\alpha_{xm}) = \delta_{lm}. \tag{2.85}$$

According to (2.72) for $l \neq m$, $\alpha_{xl}\alpha_{xm} = 1$, and the condition (2.83) is fulfilled automatically. For $l = m$ we have:

$$a_y = (1 + \alpha_x^2)^{-1/2}. \tag{2.86}$$

Now we can solve Equation (2.58), presenting the electric field as the sum of the eigenmodes (2.60). Substituting (2.60) into (2.58) we have:

$$\mathcal{D}_{\alpha\beta} E_\beta = \sum_m \left\{ k^2(\delta_{\alpha\beta} - n_\alpha n_\beta)a_{\beta m} - \frac{\omega^2}{c^2}\varepsilon_{\alpha\beta} a_{\beta m} \right\} E_m = -\frac{4\pi i\omega}{c^2} j_{\mathrm{h}\alpha}. \tag{2.87}$$

Multiplying (2.87) by a complex conjugate polarization vector $a_{\alpha l}^*$ and using the relations (2.83), (2.84) and (2.86) we find that:

$$\mathcal{D}_l E_l \equiv \left(k^2 - \frac{\omega^2}{c^2} N_l^2\right) E_l = -\frac{4\pi i\omega}{c^2}\left(\vec{j}_{\mathrm{h}}\left(\omega, \vec{k}\right)\vec{a}_l^*\right) \tag{2.88}$$

where $N_l^2 = \varepsilon_{\alpha\beta} a_{\beta l} a_{\alpha l}^*$ is determined by (2.73). Expression (2.88) is the spectral form of the basic wave excitation equation in CM theory and, as such, is crucial.

2.5 Wave eigenmodes in a CM. Properties of reflecting mirrors in a CM

Now we use the relations of the previous section to analyse the CM wave eigenmodes. In principle any normal plasma wave which can be amplified by an anisotropic particle distribution in the process of a cyclotron interaction can serve as a wave eigenmode in a CM. Among these eigenmodes are circularly polarized electromagnetic waves with a frequency $\omega \lesssim |\omega_{B\alpha}|$, the gyrofrequency of a particle, and a wave vector $\vec{k}$ nearly along $\vec{B}$. For an electron CM (when the energetic particle component consists of electrons) such waves are whistler-mode waves; for an ion CM such waves are Alfvén waves. These modes have some advantages over other waves.

First, the group velocity for both these waves is close (within $\sim 20°$) to the magnetic field for practically any direction of the wave vector $\vec{k}$. Thus, the wave energy is guided along the ambient magnetic field – it does not leak out of the magnetic flux tube. Secondly, even weak plasma density gradients across the magnetic field provide ducting of the rays along the magnetic field. The CI of whistler and Alfvén waves possesses the very important property that, if a CI appears at the central cross-section of a magnetic trap, i.e. in the equatorial plane of the magnetosphere, it takes place all along the magnetic flux tube. These two properties make the whistler and Alfvén wave modes the preferred eigenmodes in CMs in space plasmas.

Another important property of these modes is that, for both the whistler and Alfvén waves, their frequency $\omega < |\omega_{B\alpha}|$. This enables the absolute instability to appear in space-limited systems without reflecting mirrors.

In this section we analyse the propagation of the whistler and Alfvén waves in a space plasma. The refractive index and polarization of these waves can be found from (2.77) and (2.78). In the case of purely longitudinal propagation, $\chi = 0$, we obtain:

$$N_{1,2}^2 = \varepsilon \pm g, \quad \alpha_x = \pm 1, \quad \alpha_z = 0. \tag{2.89}$$

For a plasma with a single ion species, with $|e| = e_i$, the component g (2.55) can be written as:

$$g = \frac{\omega_{pe}^2(\omega_{Be}\,\omega)}{\left(\omega^2 - \omega_{Be}^2\right)\left(\omega^2 - \omega_{Bi}^2\right)}, \qquad \frac{\omega_{pe}^2}{|\omega_{Be}|} = \frac{\omega_{pi}^2}{\omega_{Bi}}. \tag{2.90}$$

For whistler waves (subscript 'w')

$$\omega_{Bi} \ll \omega < |\omega_{Be}|, \quad g > 0 \tag{2.91}$$

and we must take the upper sign in (2.89)

$$N_{\mathrm{w}}^2 = 1 + \frac{\omega_{pe}^2}{\omega(|\omega_{Be}| - \omega)}, \quad \alpha_x = 1. \tag{2.92}$$

For CMs in space the second term in (2.92) is much greater than unity. The polarization of the wave is circular.

The electric field for the whistler wave has the components (see (2.68))

$$\vec{E}_{\mathrm{w}} = A \exp i(\omega t - kz)\, a_y(i\vec{x}_0 + \vec{y}_0 + 0.\vec{z}_0), \quad a_y = \frac{1}{\sqrt{2}} \tag{2.93}$$

where A is the amplitude. As discussed in Section 2.6, the electron velocity in the magnetic field is $\vec{v}_e = v_\perp(\vec{y}_0 - i\vec{x}_0)\exp(-i\omega_B t)$, so that $\vec{E}_w$ and $\vec{v}_e$ rotate in the same direction around the magnetic field $\vec{B}$.

For ion CMs we have the Alfvén electromagnetic branch with

$$\omega < \omega_{Bi}, \quad g < 0. \tag{2.94}$$

In this case, taking the upper sign in (2.89):

$$N_{\mathrm{A}}^2 = 1 + \frac{\omega_{pi}^2}{\omega_{Bi}(\omega_{Bi} - \omega)}, \quad \alpha_x = -1. \tag{2.95}$$

The electric field vector of the Alfvén wave (subscript ‘A’) rotates in the same direction as the ion. The second term in (2.95) is much greater than unity for space CM conditions.

The relations (2.92) and (2.95) can be generalized for the case of oblique propagation, when $\chi \neq 0$, using (2.77) and (2.78). For whistler waves (hereafter, $\omega_B = |\omega_{Be}|$) we have:

$$N_{\mathrm{w}}^2 = \frac{\omega_{pe}^2}{\omega(\omega_B \cos\chi - \omega)} \gg 1. \tag{2.96}$$

For Alfvén waves:

$$N_{\mathrm{A}}^2 = \frac{\omega_{pi}^2/\omega_{Bi}^2}{\cos^2\chi} \frac{1 + (\omega/\omega_{Bi})\cos\chi}{1 - (\omega/\omega_{Bi})^2}. \tag{2.97}$$

For oblique propagation the polarization of the wave is determined by (2.78). For whistler waves we have:

$$\alpha_{x\mathrm{w}} = 1 + \frac{\omega}{2\omega_B} \frac{\sin^2\chi}{\cos\chi}. \tag{2.98}$$

Thus, the correction to the circular polarization for oblique whistler waves is rather small, especially in the case when $(\omega/\omega_B) \ll 1$. For Alfvén waves this change is crucial. According to (2.78), we have ($\omega/\omega_{Bi} \ll 1$):

$$\alpha_{x\mathrm{A}} = -1 - \frac{\omega_{Bi}}{2\omega} \frac{\sin^2\chi}{\cos\chi} \tag{2.99}$$

so that the circular polarization of Alfvén waves is retained only in the relatively small range of χ values given by:

$$\sin^2 \chi \ll \frac{2\omega}{\omega_{Bi}}. \tag{2.100}$$

We now briefly discuss the group velocity and the phase trajectories $\vec{k}(\vec{r})$ of whistler and Alfvén waves.

As is well known, a packet of electromagnetic waves propagates with the group velocity (see below (2.131)):

$$\vec{v}_{\mathrm{g}} = \frac{\partial \omega}{\partial \vec{k}}. \tag{2.101}$$

This expression was analysed in many papers for whistler and Alfvén waves. For example, the group velocity of Alfvén waves with the dispersion relation (2.97) is almost parallel to the magnetic field in the limit $\omega/\omega_{Bi} \ll 1$. Analysing the group velocity for whistler-mode waves by using the dispersion relation (2.96), we obtain Storey's theorem (Storey, 1953) that the angle between $\vec{v}_{\mathrm{g}}$ and $\vec{B}$ is less than $19°29'$.

Under magnetospheric conditions, the geometrical optics approximation can be used, as a rule, when the inequality is fulfilled:

$$ka \ll 1, \tag{2.102}$$

where k is the wave vector and a is the characteristic scale of the magnetic field and plasma density inhomogeneity. In this approach the energy of electromagnetic waves propagates along group trajectories, which are determined by the following system of equations (Budden, 1964)

$$\frac{d\vec{r}}{dt} = \frac{\partial \omega}{\partial \vec{k}} \equiv \vec{v}_{\mathrm{g}} \tag{2.103}$$

$$\frac{d\vec{k}}{dt} = -\frac{\partial \omega}{\partial \vec{r}} \tag{2.104}$$

where $\omega(\vec{k}, \vec{r})$ is the dispersion relation, including the medium inhomogeneity as a parameter. The propagation of whistler and Alfvén waves in the frame of (2.103)–(2.104) was analysed in numerous papers (see, for example, Smith *et al.*, 1960; Helliwell, 1965; Kimura, 1966; Inan and Bell, 1977; Jiricek and Triska, 1982; Titova *et al.*, 1985; Maltseva and Molchanov, 1987).

Figure 2.6, taken from Lyons *et al.* (1972), shows typical ray paths for unducted whistler-mode waves in the real magnetosphere. It is important for CM operation to have a wave packet's multiple crossing of the equatorial plane with the $\vec{k}$ direction

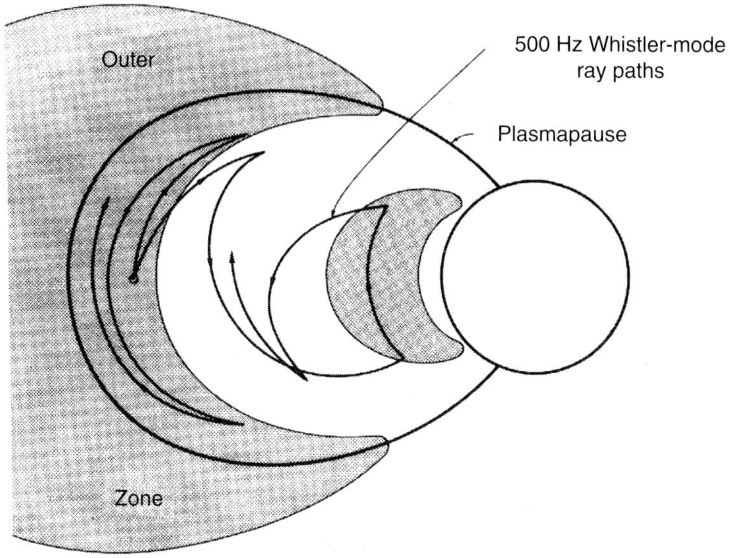

Figure 2.6 Examples of 500 Hz whistler-mode ray paths obtained using a Stanford ray-tracing program (Kimura, 1966), which illustrate how wave energy generated at a point on the magnetic equator in the outer region of the plasmasphere can propagate across field lines so as to fill the plasmasphere with waves (from Lyons *et al.*, 1972). (Copyright American Geophysical Union, reproduced with permission.)

along the magnetic field $\vec{B}_L$ and the same ray path. Such conditions occur for magnetic field-aligned inhomogeneities of a background plasma. For magnetospheric conditions these are at the plasmapause and in ducts (Smith, 1961; Helliwell, 1965; Angerami, 1970; Walker, 1971; Laird and Nunn, 1975; Inan and Bell, 1977; Ondoh, 1976; Semenova and Trakhtengerts, 1980; Karpman and Kayfman, 1982; Lemaire and Gringauz, 1998; Kondratyev *et al.*, 1999; Pasmanik and Trakhtengerts, 2001).

Some examples of phase trajectories with $\vec{k}$ close to the direction of $\vec{B}$ are shown in Fig. 2.7. The quality of a magnetospheric resonator for the cyclotron (whistler and Alfvén mode) waves is determined by partial reflection of these waves from the conjugate ionospheres. Analysis of this for whistler-mode waves can be found in the books by Alpert *et al.* (1967) and Tverskoy (1968) and in papers by Tsuruda (1973) and Bespalov *et al.* (2003). The reflection coefficient R for whistler-mode waves varies smoothly with the frequency ω. The important feature of the reflection coefficient for Alfvén waves is its non-monotonic dependence on ω. An example of such a dependence $R(\omega)$ in the frequency range corresponding to the proton cyclotron instability is shown in Fig. 2.8. The matter is important because the ionospheric F-layer serves as a resonator (Polyakov and Rapoport, 1981; Belyaev *et al.*, 1990) in this frequency range, with a scale along the magnetic field which is comparable with the Alfvén mode wavelength: $kl \sim 1$. As we shall see in Chapter 9, new generation regimes in the proton cyclotron maser are connected with this feature.

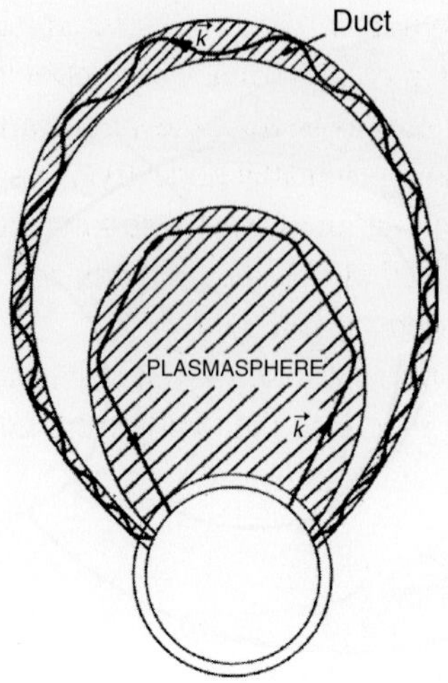

Figure 2.7 Diagram illustrating two examples of phase trajectories (ray paths) with almost field-aligned directions of the wave vector: (a) in a duct of enhanced plasma density; (b) bounded by the plasmapause.

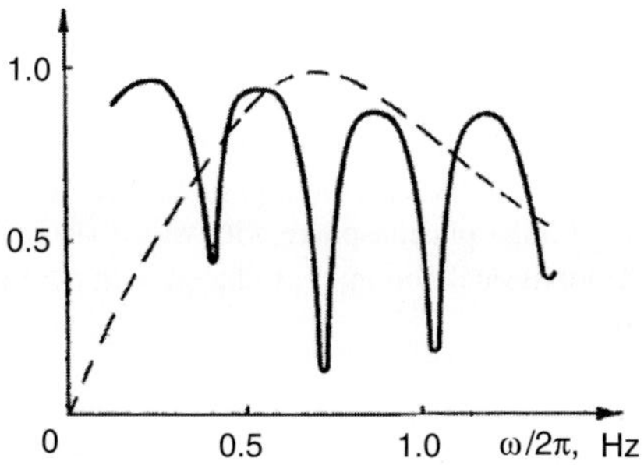

Figure 2.8 Diagram illustrating examples of the frequency dependence of the modulus of the reflection coefficient R from the ionosphere (solid line) and the amplification Γ (dashed line) for Alfvén waves.

2.6 General equations of CM theory

As has already been mentioned, the theory of a CM is based on the system of self-consistent equations, which include the kinetic equation for the distribution function of charged particles and the field equations for the electromagnetic waves. Further, we make some realistic simplifications.

First, we suppose that the ratio of the hot plasma density n_h to cold plasma density n_c is much less than unity:

$$n_\mathrm{h}/n_\mathrm{c} \ll 1. \tag{2.105}$$

Under this condition, the growth rate γ of cyclotron waves is much less than the wave frequency ω:

$$\gamma/\omega \ll 1. \tag{2.106}$$

Then we can consider the wave eigenmodes in a CM as wave packets with slowly changing amplitudes and phases.

Secondly, we neglect nonlinear wave–wave interactions and nonlinear changes of the dispersion relation in a cold plasma. These effects can be important in space plasmas, but their consideration is beyond the scope of this book.

Thirdly, only electromagnetic whistler and Alfvén waves with wave vectors $\vec{k} \parallel \vec{B}_0$ will be considered as eigenmodes in a CM. These waves are really the most important for CM operation. Some qualitatively new effects can be connected with oblique waves; these will be discussed separately.

We now discuss the equation of motion for charged particles – electrons and ions – moving in a homogeneous magnetic field and in the field of a circularly polarized wave. This is written as:

$$\frac{d\vec{v}}{dt} = \frac{e_\alpha}{m_\alpha}\left[\vec{v} \wedge \vec{B}\right] + \frac{e_\alpha}{m_\alpha}\vec{E} \tag{2.107}$$

where m_α and e_α are the particle mass and charge respectively, $\vec{B} = \vec{B}_0 + \vec{B}_\sim$, where $\vec{B}_\sim$ is the wave magnetic field ($\vec{B}_\sim = [\vec{N} \wedge \vec{E}_\sim]$), the ambient electric field in a plasma $\vec{E}_0 = 0$, and the refractive index $\vec{N} = \vec{k}c/\omega$. In the three Cartesian directions we have, for electrons ($e_\alpha = -e$, $e > 0$):

$$\frac{dv_x}{dt} = -v_y\omega_B - \frac{e}{m}(1 - N_\mathrm{w}\beta_\parallel)\,\mathrm{Re}\,E_x \tag{2.108}$$

$$\frac{dv_y}{dt} = v_x\omega_B - \frac{e}{m}(1 - N_\mathrm{w}\beta_\parallel)\,\mathrm{Re}\,E_y \tag{2.109}$$

$$\frac{dv_\parallel}{dt} = -\frac{e}{cm}\left(\left[\vec{v} \wedge \vec{B}_\sim\right]\right)_\parallel = -\frac{eN_\mathrm{w}}{mc}(v_x\,\mathrm{Re}\,E_x + v_y\,\mathrm{Re}\,E_y) \tag{2.110}$$

where $\vec{E}(E_x, E_y)$ is the electric field, introduced by the relations (2.52) and (2.60), $\omega_B = eB/mc$ is the gyrofrequency of energetic electrons, and $\beta_\parallel = v_\parallel/c$.

According to Sections 2.4 and 2.5, the electric field $\vec{E}_\sim$ of the quasi-monochromatic wave $\vec{E}(z, t)$ is written as:

$$\vec{E}_\sim = A\vec{a}\exp\{i\Theta\}, \quad \Theta = \omega t - kz + \Theta_0. \tag{2.111}$$

The wave phase Θ includes, in general, nonlinear terms due to interactions with the ensemble of energetic electrons. The polarization vector $\vec{a}(i\alpha_x, 1, i\alpha_z)$ has the following components for the whistler wave (see (2.68), (2.86) and (2.93)):

$$\vec{a} = a_y(i\vec{x}_0 + \vec{y}_0), \quad a_y = 1/\sqrt{2} \tag{2.112}$$

where $\vec{x}_0$ and $\vec{y}_0$ are unit vectors. We put

$$\vec{v}_\perp = v_x\vec{x}_0 + v_y\vec{y}_0 = \mathrm{Re}\,v_\perp(-i\vec{x}_0 + \vec{y}_0)\exp(-i\varphi) \tag{2.113}$$

and $\psi = \Theta - \varphi$. Substituting (2.113) into the system (2.108)–(2.110), we obtain:

$$\frac{dv_\perp}{dt} = -\frac{e}{m}(1 - N_w\beta_\parallel)\,\mathrm{Re}\,(A_y \exp i\psi) \tag{2.114}$$

$$\frac{d\psi}{dt} = -\omega_B - kv_\parallel + \omega - \frac{e}{mv_\perp}(1 - N_w\beta_\parallel)\,\mathrm{Im}\,(A_y \exp i\psi) \tag{2.115}$$

where $A_y = Aa_y$, $a_y = 1/\sqrt{2}$, A is the electric field amplitude.

In an inhomogeneous magnetic field, ω_B and N_ω depend on z, and there are additional terms to the right-hand sides of Equations (2.116)–(2.117) according to relations (2.12)–(2.13). Therefore:

$$\frac{dv_\perp}{dt} = \frac{v_\perp v_\parallel}{2\omega_B}\frac{d\omega_B}{dz} - \frac{e}{m}(1 - N_w\beta_\parallel)\,\mathrm{Re}\,(A_y \exp i\psi) \tag{2.116}$$

$$\frac{dv_\parallel}{dt} = -\frac{1}{2}J_\perp\frac{d\omega_B}{dz} - \frac{eN_w}{mc}v_\perp(\mathrm{Re}\,A_y \exp i\psi) \tag{2.117}$$

$$\frac{d\psi}{dt} = -\omega_B(z) - k_w(z)\,v_\parallel + \omega - \frac{e}{mv_\perp}(1 - N_w\beta_\parallel)(\mathrm{Im}\,A_y \exp i\psi). \tag{2.118}$$

We now introduce new variables into Equations (2.116)–(2.118), the electron's kinetic energy and first adiabatic invariant

$$W = m\,\frac{v_\perp^2 + v_\parallel^2}{2}, \quad J_\perp = \frac{mv_\perp^2}{2B} \tag{2.119}$$

and a new time variable, according to the relation:

$$dt = \frac{dz}{(2/m)^{1/2}(W - J_\perp B)^{1/2}}, \quad t - t_0 = \int_{z_0}^{z} \frac{d\xi}{(2/m)^{1/2}(W - J_\perp B)^{1/2}}. \tag{2.120}$$

The 'new time' is appropriate because ω_B and k_w depend on z, and it is straightforward to write the relation $t_0(t, z, z_0)$ with the help of (2.120) in order to solve the kinetic equation. Substituting (2.119) and (2.120) into (2.116)–(2.118), we obtain:

$$\frac{dW}{dz} = -e\left(\frac{J_\perp B}{W - J_\perp B}\right)^{1/2}\mathrm{Re}\,(A_y \exp i\psi) \tag{2.121}$$

$$B\frac{dJ_\perp}{dz} = -e\left(\frac{J_\perp B}{W - J_\perp B}\right)^{1/2}(1 - N_w\beta_\parallel)\,\mathrm{Re}\,(A_y \exp i\psi) \tag{2.122}$$

$$\frac{d\psi}{dz} = \left(\omega - \omega_B(z) - k_w(z)\,v_\parallel(z)\right)v_\parallel^{-1} - \frac{e(1 - N_w\beta_\parallel)}{mv_\perp v_\parallel}\,\mathrm{Im}\,(A_y \exp i\psi). \tag{2.123}$$

These equations are still valid if the wave amplitude A and the frequency ω are changing slowly in space and time. If there are several waves, we have to change $A\exp\{i\Theta\} \to \sum_k A_k \exp\{i\Theta_k\}$.

Next we consider the wave equation. The excitation of electromagnetic waves in a CM by a current of energetic particles is analysed using Equation (2.88). Taking into account the inequality (2.105), we apply an iterative procedure to solve (2.88). In the zeroth approximation for the parameter $\delta = n_{\rm h}/n_{\rm c}$ we have the dispersion relation for whistler waves:

$$\mathcal{D}_l^{(0)}(\omega, \vec{k}) = 0, \quad k^2c^2/\omega^2 = a_\alpha\, \varepsilon_{\alpha\beta}\, a_\beta^* \equiv N_{\rm w}^2(\omega, \vec{a}\,). \tag{2.124}$$

In the first approximation we expand $\mathcal{D}_l(\omega k)$ near a solution to (2.124):

$$\mathcal{D} = \mathcal{D}_l^0 + \frac{\partial \mathcal{D}_l^0}{\partial \omega}(\omega - \omega_{\rm w}) + \frac{\partial \mathcal{D}_l^0}{\partial \vec{k}}\left(\vec{k} - \vec{k}_{\rm w}\right) \tag{2.125}$$

where $\omega_{\rm w}$ and $\vec{k}_{\rm w}$ satisfy (2.124). With help of (2.124) and (2.125), Equation (2.88) can be rewritten as

$$\left[\frac{\partial \mathcal{D}_l^0}{\partial \omega}(\omega - \omega_{\rm w}) + \frac{\partial \mathcal{D}_l^0}{\partial \vec{k}}(\vec{k} - \vec{k}_{\rm w})\right] E_l(\omega, \vec{k}\,) = -i\,\frac{4\pi\omega}{c^2}\left\{\vec{j}_{\rm h}(\omega, \vec{k}\,)\,\vec{a}_l^*\right\}. \tag{2.126}$$

The physical meaning of (2.126) is that a monochromatic wave due to $\vec{j}_{\rm h}$, the current of hot particles, is an evolving wave packet with a slowly changing amplitude. This evolution can be determined, keeping the definitions (2.111), by the relation:

$$\begin{aligned} &\int E_l(\omega, \vec{k}) \exp i(\omega t - \vec{k}\vec{r} - \omega_{\rm w} t + \vec{k}_{\rm w}\vec{r}\,)\, d\vec{k}\, d\omega \\ &\quad = A \exp i(\Theta - \Theta_{\rm w}) \equiv \mathcal{A}(\vec{r}, t) \end{aligned} \tag{2.127}$$

where A and $\Delta\Theta = \Theta - \Theta_{\rm w}$ are real functions, which change slowly with $\vec{r}$ and t, $\Theta_{\rm w} = \omega_{\rm w} t - \vec{k}_{\rm w}\cdot\vec{r}$. Now we write (2.126), taking (2.127) into account, in the form:

$$\begin{aligned} &\int d\omega\, d\vec{k} \exp(i\Delta\omega t - i\Delta\vec{k}\vec{r}\,)\left\{\frac{\partial \mathcal{D}}{\partial \omega}\Delta\omega + \frac{\partial \mathcal{D}}{\partial \vec{k}}\Delta\vec{k}\right\} E_l(\omega, \vec{k}) \\ &\quad = -\frac{4\pi i}{c^2}\int \omega_{\rm w}\, d\omega\, d\vec{k}\{\vec{j}_{\rm h}(\omega, \vec{k}\,)\,\vec{a}_l^*\} \exp(i\Delta\omega t - i\Delta\vec{k}\vec{r}\,) \end{aligned} \tag{2.128}$$

where $\Delta\omega = \omega - \omega_{\rm w}$, $\Delta\vec{k} = \vec{k} - \vec{k}_{\rm w}$. The integral on the right-hand side of (2.128) is transformed as:

$$\begin{aligned} &\int d\omega\, d\vec{k}\{\vec{j}_{\rm h}(\vec{k}, \omega)\,\vec{a}_l^*\} \exp(i\omega t - i\vec{k}\vec{r} - i\Theta_{\rm w}) = \{\vec{j}_{\rm h}(\vec{r}, t)\,\vec{a}^*\} \exp(-i\Theta_{\rm w}) \\ &\quad = -ea_y \int \mathcal{F} v_\perp\, d^3\vec{v} \exp\{-i\Theta_{\rm w} + i\varphi\} \end{aligned} \tag{2.129}$$

where we have used the relation:

$$(\vec{v}_\perp \vec{a}^*) = \frac{v_\perp}{2}\left[(\vec{y}_0 - i\vec{x}_0)e^{-i\varphi} + (\vec{y}_0 + i\vec{x}_0)e^{i\varphi}\right](\vec{y}_0 - i\vec{x}_0)a_y$$
$$= a_y v_\perp \exp\{i(\varphi - \Theta_\mathrm{w})\}. \tag{2.130}$$

The left-hand side of (2.128) can be written as

$$-i\left(\frac{\partial \mathcal{D}_l^{(0)}}{\partial \omega}\right)\left(\frac{\partial}{\partial t} + \vec{v}_\mathrm{g}\frac{\partial}{\partial \vec{r}}\right) A\exp(i\Delta\Theta) \equiv -i\left(\frac{\partial \mathcal{D}_l^{(0)}}{\partial \omega}\right)\left(\frac{\partial \mathcal{A}}{\partial t} + \vec{v}_\mathrm{g}\frac{\partial \mathcal{A}}{\partial \vec{r}}\right) \tag{2.131}$$

where $\mathcal{D}_l^{(0)} = k^2 - \dfrac{\omega^2}{c^2} N_\mathrm{w}^2$ for whistler waves, the group velocity $\vec{v}_\mathrm{g} = -\,(\partial \mathcal{D}_l^{(0)}/\partial k)/(\partial \mathcal{D}_l^{(0)}/\partial \omega)$, $\Delta\Theta = \Theta - \Theta_\mathrm{w}$, and $\mathcal{A}$ is determined by the relation (2.127).

We thus have the equation for the wave electric field complex amplitude $\mathcal{A}$ in the final form:

$$\frac{\partial \mathcal{A}}{\partial t} + \vec{v}_\mathrm{g}\frac{\partial \mathcal{A}}{\partial \vec{r}} = \frac{4\pi e\omega a_y}{(\partial \omega^2 N_\mathrm{w}^2/\partial \omega)\, c^2}\int d^3\vec{v}\,\mathcal{F} v_\perp \exp(i\varphi - i\Theta_\mathrm{w}). \tag{2.132}$$

For real amplitude A and phase $\Delta\Theta = \Theta - \Theta_\mathrm{w}$ we have:

$$\frac{\partial A}{\partial t} + \vec{v}_\mathrm{g}\frac{\partial A}{\partial \vec{r}} = \frac{4\pi e\omega a_y}{(\partial \omega^2 N_\mathrm{w}^2/\partial \omega)\, c^2}\int d^3\vec{v}\,\mathcal{F} v_\perp \cos\psi \tag{2.133}$$

$$A\left(\frac{\partial \Delta\Theta}{\partial t} + \vec{v}_\mathrm{g}\frac{\partial \Delta\Theta}{\partial \vec{r}}\right) = \frac{4\pi e\omega a_y}{(\partial \omega^2 N_\mathrm{w}^2/\partial \omega)\, c^2}\int d^3\vec{v}\,\mathcal{F} v_\perp \sin\psi. \tag{2.134}$$

For ions with charge number Z, it is necessary to change e to Ze in Equations (2.132)–(2.134) and to take Alfvén modes as the corresponding wave eigenmodes.

2.7 Summary

The equations of motion (2.116)–(2.118), the kinetic equation (2.34) in the case of a homogeneous plasma and (2.36) and (2.121)–(2.123) for an inhomogeneous medium, together with the wave equations (2.132)–(2.134), are the complete self-consistent set of equations describing wave–particle interactions in a CM. This system of equations is what is initially required to construct CM theory.

Chapter 3

Linear theory of the cyclotron instability (CI)

In this chapter, we consider the linear theory of the cyclotron instability. As applied to the Earth's radiation belts this question was considered in many papers (see Trakhtengerts, 1963; Kennel and Petschek, 1966; Tverskoy, 1968; Ashour-Abdalla, 1972; Cuperman and Landau, 1974; Cuperman and Salu, 1974; Huang *et al.*, 1983; Church and Thorne, 1983; Gary, 1993; Sazhin, 1993).

The topics of our interest will be

- the temporal growth rate γ due to the cyclotron instability (CI) for three different velocity distributions:
  - (i) a smooth function, representing the energetic charged particles observed in the magnetosphere,
  - (ii) a delta-function, representing a mono-energetic beam as occurs in a laboratory plasma machine, and
  - (iii) a step between energetic charged particles which are in cyclotron resonance with waves and those which have insufficient energy ever to be in resonance for electrons; these waves are whistler-mode waves whereas, for ions, they are Alfvén (magnetohydrodynamic) waves;
- the crucial role of the spatially inhomogeneous (e.g. dipolar) magnetic field along a particular geomagnetic flux tube.

These factors determine the cyclotron amplification the whistler and Alfvén waves in the magnetosphere.

3.1 The dispersion equation for whistler and Alfvén waves, taking the hot plasma fraction into account

The linear theory of a CM considers the necessary conditions for switching on the cyclotron instability (CI), the growth rate and the frequency spectrum of unstable waves at the initial stage of the instability. This theory is based upon an analysis of the dispersion equation, obtained from (2.133)–(2.134) by substituting the electric current of energetic electrons in a linear expansion of the whistler wave amplitude. Following this approach, we write the solution of the system (2.116)–(2.118) in the following form:

$$\psi = (\omega - \omega_B - kv_{\|})t + \Theta_0 - \varphi_0, \tag{3.1}$$

$$v_{\|} = v_{\|0} - \frac{eN_{\mathrm{w}}v_{\perp}}{mc\Delta}\,\mathrm{Im}\,A_y \exp(i\psi) = v_{\|0} - a, \tag{3.2}$$

$$v_{\perp} = v_{\perp 0} + \frac{1 - N_{\mathrm{w}}\beta_{\|}}{N_{\mathrm{w}}\beta_{\perp}}a = v_{\perp 0} - b, \quad \Delta = \omega - \omega_B - kv_{\|}. \tag{3.3}$$

The solution for the distribution function can be written (see 2.42) as:

$$\mathcal{F}(\vec{v}, \vec{r}, t) = \mathcal{F}_0(v_{\|} + a, v_{\perp} + b) \tag{3.4}$$

where we have supposed that the initial function $\mathcal{F}_0$ does not depend on φ. Using the inequalities:

$$|a| \ll v_{\|}, \quad |b| \ll v_{\perp}, \tag{3.5}$$

we can put $\mathcal{F}_0$ (3.4) in the form of a Taylor series expansion

$$\mathcal{F} = \mathcal{F}(v_{\|}, v_{\perp}) + a\frac{\partial \mathcal{F}}{\partial v_{\|}} + b\frac{\partial \mathcal{F}}{\partial v_{\perp}} \tag{3.6}$$

where the subscript ‘0’ has been omitted.

Now we use (3.6) to calculate the current on the right-hand side of (2.133), putting (3.2) in the form:

$$a = \frac{eN_{\mathrm{w}}\, v_{\perp}}{mc\Delta\,\sqrt{2}}\,\mathrm{Im}(\mathcal{A}\exp\{-i\varphi + i\Theta_{\mathrm{w}}\}), \tag{3.7}$$

where we put $A_y = Aa_y = A/\sqrt{2}$. Substituting (3.6) together with (3.7) into (2.132) and integrating over φ we obtain:

$$\frac{\partial \mathcal{A}}{\partial t} + \vec{v}_{\mathrm{g}}\frac{\partial \mathcal{A}}{\partial \vec{r}} = -\frac{2\pi^2\omega e^2}{mc}\frac{N_{\mathrm{w}}\,\mathcal{A}\,i}{\partial N_{\mathrm{w}}^2\omega^2/\partial\omega} \times \int_{-\infty}^{\infty} dv_{\|}\int_0^{\infty} v_{\perp}^3\,dv_{\perp} \times \left(\frac{\partial \mathcal{F}}{\partial v_{\|}} + \frac{1 - N_{\mathrm{w}}\,\beta_{\|}}{N_{\mathrm{w}}\,\beta_{\perp}}\frac{\partial \mathcal{F}}{\partial v_{\perp}}\right)(\omega - kv_{\|} - \omega_B)^{-1}. \tag{3.8}$$

In the case of an infinite plasma, $\frac{\partial \mathcal{A}}{\partial \vec{r}} = 0$ and $\frac{\partial \mathcal{A}}{\partial t} = i\,(\omega - \omega_{\mathrm{W}})\,A$; here ω and ω_{W} is the frequency of a whistler wave with (ω) and without (ω_{W}) $\vec{j}_h$ taken into account. Thus we have the following equation:

$$\omega - \omega_{\mathrm{W}} = -\frac{4\pi^2 \omega e^2}{mc} \frac{N_{\mathrm{W}}}{(\partial N_{\mathrm{W}}^2 \omega^2/\partial \omega)} \int_{-\infty}^{\infty} dv_{\|} \int_0^{\infty} v_{\perp}^3\, dv_{\perp} \times \left(\frac{\partial \mathcal{F}}{\partial v_{\|}} + \frac{1 - N_{\mathrm{W}} \beta_{\|}}{N_{\mathrm{W}} \beta_{\perp}} \frac{\partial \mathcal{F}}{\partial v_{\perp}} \right) (\omega - k v_{\|} - \omega_B)^{-1}. \tag{3.9}$$

The dispersion equation in the case of ions i has the same form as (3.9), making the substitution $m \to m_i$, and $N_{\mathrm{W}} \to N_{\mathrm{A}}$. Using (2.92) for N_{W}^2 and (2.95) for N_{A}^2 we obtain:

$$\frac{\partial N_{\mathrm{W}}^2 \omega^2}{\partial \omega} = \frac{\omega_B\, \omega}{\omega_B - \omega} N_{\mathrm{W}}^2 \tag{3.10}$$

and

$$\frac{\partial N_{\mathrm{A}}^2 \omega^2}{\partial \omega} = \frac{(2\omega_{Bi} - \omega)\, \omega}{\omega_{Bi} - \omega} N_{\mathrm{A}}^2. \tag{3.11}$$

3.2 The growth rate for a smooth particle distribution

We consider three examples, which differ significantly from each other. The difference is due to different forms of the distribution functions taken. The first example is an energetic charged particle distribution function with a wide velocity spread. It is typical of space plasma sources of energetic particles. The other extreme case, typical of laboratory devices, is a delta-function in velocity space. The third example relates to a specific step deformation of the distribution function, which appears during the process of wave–particle interactions with smooth particle distributions. This serves to connect noise-like and discrete electromagnetic emissions.

A classical example of a smooth distribution function is a particle distribution with different temperatures along ($\|$) and perpendicular to ($\perp$) the ambient magnetic field, i.e. with a temperature anisotropy:

$$\mathcal{F} = \frac{n_{\mathrm{h}}\, m^{3/2}}{(2\pi)^{3/2} T_{\perp} T_{\|}^{1/2}} \exp\left\{ -\frac{m v_{\perp}^2}{2T_{\perp}} - \frac{m v_{\|}^2}{2T_{\|}} \right\}. \tag{3.12}$$

Thus $\mathcal{F}$ satisfies the normalization condition:

$$\int \mathcal{F} d^3\vec{v} = \int_0^{\pi} d\varphi \int_{-\infty}^{\infty} dv_{\|} \int_0^{\infty} v_{\perp}\, dv_{\perp}\, \mathcal{F} = n_{\mathrm{h}} \tag{3.13}$$

where $n_{\rm h}$ is the hot charged particle number density, and the temperatures $T_\parallel$ and $T_\perp$ are everywhere in this book given in units of energy. For a smoothly varying $\mathcal{F}$, it is possible for the integration of (3.9) over $v_\parallel$ to use the following asymptotic relation:

$$\left.\frac{1}{\omega - k\,v_\parallel - \omega_B + i\nu}\right|_{\nu\to 0} \equiv \left.\frac{1}{\Delta + i\nu}\right|_{\nu\to 0} = \frac{P}{\Delta} + i\pi\,\delta(\Delta) \tag{3.14}$$

where $\Delta = \omega - k\,v_\parallel - \omega_B$, and the symbol P means the principal part of the integral. The expression for the temporal growth rate γ, after substitution of (3.14) into (3.9), has the form (with $\omega = \mathrm{Re}\,\omega - i\gamma$):

$$\gamma = \frac{2\pi^3\,e^2\,\omega\,N_{\rm w}}{mc\,(\partial N_{\rm w}^2\omega^2/\partial\omega)_0}\int_0^\infty\int_{-\infty}^\infty dv_\perp\,v_\perp^3\,dv_\parallel\left(\frac{\partial\mathcal{F}}{\partial v_\parallel} + \frac{\omega_B}{N_{\rm w}\omega\beta_\perp}\frac{\partial\mathcal{F}}{\partial v_\perp}\right)$$
$$\times\ \delta(\omega - k\,v_\parallel - \omega_B). \tag{3.15}$$

Using (3.12), we find that:

$$\frac{\partial\mathcal{F}}{\partial v_\parallel} + \frac{\omega_B}{k\,v_\perp}\frac{\partial\mathcal{F}}{\partial v_\perp} = -m\left(\frac{v_\parallel}{T_\parallel} + \frac{\omega_B}{kT_\perp}\right)\mathcal{F}. \tag{3.16}$$

Substituting (3.10) and (3.16) into (3.15), we obtain:

$$\gamma = \frac{n_{\rm h}}{n_{\rm c}}\frac{\alpha\sqrt{\pi}}{k\,v_{T_\parallel}}\frac{(\omega_B-\omega)^2}{2\omega_B}\left[\omega_B\left(1-\alpha^{-1}\right) - \omega\right]\exp\left\{-\frac{(\omega_B-\omega)^2}{k^2\,v_{T_\parallel}^2}\right\}, \tag{3.17}$$

where $n_{\rm c}$ is the cold plasma density, $\alpha = T_\perp/T_\parallel$ is the temperature anisotropy factor, and $\gamma > 0$ corresponds to the instability. It is seen from (3.17) that the instability ($\gamma > 0$) arises when

$$\omega < \omega_* = \omega_B(1 - \alpha^{-1}). \tag{3.18}$$

The relation (3.17) can be written as:

$$\tilde{\gamma} \equiv \frac{\gamma}{\gamma_0} = \alpha\,\frac{(1-\tilde{\omega})^{5/2}}{\sqrt{\tilde{\omega}\beta_*}}\,(y - \tilde{\omega})\exp\left\{-\frac{(1-\tilde{\omega})^3}{\beta_*\tilde{\omega}}\right\}, \tag{3.19}$$

where $\tilde{\omega} = \omega/\omega_B$, $\beta_* = \dfrac{\omega_{\rm p}^2\,v_{\parallel T}^2}{\omega_B^2\,c^2} = \dfrac{8\pi n_{\rm c} T_\parallel}{B^2}$ (with T in energy units) characterizes the ratio of the kinetic plasma pressure to the magnetic pressure, but with the hot plasma density $n_{\rm h}$ replaced by the cold plasma density $n_{\rm c}$, $y = 1 - \alpha^{-1}$, and

$$\gamma_0 = \frac{\sqrt{\pi}}{2}\frac{n_{\rm h}}{n_{\rm c}}\,\omega_B. \tag{3.20}$$

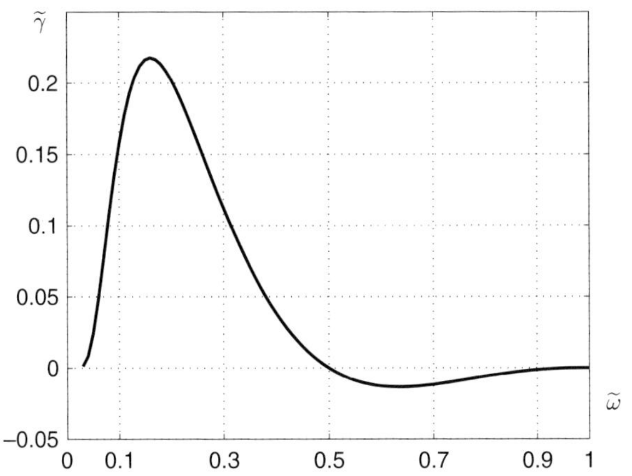

Figure 3.1 Diagram illustrating the dependence of normalized growth rate $\tilde{\gamma} = (\gamma/\gamma_0)$ on dimensionless frequency $\tilde{\omega} = (\omega/\omega_{BL})$, ω_{BL} is the electron gyrofrequency at the equator; the parameters y and β_* are chosen to be $y = 0.5$ and $\beta_* = 4$.

The dependence of $\tilde{\gamma}$ as a function of $\tilde{\omega}$ is shown in Fig. 3.1. There is a maximum value of $\gamma_{\rm m}$, whose position along the $\tilde{\omega}$-axis is determined by two parameters: α (or y) and β_*.

For the case $\tilde{\omega} \equiv \omega/\omega_B \ll 1$ we have from (3.19):

$$\tilde{\gamma} = \alpha x \left(y - \frac{1}{\beta_* x^2} \right) \exp\{-x^2\}. \tag{3.21}$$

The maximum of $\tilde{\gamma}$ is reached in this case for $x = x_{\rm m}$, where

$$x_{\rm m}^2 = \frac{y\beta_* + 2}{4y\beta_*} + \sqrt{\frac{(y\beta_* + 2)^2}{16y^2\beta_*^2} + \frac{1}{2\beta_* y}} \tag{3.22}$$

which corresponds to

$$\tilde{\omega}_{\rm m} = 2y \left[1 + \frac{y\beta_*}{2} + \sqrt{\left(1 + \frac{y\beta_*}{2}\right)^2 + 2y\beta_*} \right]^{-1}. \tag{3.23}$$

The expressions (3.21)–(3.23) are simplified in two limiting cases, when $\beta_* y \gg 2$ or $\beta_* y \ll 2$. In the case of a dense cold plasma, $\beta_* y \gg 2$, we have:

$$\tilde{\omega}_{\rm m} = \frac{2}{\beta_*} \quad \text{and} \quad \gamma_{\rm m} \approx 0.43(\alpha - 1)\gamma_0. \tag{3.24}$$

In the opposite case ($y\beta_* \ll 2$) the frequency $\omega_{\rm m}$ is close to the maximum frequency ω_* and $\gamma_{\rm m}$ is exponentially small:

$$\tilde{\omega}_{\rm m} = \frac{y}{1 + \beta_* y} \quad \text{and} \quad \gamma_{\rm m} = \gamma_0\, \alpha \beta_*^{1/2} y^{3/2} \exp\left\{ -\frac{1}{\beta_*^2 y^2} \right\}. \tag{3.25}$$

The formulae (3.24) and (3.25) both illustrate the very important dependence of the CI on the cold plasma density n_c. In the case of a smooth distribution function the cyclotron instability is effective only in a rather dense cold plasma.

3.3 The distribution function with step and delta-function features. Hydrodynamic stage of the instability

Now we consider a qualitatively different situation, when the distribution function of energetic particles has a small velocity dispersion and can be approximated by a delta-function:

$$\mathcal{F} = \frac{n_h}{2\pi v_{\perp 0}}\,\delta(v_\perp - v_{\perp 0})\,\delta(v_\| - v_{\|0}). \tag{3.26}$$

Such a well-organized helical beam is used in man-made microwave sources, which find many uses and applications nowadays in different electronic devices.

In the Earth's magnetosphere such beams can arise from the interaction of quasi-monochromatic electromagnetic waves from a ground-based VLF transmitter with radiation belt electrons which give rise to so-called triggered emissions (see Chapter 6). Sometimes well-organized electron beams are also observed in auroral regions. The specific feature of space CMs is the presence of a dense cold plasma, which changes the excitation conditions in the CM. To obtain the growth rate in this case, we substitute the function (3.26) into the general dispersion equation (3.9). As a result we have:

$$\omega - \omega_{\mathrm{w}} = -\frac{\omega_{\mathrm{ph}}^2}{2}\left(1 - \frac{\omega}{\omega_B}\right)\left[\frac{\beta_{\perp 0}^2\omega}{2\,(\omega_B - \omega + kv_{\|0})^2} - \frac{(\omega_B/\omega)}{N_\omega^2\,(\omega_B - \omega + kv_{\|0})}\right] \tag{3.27}$$

where the hot plasma frequency is given by $\omega_{\mathrm{ph}}^2 = \dfrac{4\pi e^2 n_{\mathrm{h}}}{m_e}$, $\beta_{\perp 0} = v_{\perp 0}/c$, and $\omega_{\mathrm{w}}(k)$ is the solution of the whistler dispersion equation (2.49). Equation (3.27) is not applicable when $\beta_{\perp 0} \to 0$; in this case it is necessary to take into account the velocity dispersion across the magnetic field.

Defining

$$\Omega = \omega - kv_\| - \omega_B, \quad \Delta_{\mathrm{w}} = \omega_{\mathrm{w}} - kv_\| - \omega_B, \quad \nu = \frac{\omega_{\mathrm{ph}}^2}{4}\left(1 - \frac{\omega}{\omega_B}\right), \tag{3.28}$$

(3.27) becomes

$$\Omega^3 - \Delta_{\rm w}\Omega^2 + \frac{2\nu\,\omega_B}{N_\omega^2\,\omega}\,\Omega + \nu\beta_{\perp 0}^2\omega = 0. \tag{3.29}$$

The maximum growth rate occurs when $\Delta_{\rm w}$ in Equation (3.29) is zero, i.e. exactly at the Doppler shifted cyclotron resonance condition.

For a weak beam, when the following inequality is satisfied,

$$\frac{n_{\rm h}}{n_{\rm c}} \ll \frac{27}{16}\,(N_{\rm w}\,\beta_{\perp 0})^4\left(1-\frac{\omega}{\omega_B}\right)^{-2}, \tag{3.30}$$

the solution of (3.29) has the simple form:

$$\Omega_1 = P^{1/3}, \quad \Omega_{2,3} = -\frac{P^{1/3}}{2}\left(1 \pm i\sqrt{3}\right), \tag{3.31}$$

$$P = \frac{\omega_{\rm ph}^2}{2}\,\beta_{\perp 0}^2\left(1-\frac{\omega}{\omega_B}\right)\omega. \tag{3.32}$$

One solution (Ω_2) corresponds to instability. It is important to note that, in the case of a well-organized beam (3.26), there are three waves (3.31) in comparison with just one wave for the case of a smooth distribution function. As we shall see later, the presence of more than one wave in the system is a necessary condition for the absolute instability in space-limited systems. Besides, the growth rate (3.31) is proportional to $(n_{\rm h}/n_{\rm c})^{1/3}$:

$$\gamma_\delta = \frac{\sqrt{3}}{2}\left(\frac{n_{\rm h}}{n_{\rm c}}\right)^{1/3}\left[N_{\rm w}\,\beta_{\perp 0}\,\frac{\omega}{\omega_B}\left(1-\frac{\omega}{\omega_B}\right)\right]^{2/3}\omega_B. \tag{3.33}$$

This is the so-called hydrodynamical stage of the instability when all beam particles interact with one monochromatic wave; the subscript δ in (3.33) means that $\mathcal{F}$ is a delta-function.

The mono-energetic beam in the form (3.26) may be a rather rare phenomenon in space plasmas. Now we shall consider the second example of a non-smooth distribution function with a jump-like feature or step. This is now believed to be very important in natural space plasmas. We put the distribution function in the form (for $k > 0$, $v_\parallel < 0$):

$$\mathcal{F} = \mathcal{F}_{\rm sm}(v_\perp, v_\parallel)\,{\rm He}(v_0 + v_\parallel) + \mathcal{F}_{\rm sm1} \tag{3.34}$$

where $\mathcal{F}_{\rm sm}$ and $\mathcal{F}_{\rm sm1}$ are some smooth functions of $v_\perp$ and $v_\parallel$ (for example (3.12)), and ${\rm He}(x)$ is the Heaviside operator, equal to zero if $x < 0$, and unity if $x \geq 1$. Such a distribution function, shown in Fig. 3.2, appears when a noise-like emission interacts with the smooth distribution function. The value v_0 in (3.34) marks the boundary between resonant and non-resonant particles. The resonant beam electrons satisfy

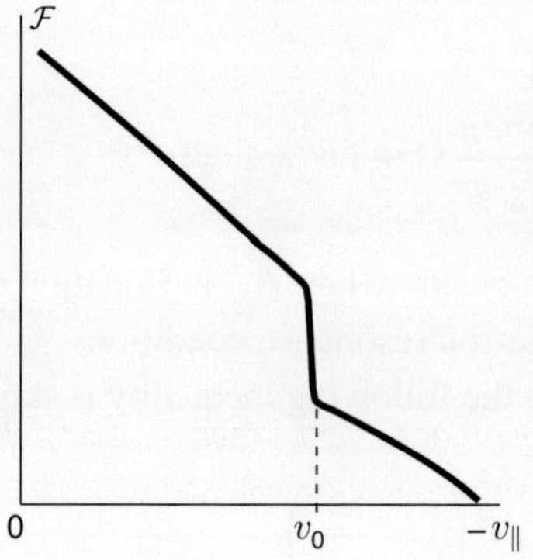

Figure 3.2 The distribution function $\mathcal{F}$ of the radiation belt electrons in the form of a step-like deformation; the velocity of the step v_0 satisfies the cyclotron resonance condition $\omega_{\rm m} = k_{\rm m} v_0 + \omega_{BL}$, where $\omega_{\rm m}$ and $k_{\rm m} > 0$ are the frequency and the wave vector of a whistler wave with the maximum growth rate.

the cyclotron resonance condition

$$\omega_{\rm m} = \omega_B - k_{\rm m}\, v_0, \tag{3.35}$$

where $\omega_{\rm m}$ is the maximum frequency in the spectrum of the noise-like emission. These electrons diffuse into the loss cone, and are lost from the magnetosphere. The distribution function of non-resonant particles conserves its form, and therefore leads to the formation of (3.34). This process will be considered in detail in Chapters 7 and 9.

To obtain the growth rate, we substitute (3.34) into (3.9), as in the previous case (3.26). For simplicity we put $\mathcal{F}_{\rm sm1} = 0$ and $\mathcal{F}_{\rm sm}$ equal to (3.12); we shall not consider the weak instability due to the smooth part of the distribution function $\mathcal{F}_{\rm sm}$ since it is insignificant. In this approximation we have:

$$\frac{\partial \mathcal{F}}{\partial v_\parallel} \approx +\mathcal{F}_{\rm sm}(v_\perp, v_\parallel)\,\delta(v_0 + v_\parallel), \qquad \frac{\partial \mathcal{F}}{\partial v_\perp} \simeq 0 \tag{3.36}$$

where we must account for the fact that $v_\parallel < 0$ (see Fig. 3.2). Substituting (3.36) into (3.9), we obtain:

$$\omega - \omega_{\rm w}(k_{\rm w}) = \frac{4\pi^2 e^2\,(\omega_B - \omega)}{mc\,\omega_B\,N_{\rm w}} \int_0^\infty \frac{\mathcal{F}_{\rm sm}(v_\perp, v_0)\,v_\perp^3\,dv_\perp}{\omega + kv_0 - \omega_B}\,. \tag{3.37}$$

Using (3.12) and the definitions given as (3.28) we find that

$$(\Omega - \Delta_{\rm w})\,\Omega = -\frac{n_{\rm h}}{n_{\rm c}}\,\frac{\omega_B}{2\sqrt{\pi}}\,\frac{\omega_p\,\beta_{\perp T}^2}{\beta_{\parallel T}}\left(1 - \frac{\omega}{\omega_B}\right)^{3/2}\left(\frac{\omega}{\omega_B}\right)^{1/2} \times \exp\left\{-\frac{v_0^2}{v_{\parallel T}^2}\right\} \equiv \frac{n_{\rm h}}{n_{\rm c}}\,\eta(\omega) \tag{3.38}$$

where $\beta_{\perp T} = \left(\frac{2T_\perp}{mc^2}\right)^{1/2}$ and $\beta_{\| T} = \left(\frac{2T_\|}{mc^2}\right)^{1/2}$. The solution of (3.38) is

$$\Omega = -\frac{\Delta_{\mathrm{w}}}{2} \pm \sqrt{\frac{\Delta_{\mathrm{w}}^2}{4} - \frac{n_{\mathrm{h}}}{n_{\mathrm{c}}}\,\eta(\omega)}\,. \tag{3.39}$$

The maximum growth rate is achieved when $\Delta_{\mathrm{w}} = 0$. Thus, we have two wave modes, one stable and one unstable, which are determined by the following relations:

$$\Omega_{1,2} = \omega_{1,2} + kv_0 - \omega_B = \omega_{1,2} - \omega_{\mathrm{w}}(k) = \pm i\left(\frac{n_{\mathrm{h}}}{n_{\mathrm{c}}}\right)^{1/2} \eta^{1/2}(\omega). \tag{3.40}$$

The growth rate in this case with a step in the distribution function is equal to:

$$\begin{aligned} \gamma_{\mathrm{st}} &= \left(\frac{n_{\mathrm{h}}}{n_{\mathrm{c}}}\right)^{1/2} \eta^{1/2}(\omega), \\ \eta &= \frac{\omega_B}{2\sqrt{\pi}} \frac{\omega_p \beta_{\perp T}^2}{\beta_{\| T}} \left(1 - \frac{\omega}{\omega_B}\right)^{3/2} \left(\frac{\omega}{\omega_B}\right)^{1/2} \exp\left\{-\frac{v_0^2}{v_{T\|}}\right\}, \end{aligned} \tag{3.41}$$

together with $\omega - \omega_B = kv_{\mathrm{st}\,\|} \equiv -kv_0$.

We summarize the results of Sections 3.2 and 3.3 in Table 3.1.

Thus, as the organization of the distribution function grows from a smooth distribution through a step-like deformation into a delta-function, qualitative changes take place in the number of wave modes that exist and in the dependence on the small parameter $n_{\mathrm{h}}/n_{\mathrm{c}}$. As will be shown later, this also leads to qualitatively new effects in CM generation.

Problem

Obtain the dispersion relation for the distribution function (1) with a step deformation (3.34), and (2) with a delta-function (3.26), without using the Taylor series expansion (3.1).

Table 3.1

Type of distribution function	Number of wave modes	Growth rate
Smooth	1	$\sim n_{\mathrm{h}}/n_{\mathrm{c}}$
Step (jump-like) deformation	2	$\sim (n_{\mathrm{h}}/n_{\mathrm{c}})^{1/2}$
Delta-function	3	$\sim (n_{\mathrm{h}}/n_{\mathrm{c}})^{1/3}$

Solution

(1) The current derived from Equation (2.132) can be written as:

$$e \int v_\perp^2 \, dv_\perp \, dv_\parallel \, d\varphi \, \exp\{i\,(\varphi - \Theta_\mathrm{w})\}$$
$$\times\, \mathcal{F}_0(v_\parallel + a,\; v_\perp + b)\,\mathrm{He}(v_0 + a + v_\parallel), \tag{3.42}$$

where

$$a = \frac{eN_\mathrm{w}\, v_\perp}{mc\,\Delta\,\sqrt{2}}\,\mathrm{Im}\,A_0\, e^{-i\varphi + i\Theta} = \frac{eN_\mathrm{w} v_\perp}{mc\,\Delta\,\sqrt{2}}\,\mathrm{Im}\,A\, e^{-i\varphi + i\Theta_\mathrm{w}},$$
$$\tag{3.43}$$
$$b = \frac{1 - N_\mathrm{w}\,\beta_\parallel}{N_\mathrm{w}\,\beta_\parallel}\,a, \qquad \Delta = \omega + kv_0 - \omega_B.$$

We have used the expression (3.34) ($\mathcal{F}_1 = 0$) and the formulae (3.3)–(3.5), and have taken into account that $a_y = 1/\sqrt{2}$. As in (3.36) we take into account only the jump-like feature, so that (3.42) can be rewritten in the form:

$$e \int_0^\infty v_\perp^2 \, dv_\perp \int_0^{2\pi} d\varphi \, e^{i\varphi - i\Theta_\mathrm{w}} \int_{-a-v_0}^\infty dv_\parallel \, \mathcal{F}_0(v_\parallel, v_\perp). \tag{3.44}$$

Performing the integration in (3.44) over φ by parts, we find (to first order in A)

$$ie \int_0^\infty v_\perp^2 \, dv_\perp \int_0^{2\pi} e^{i\varphi - i\Theta_\mathrm{w}} \left(\frac{\partial a}{\partial \varphi}\right)_{v_0} \mathcal{F}_0(v_0, v_\perp)\, d\varphi$$
$$= -\frac{ie^2\, N_\mathrm{w}\, A\,\pi}{\sqrt{2}\,mc\,\Delta} \int_0^\infty \mathcal{F}_0(v_0, v_\perp)\, v_\perp^3 \, dv_\perp. \tag{3.45}$$

The substitution of (3.45) into (2.132) gives the dispersion equation as (3.37).

(2) In the case of (3.26) the current in Equation (2.132) can be written as ($v_0 > 0$)

$$\frac{en_\mathrm{h}}{2\pi} \int_0^{2\pi} d\varphi \int_0^\infty \frac{v_\perp^2 \, dv_\perp}{v_{\perp 0}}$$
$$\times \int_{-\infty}^\infty dv_\parallel \, e^{i\varphi - i\Theta_\mathrm{w}}\, \delta(v_\parallel + a + v_0)\, \delta(v_\perp + b - v_{\perp 0}). \tag{3.46}$$

After integration of (3.46) over $v_\parallel$ and $v_\perp$, we have to first order in a and b:

$$\frac{e\,n_{\rm h}}{2\pi\,v_{\perp 0}}\int_0^{2\pi} d\varphi\, e^{i\varphi - i\Theta_{\rm w}}(v_{\perp 0}-b)^2\left[1+\left(\frac{\partial a}{\partial v_\parallel}\right)_{v_0}\right]^{-1}$$

$$\approx -\frac{n_{\rm h}\,e}{2\pi}\int_0^{2\pi} e^{i\varphi - i\Theta_{\rm w}}\left[2b + v_{\perp 0}\left(\frac{\partial a}{\partial v_\parallel}\right)_{v_0}\right] d\varphi. \tag{3.47}$$

Substituting (3.42) into (3.47) and integrating over φ, we obtain from (2.132) the dispersion equation (3.27).

3.4 CI for an inhomogeneous magnetic field. Wave amplification

In a non-uniform magnetic field we use the equations of motion of a charged particle in the form (2.124)–(2.126). In the linear approximation we have:

$$\psi^{(0)} \cong (\Theta-\varphi)^{(0)} = \int_{z_0}^{z}\frac{\omega-\omega_B-kv_\parallel}{v_\parallel}\,dz' + \omega t_0 - \varphi_0, \tag{3.48}$$

$$W = W_0 - e\int_{z_0}^{z} dz'\left(\frac{J_\perp B}{W-J_\perp B}\right)^{1/2}\mathrm{Re}\left(\frac{A\,e^{i\psi^{(0)}(z')}}{\sqrt{2}}\right), \tag{3.49}$$

$$J_\perp = J_{\perp 0} - e\int_{z_0}^{z}\frac{dz'}{B}\left(\frac{J_\perp B}{W-J_\perp B}\right)^{1/2}(1-N_{\rm w}\beta_\parallel)\,\mathrm{Re}\left(\frac{A\,e^{i\psi^{(0)}(z')}}{\sqrt{2}}\right), \tag{3.50}$$

where

$$v_\parallel = \pm\sqrt{\frac{2}{m}\,(W-J_\perp B)}, \quad t_0 = t - \int_{z_0}^{z}\frac{dz'}{v_\parallel(z')}, \quad \varphi = \varphi_0 + \int_{z_0}^{z}\frac{\omega_B}{v_\parallel}\,dz'. \tag{3.51}$$

The lower limit in the integrals (3.49)–(3.50) can be chosen as $z = \mp\ell$ for particles moving along the $\pm z$-direction (see Fig. 3.3); here the coordinates $z = \pm\ell$ correspond to the ends of the magnetic trap.

For a dipolar representation of the geomagnetic field in the magnetosphere, the trap is formed between two points (in the northern and southern hemispheres) with the same value of B. The cyclotron interaction in a non-uniform magnetic field takes place at two points, as shown in Fig. 3.3. Here the closed line is the trajectory of a particle in the absence of wave–particle interactions. This is described by the relation:

$$W_\parallel + J_\perp B(z) = W = \text{const.} \tag{3.52}$$

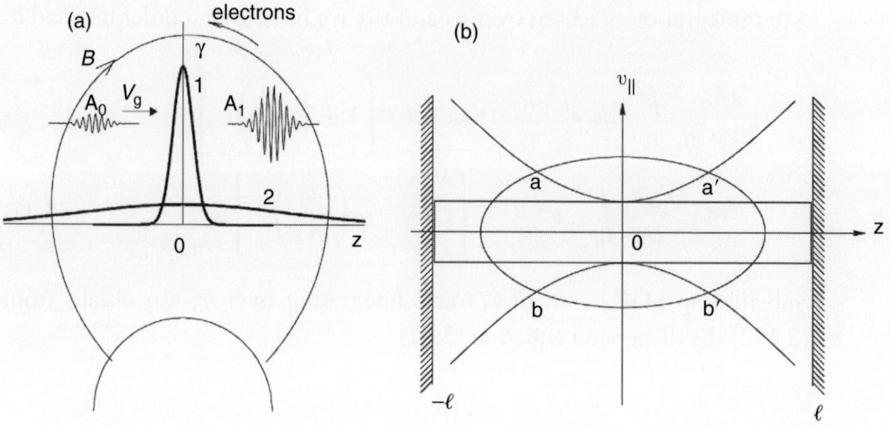

Figure 3.3 (a) The dependence of the local growth rate $\gamma(z)$ on the z-coordinate, the distance along the magnetic field line from the equator. The one-hop amplification, $C = \ln(A_1/A_0)$, describes the cyclotron instability for a non-uniform magnetic field, and A_1 and A_0 are the whistler wave amplitudes after and before crossing the equatorial plane (Copyright American Geophysical Union, reproduced with permission.) (Trakhtengerts *et al.*, 1996) (b) Diagram showing the points (a, a′) and (b, b′) of cyclotron interaction where the curves of cyclotron resonance (thick lines) cross the electron trajectory (closed line, like an ellipse).

The parabolic curve in Fig. 3.3(b) corresponds to the cyclotron resonance condition

$$v_\parallel = v_\mathrm{R} = \frac{\omega - \omega_B(z)}{k(z)} . \tag{3.53}$$

The cyclotron interaction occurs where these curves cross. Expressions (3.49) and (3.50), after substitution of (3.48) and (3.51), take the form:

$$W_0 = W + w, \quad w = e\, e^{i\psi} \int_{z_0}^{z} dz' \left(\frac{J_\perp B}{w - J_\perp B} \right)^{1/2} \times \mathrm{Re}\left(\frac{A}{\sqrt{2}} \exp\left\{ i \int_{z'}^{z} (\Delta / v_\parallel)\, d\xi \right\} \right), \tag{3.54}$$

$$J_{0\perp} = J_\perp + i_\perp, \quad i_\perp = e\, e^{i\psi} \int_{z_0}^{z} dz' \left(\frac{J_\perp B}{w - J_\perp B} \right)^{1/2} \frac{1 - N_\mathrm{w}\beta_\parallel}{B} \times \mathrm{Re}\left(\frac{A}{\sqrt{2}} \exp\left\{ i \int_{z'}^{z} \frac{\Delta}{v_\parallel}\, d\xi \right\} \right). \tag{3.55}$$

Now, following (3.4) we can write:

$$\mathcal{F}(W, J_\perp, \psi, z, t) = \mathcal{F}_0(W + w, J_\perp + i_\perp) \tag{3.56}$$

where $\mathcal{F}_0$ is the initial distribution function. Similarly to (3.6), we present (3.56) as a Taylor series expansion in the small parameters w/W and $i_\perp/J_\perp$:

$$\mathcal{F} = \mathcal{F}_0(W, J_\perp) + \frac{\partial \mathcal{F}_0}{\partial W} w + \frac{\partial F_0}{\partial J_\perp} i_\perp + \cdots \tag{3.57}$$

We now take into account that:

(i) for cyclotron resonance

$$\frac{1 - N_{\text{w}} \beta_\parallel}{B} = \frac{e}{mc\omega}; \tag{3.58}$$

(ii) the integral

$$\begin{aligned}
&\int_0^{2\pi} d\varphi \exp\left\{i\varphi - i\Theta^{(0)}\right\} \\
&\quad \times \operatorname{Re}\left(\int_{z_0}^{z} dz' \left(\frac{J_\perp B}{W - J_\perp B}\right)^{1/2} \frac{A}{\sqrt{2}} \exp\left\{i\left(\psi + i\int_{z'}^{z} (\Delta/v_\parallel)\, d\xi\right)\right\}\right) \\
&= \pi \int_{z_0}^{z} dz' \left(\frac{J_\perp B}{W - J_\perp B}\right)^{1/2} A \exp\left\{i\int_{z'}^{z} (\Delta/v_\parallel)\, d\xi\right\};
\end{aligned} \tag{3.59}$$

(iii) an element of phase space $d^3\vec{v}$ is given by

$$d^3\vec{v} = \frac{B}{m^2} \frac{dW\, dJ_\perp\, d\varphi}{\sqrt{\dfrac{2}{m}(W - J_\perp B)}}. \tag{3.60}$$

Calculating the right-hand part of (2.132) together with (3.54)–(3.60), we obtain:

$$\begin{aligned}
&\frac{1}{\mathcal{A}}\left(\frac{\partial \mathcal{A}}{\partial t} + \vec{v}_{\text{g}} \frac{\partial \mathcal{A}}{\partial \vec{r}}\right) \\
&\quad = \frac{2\pi^2 \omega e^2}{(\partial N_{\text{w}}^2 \omega^2/\partial\omega)\, m^2} \int \frac{J_\perp^{1/2} B^{3/2}\, dW\, dJ_\perp}{\sqrt{W - J_\perp B}} \\
&\quad \times \left(\frac{\partial \mathcal{F}_0}{\partial W} + \frac{e}{mc\omega} \frac{\partial \mathcal{F}_0}{\partial J_\perp}\right) \int_{z_0}^{z} dz' \left(\frac{J_\perp B}{W - J_\perp B}\right)^{1/2} \exp\left\{i\int_{z'}^{z} \frac{\Delta}{v_\parallel}\, d\xi\right\}.
\end{aligned} \tag{3.61}$$

It is important to compare the local growth rate in the inhomogeneous magnetic field, which is determined as the real part of the right-hand side of expression (3.61), with the growth rate γ in a homogeneous plasma (3.15), written using the variables

$(W, J_\perp)$ as:

$$\gamma = \frac{2\pi^3 \omega e^2 B^2}{(\partial N_{\rm w}^2 \omega^2/\partial\omega)\, m^2} \int \frac{J_\perp\, dJ_\perp\, dW}{W - J_\perp B} \left(\frac{\partial \mathcal{F}_0}{\partial W} + \frac{e}{mc\omega}\frac{\partial \mathcal{F}_0}{\partial J_\perp}\right) \delta\left(\frac{\Delta}{v_\|}\right). \tag{3.62}$$

It can be seen that the right-hand side of (3.61) differs from (3.62) by the presence of the integral over z' rather than the Dirac function, $\delta(\Delta/v_\|)$.

The character of the instability changes in the case of a non-uniform magnetic field. For a smooth distribution function the CI is convective; a wave packet propagating through the instability region is amplified up to a limiting value. That value is determined from the wave equation (3.61) by integrating over z in the stationary case $(\partial/\partial t = 0)$:

$$\Gamma = \ln\left[\frac{v_{\rm g}(\ell/2)\, A(\ell/2)}{v_{\rm g}(-\ell/2)\, A(-\ell/2)}\right] \tag{3.63}$$

where $\pm\ell/2$ are the lengths of the magnetic field line from the equator to its conjugate ionospheric feet, and $A = \mathrm{Re}\,\mathcal{A}$. Taking into account the right-hand side of Equation (3.61), we have for a wave propagating in the $+z$-direction:

$$\Gamma^+ = \frac{\pi^2 e^2}{cm^2} \int_{-\ell/2}^{\ell/2} \frac{dz}{N_{\rm w}} \int \frac{B^{3/2} J_\perp^{1/2}\, dJ_\perp\, dW}{(W - J_\perp B)^{1/2}} \left(\frac{\partial \mathcal{F}_0}{\partial W} + \frac{e}{mc\omega}\frac{\partial \mathcal{F}_0}{\partial J_\perp}\right)$$
$$\times \int_{-\ell/2}^{z} dz' \left(\frac{J_\perp B}{W - J_\perp B}\right)^{1/2} \mathrm{Re}\left(\exp\left\{i\int_{z'}^{z} \frac{\Delta}{v_\|}\, d\xi\right\}\right). \tag{3.64}$$

The integrals over z and z' in (3.64) can be calculated by the method of stationary phase because the phase $\displaystyle\int_{z'}^{z} \frac{\Delta}{v_\|}\, d\xi$ is rapidly changing everywhere except in the vicinity of the resonance, at $\Delta = 0$. Thus, the expression (3.64) is written as

$$\Gamma^+ = \mathrm{Re} \int_{-\ell/2}^{\ell/2} dz \exp\left\{i\int_0^z \frac{\Delta}{v_\|}\, d\xi\right\} Q_1(z)$$
$$\int_{-\ell/2}^{z} dz' \exp\left\{i\int_0^{z'} \frac{\Delta}{v_\|}\, d\xi\right\} Q_2(z')$$
$$= Q_1(z_{\rm st})\, Q_2(z_{\rm st})\, \mathrm{Re} \int_{-\ell/2}^{\ell/2} dz \exp\left\{i\int_0^z \frac{\Delta}{v_\|}\, d\xi\right\} \int_{-\ell/2}^{z} dz'$$
$$\times \exp\left\{-i\int_0^{z'} \frac{\Delta}{v_\|}\, d\xi\right\} \tag{3.65}$$

where the slowly changing functions Q_1 and Q_2 are taken at the stationary point $z_{\rm st}(W, J_\perp)$, which is the root of the equation:

$$\Delta = \omega - \omega_B(z_{\rm st}) - k\left(\frac{2}{m}\right)^{1/2}\Big(W - J_\perp B(z_{\rm st})\Big)^{1/2} = 0. \tag{3.66}$$

Using integration by parts in (3.65), we obtain the relation:

$$2{\rm Re}\int_{-\ell/2}^{\ell/2} dz\,\exp\left\{i\int_0^z \frac{\Delta}{v_\|}\,d\xi\right\}\int_{-\ell/2}^{z} dz'\,\exp\left\{-i\int_0^{z'}\frac{\Delta}{v_\|}\,d\xi\right\}$$
$$= \left|\int_{-\ell/2}^{\ell/2}\exp\left\{i\int_0^z\frac{\Delta}{v_\|}\,d\xi\right\}dz\right|^2. \tag{3.67}$$

In the final form we have:

$$\Gamma^+ = \frac{\pi^2 e^2}{cm^2}\int_{W_{\rm RL}}^{\infty}\int_0^{W-W_{\rm RL}}\left[\frac{J_\perp B^2\,dW\,dJ_\perp}{N_{\rm w}(W - J_\perp B)}\right.$$
$$\left.\times\left\{\frac{\partial\mathcal{F}_0}{\partial W} + \frac{e}{mc\omega}\frac{\partial\mathcal{F}_0}{\partial J_\perp}\right\}I\right]_{z_{\rm st}(W,J_\perp,\omega)} \tag{3.68}$$

where we have defined through I the right-hand side of the equality (3.67); the integration limits in (3.68) are determined from the condition that the lowest (minimum) resonant energy $W_{\rm RL} = m\left(v_{\rm RL}^2/2\right)$ is achieved at the equator, and the maximum value of $J_\perp$ is determined from the cyclotron resonance condition on the equator for a fixed value of the electron energy W.

Applying the stationary phase method, we find that

$$I = \left|\frac{\pi v_\|}{\partial\Delta/\partial z}\right|_{z_{\rm st}} = \pi\int_{-\infty}^{\infty} dz\,\delta(\Delta/v_\|), \tag{3.69}$$

where $\delta(x)$ is the Dirac function, and $\Delta(z)$ is determined by the relation (3.66). The result (3.69) coincides with Γ, calculated from the formulae (3.62) for a homogeneous medium according to the relation

$$\Gamma^+ = \int_{-\ell/2}^{\ell/2}\frac{dz}{v_{\rm g}}\,\gamma(z). \tag{3.70}$$

Let us consider a particular example, when $\mathcal{F}_0(J_\perp, W)$ has the realistic form (Bespalov and Trakhtengerts, 1986):

$$\mathcal{F}_0 = \mathcal{F}_\nu \equiv c_\nu J_\perp^\nu \exp\left\{-\frac{W}{W_0}\right\} \tag{3.71}$$

where $\nu > 0$

$$c_\nu = n_\mathrm{h}\left[\left(\frac{W_0}{B}\right)^\nu W_0^{3/2}\left(\frac{2\pi}{m}\right)^{3/2}\Gamma(\nu+1)\right]^{-1} \tag{3.72}$$

is the normalization constant $\left(\int \mathcal{F}_0\, d^3\vec{v} = n_h\right)$, and $\Gamma(\nu+1)$ is the gamma-function. Leaving the reader with the opportunity of calculating Γ^+, we write the final result for $\Gamma^+_{\max}$ and ω_m in the case of a dense plasma $\beta_* \gg 1$:

$$\omega_\mathrm{m} = \omega_{B\mathrm{L}}/\beta_*, \quad \Gamma^+_{\max} \approx \left(\frac{n_\mathrm{hL}}{n_\mathrm{cL}}\right)\frac{4.3\omega_{B\mathrm{L}}\, a\nu}{v_\mathrm{ph}} \approx \frac{10a\omega_{B\mathrm{L}}}{v_\mathrm{ph}}\gamma_\mathrm{m} \tag{3.73}$$

where $v_\mathrm{ph} = c/N_\mathrm{w}$, a is the scale in the parabolic approximation of the geomagnetic field (2.8), and γ_m is determined by the expression (3.24) with the replacement of $(\alpha - 1)$ by ν. The frequency dependence of $\Gamma^+(\omega)$ is similar to the dependence of γ on ω, Fig. 3.1, but the maximum of Γ^+ is shifted to lower frequencies in comparison with the maximum in $\gamma(\omega)$ (see (3.24) and (3.73)). This shift is due to the frequency dependence of the whistler wave group velocity $v_\mathrm{g}(\omega)$.

3.5 Amplification for the distribution function with a step

It has been seen from Section 3.3 that the character of the cyclotron instability changes markedly if the particle distribution function is not smooth. It is interesting to investigate this question for the case of an inhomogeneous magnetic field. The amplification factor will be considered here, following the general approach given in the previous section.

We now obtain the amplification in the case of the distribution function with a jump-like feature or step. We put

$$\mathcal{F}_0 = \mathcal{F}_\nu \cdot \mathrm{He}(v_*^2 - v_{\|\mathrm{L}}^2) \tag{3.74}$$

where we consider a jump across the curve

$$v_{\|\mathrm{L}}^2 \equiv \frac{2}{m}(W - J_\perp B_\mathrm{L}) = v_*^2. \tag{3.75}$$

This is in accordance with the physical picture described in Section 3.3. For (3.74) we can use (3.68) for the calculation of Γ, as before. However, the formula (3.69) is no longer valid. As a matter of fact, $\partial\Delta/\partial z$ can be written as:

$$\left(\frac{\partial\Delta}{\partial z}\right)_{z_\mathrm{st}} = \frac{z_\mathrm{st}}{a^2}\left[2\omega_{B\mathrm{L}} + \frac{kJ_\perp B_\mathrm{L}}{\sqrt{\frac{m}{2}(W - J_\perp B)}}\right] \tag{3.76}$$

where z_{st} is the root of Equation (3.66); it is supposed, for simplicity, that $k(z) = \text{const}$.

If a cyclotron interaction takes place near the central cross-section of the magnetic trap (near the equator), from (3.66) we obtain:

$$\frac{z_{\text{st}}}{a} = \left(\frac{v_{\|\text{L}} - v_{\text{R}}}{\omega_{\text{BL}}/k + (J_\perp B_\perp)/(m v_{\|\text{L}})} \right)^{1/2} \tag{3.77}$$

where $v_{\text{R}} = (\omega_{\text{BL}} - \omega)/k(\omega)$. From (3.76)–(3.77) and (3.69)

$$I = \frac{\pi a v_\|(z_{\text{st}})}{2(v_{\|\text{L}} - v_{\text{RL}})^{1/2} \left(\dfrac{\omega_{\text{BL}}}{k} + \dfrac{v_{\perp\text{L}}^2}{2 v_\|(z_{\text{st}})} \right)^{1/2} k}. \tag{3.78}$$

For the smooth $\mathcal{F}_0$, $I \sim \dfrac{1}{\sqrt{v_{\|\text{L}} - v_{\text{R}}}}$ is integrated over $(W, J_\perp)$ and gives a finite value. However, for the distribution function (3.74), all interacting particles have one value of $v_{\|\text{L}} = v_*$, and $\Gamma^+ \sim (v_* - v_{\text{R}})^{-1/2}$. In this approach, $\Gamma^+ \to \infty$ when $|v_{\text{R}}| \to v_*$. To remove this divergence we have to use the improved stationary phase method, instead of (3.69), with a longer Taylor series expansion. We present $\Delta(\xi)$ near the stationary point z_{st} in the form:

$$\Delta = \left(\frac{\partial \Delta}{\partial z} \right)_{z_{\text{st}}} \xi + \left(\frac{\partial^2 \Delta}{\partial z^2} \right)_{z_{\text{st}}} \frac{\xi^2}{2} + \dots, \tag{3.79}$$

where $(\partial \Delta / \partial z)_{z_{\text{st}}}$ is determined by (3.76), and

$$\left(\frac{\partial^2 \Delta}{\partial z^2} \right)_{z_{\text{st}}} \approx a^{-2} \left(2\omega_{\text{BL}} + \frac{k v_{\perp\text{L}}^2}{v_{\|\text{L}}} \right). \tag{3.80}$$

We remember that $v_{\perp\text{L}}^2 = \dfrac{2}{m} J_\perp B_{\text{L}}$, and $v_{\|\text{L}} = \left(\dfrac{2}{m} \right)^{1/2} (W - J_\perp B_{\text{L}})^{1/2}$. Now we use (3.79) for a more exact calculation of the integral (3.69):

$$I = \left| \int_{-\infty}^{\infty} dx \exp \left\{ i \left(\frac{\omega_{BL}}{v_*} + \frac{k v_{\perp\text{L}}^2}{2 v_*^2} \right) \left(z_{\text{st}} x^2 + \frac{a}{3} x^3 \right) \right\} \right|^2 a^2. \tag{3.81}$$

Substituting (3.74) into (3.68) and taking (3.71) and (3.81) into account, we obtain $(\nu = 0)$:

$$\Gamma^+(v_{\text{R}}) = \frac{\omega_{\text{ph}}^2}{8\sqrt{\pi} c^2} \frac{v_{\text{RL}} v_0}{v_*^2} e^{-v_*^2/v_0^2} I(v_{\text{R}}) \tag{3.82}$$

where $\omega_{\text{ph}}^2 = \dfrac{4\pi e^2 n_{\text{h}}}{m}$, $v_0 = \left(\dfrac{2 W_0}{m} \right)^{1/2}$, and I is determined by (3.81).

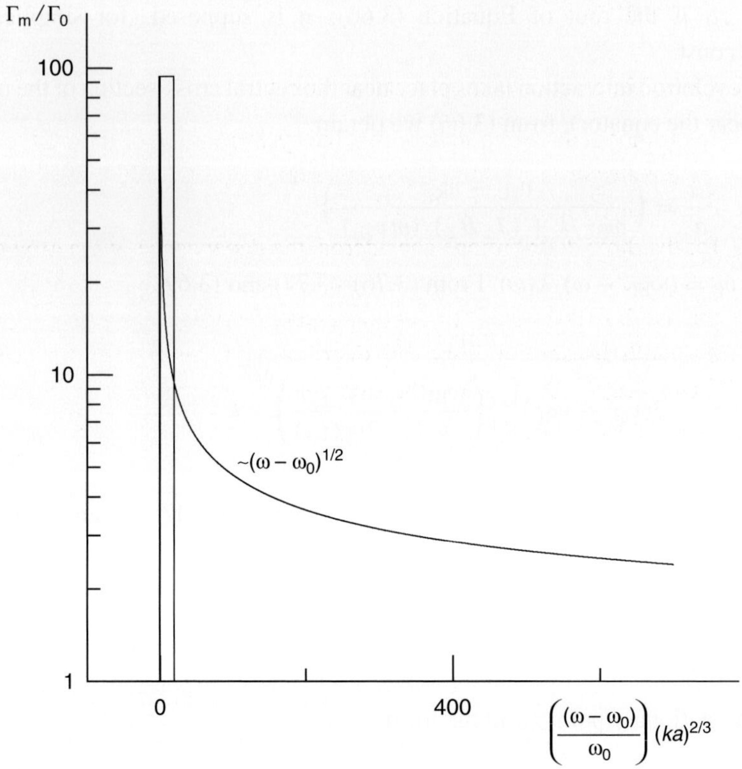

Figure 3.4 Diagram showing the dependence of the amplification Γ_m^+ on the frequency shift $\Delta\omega$ in relative units (a qualitative picture). Γ_m^+ is normalized to the value Γ_0 for the smooth distribution function; $\Delta\omega$ is measured from the step frequency ω_{st}, corresponding to $v_{RL} = v_{st}$, in units of $(ka)^{-2/3}$.

The function $I(v_R)$ has a sharp maximum I_m when $v_R = v_*$; the value I_m and its width in relation to v_R are given by

$$I_m \approx 5\left[a\left(\frac{\omega_{BL}}{v_*} + \frac{kv_0^2}{2v_*^2}\right)\right]^{-2/3} a^2 \tag{3.83}$$

$$\frac{I_m(v_R = v_*)}{I(v_R \sim 0.5\, v_*)} \sim 2(ka)^{1/3}, \qquad \frac{\Delta v_R}{v_R} \sim (ka)^{-2/3}.$$

A particular example of $I(v_R)$ is shown in Fig. 3.4 for $ka \sim 10^4$ and $v_{\perp L} \sim v_*$.

Thus, the formation of a step on the distribution function sharply increases the wave amplification in a narrow frequency band corresponding to the resonance velocity v_R, which is equal to the 'step velocity' v_*. This can explain a very interesting phenomenon, observed by satellites, namely the excitation of the narrow frequency band (discrete) emissions near the high-frequency boundary of ELF hiss. We shall discuss this phenomenon in more detail in Chapters 9 and 11.

3.6 Summary

We summarize the main results obtained as follows:

- for a smooth distribution function, the cyclotron instability is effective near the equator on a particular geomagnetic flux tube only if the cold plasma density is rather large ($n_c \gtrsim B_L^2/8\pi T_\|$);
- for the three distributions considered, the dependence of the growth on the ratio of the hot to cold plasma densities is given in Table 3.1; and
- the one-hop growth Γ^+ (across the equatorial region) is given by (3.73) for a smooth distribution; for a step distribution, Γ^+ is greatly enhanced by the factor $(ka)^{1/3} \gg 1$, when the step velocity v_* and the cyclotron resonant velocity v_R are equal.

Chapter 4

Backward wave oscillator (BWO) regime in CMs

The purpose of this chapter is:

- to explain how a backward wave oscillator (BWO) operates, from both a physical and a mathematical viewpoint;
- to demonstrate that a BWO cannot work with a smooth distribution;
- to estimate the minimum length of the generator region for the threshold condition to be met;
- to find the temporal growth rate for a step distribution and for a beam distribution; and
- to consider the effect of magnetic field inhomogeneity, e.g. as occurs along a geomagnetic flux tube.

The BWO generation regime is well-known in electronic devices, where wave generation is due to the interaction of a well-organized electron beam with man-made electrodynamic systems or with the external magnetic field (Johnson, 1955; Shevchik and Trubetskov, 1975; Wachtel and Wachtel, 1980; Ginzburg and Kuznetsov, 1981). The positive feedback inherent in such devices is due to the volume interaction of a 'beam' wave with an electromagnetic wave, the group velocities of the two interacting waves being in opposite directions. Thus, the absolute instability in the BWO generation regime does not need wave reflecting mirrors, and can be realized in length-limited systems. Such a regime is very attractive in space plasmas with their tremendous dimensions but may seem to be improbable there due to the need for having a well-organized charged particle beam. However, it turns out that such a 'beam' wave is not a rare phenomenon in space because that can appear on the step-like deformation of the distribution function (see Section 3.3). Particle beams appear in triggered ELF/VLF radio emission phenomena in the Earth's magnetosphere (see Chapter 6). The BWO generation regime seems to provide a promising explanation for certain emissions with discrete spectral forms that are observed in space plasmas (Trakhtengerts, 1995, 1999; Demekhov and Trakhtengerts, 2005; Trakhtengerts *et al.*, 2004).

4.1 Physical mechanism for BWO generation. Absolute instability in a spatially limited region

To elucidate the physical mechanism for the BWO generation regime, we consider the idealized problem of the volume interaction of any two waves in a space-limited system. The simplest equations, describing such an interaction, can be written in the form:

$$\frac{\partial \mathcal{A}}{\partial t} + v_1 \frac{\partial \mathcal{A}}{\partial z} = \alpha_1 J, \tag{4.1}$$

$$\frac{\partial J}{\partial t} + v_2 \frac{\partial J}{\partial z} = \alpha_2 \mathcal{A} \tag{4.2}$$

where $\mathcal{A}$ and J are the complex amplitudes of the interacting waves, $v_{1,2}$ are the group velocities, and the feedback interaction coefficients $\alpha_{1,2}$ can be either real or imaginary. In an infinite homogeneous medium we apply the complex Fourier transform to solve (4.1)–(4.2). Putting

$$\mathcal{A} = \mathcal{A}_\Omega \exp\{i\Omega t - i æ z\},$$
$$J = J_\Omega \exp\{i\Omega t - i æ z\} \tag{4.3}$$

we find the dispersion equation for waves of frequency Ω and wavenumber æ:

$$(\Omega - æ v_1)(\Omega - æ v_2) = -\alpha_1 \cdot \alpha_2 \equiv \alpha_{\text{int}}. \tag{4.4}$$

Such a dispersion equation is discussed by Stix (1992, p. 235). For a fixed sign of v_1, we have four possible solutions depending on the sign of v_2 and α_{int}. Amongst these, only the solution with $v_2 < 0$ and $\alpha_{\text{int}} < 0$ can grow in time; that is the condition for an absolute instability to take place.

The physical sense of this is clear. The first condition ($\alpha_{\text{int}} < 0$) means that the system possesses free energy and can be unstable. The second condition $v_1 > 0$ ($v_2 < 0$) provides the positive feedback 'connection' between the waves. This case illustrates the basic regime for BWO generation in a CM. If the system is space-limited (in the z-direction), the absolute instability can occur, but a threshold value of the generation length has to be exceeded. This can be shown for the example of the system (4.1)–(4.2), for which we must solve the boundary-value problem for

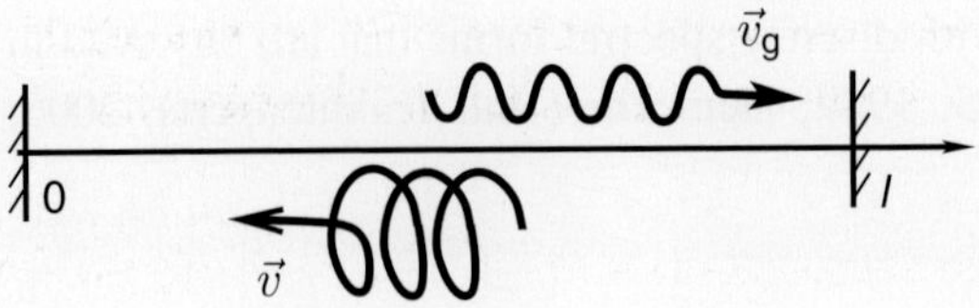

Figure 4.1 Diagram illustrating the interaction between waves and charged particles moving in opposite directions in a system of fixed length l.

the system of equations:

$$i\Omega\mathcal{A}_\Omega + v_1\frac{\partial\mathcal{A}_\Omega}{\partial z} = \alpha_1 J_\Omega, \quad \text{with} \quad \alpha_1\alpha_2 > 0, \tag{4.5}$$

$$i\Omega J_\Omega - v_2\frac{\partial J_\Omega}{\partial z} = \alpha_2\mathcal{A}_\Omega, \quad \text{with} \quad v_1, v_2 > 0 \tag{4.6}$$

with the boundary conditions (Fig. 4.1)

$$z = \ell, \quad \mathcal{A}_\Omega = \mathcal{A}_{\text{ext}}, \quad J_\Omega = 0, \tag{4.7}$$

where ℓ is the length of the system, and $\mathcal{A}_{\text{ext}}$ is the wave amplitude at the exit of the generator. It is supposed that the second wave J, propagating in the opposite direction to wave $\mathcal{A}$, comes into the generation region with zero amplitude.

For the case of constant coefficients in Equations (4.5)–(4.6), we can put $\mathcal{A}_\Omega, J_\Omega \sim \exp\{-i\text{æ}z\}$ and obtain the dispersion equation (4.4). This now determines the eigenvalues $\text{æ}_{1,2}$ of the system (4.5)–(4.6); we find these from (4.4):

$$\text{æ}_{1,2} = \frac{\Omega}{2}\left(\frac{1}{v_1} - \frac{1}{v_2}\right) \pm \sqrt{\frac{\Omega^2}{4}\left(\frac{1}{v_1} + \frac{1}{v_2}\right)^2 + \frac{\alpha_{\text{int}}}{v_1 v_2}}. \tag{4.8}$$

The solution of (4.5)–(4.6) takes the form:

$$\mathcal{A}_\Omega = C_1\exp\{-i\text{æ}_1 z\} + C_2\exp\{-i\text{æ}_2 z\}, \tag{4.9}$$

$$J_\Omega = \alpha_1^{-1}\left(i\Omega\mathcal{A}_\Omega + v_1\frac{\partial\mathcal{A}_\Omega}{\partial z}\right), \tag{4.10}$$

where the constants $C_{1,2}$ are found from the boundary conditions (4.7).

First, we put the origin of the coordinate system at the point $z = \ell$. Substituting (4.7) into (4.9)–(4.10), we obtain:

$$C_1 + C_2 = \mathcal{A}_{\text{ext}}, \tag{4.11}$$

$$\Omega(C_1 + C_2) - v_1(C_1\text{æ}_1 + C_2\text{æ}_2) = 0. \tag{4.12}$$

The solution of (4.11)–(4.12) is

$$C_1 = \frac{\Omega/v_1 - \text{æ}_2}{\text{æ}_1 - \text{æ}_2}\mathcal{A}_{\text{ext}}, \quad C_2 = \frac{\text{æ}_1 - \Omega/v_1}{\text{æ}_1 - \text{æ}_2}\mathcal{A}_{\text{ext}}. \tag{4.13}$$

The absolute instability occurs if the transmission coefficient of the device $T \equiv \mathcal{A}_{\text{ext}}/\mathcal{A}_{\text{ent}}$ goes to infinity, $\mathcal{A}_{\text{ent}}$ being the amplitude at the entrance to the generator: $\mathcal{A}_{\text{ent}} = \mathcal{A}(z = 0)$. Thus, we have the dispersion relation for complex Ω:

$$T^{-1} = \left(C_1 \exp\{i\text{æ}_1 \ell\} + C_2 \exp\{i\text{æ}_2 \ell\}\right) = 0. \tag{4.14}$$

Substituting C_1 and C_2 (4.13) into (4.14), we obtain:

$$\frac{\text{æ}_2 - \Omega/v_1}{\text{æ}_1 - \Omega/v_1} = \exp\left\{ i2\ell \sqrt{\frac{\Omega^2}{4}\left(\frac{1}{v_1} + \frac{1}{v_2}\right)^2 + \frac{\alpha_{\text{int}}}{v_1 v_2}} \right\}. \tag{4.15}$$

The minimum instability threshold is achieved when $\Omega = 0$. Then from (4.15) for $\Omega = 0$:

$$\exp\left\{ \frac{2i\alpha_{\text{int}}^{1/2}}{\sqrt{v_1 v_2}} \ell \right\} = -1, \quad \left[\left(\frac{\alpha_{\text{int}}}{v_1 v_2} \right)^{1/2} \ell \right]_{\text{thr}} = \frac{\pi}{2}. \tag{4.16}$$

If $\left(\frac{\alpha_{\text{int}}}{v_1 v_2} \right)^{1/2} \ell$ exceeds the threshold value, Equation (4.15) serves to find the growth rate γ_{BWO}. Near the threshold (4.16), γ_{BWO} is equal to:

$$\gamma_{\text{BWO}} = \left(\frac{\ell}{v_1} + \frac{\ell}{v_2} \right)^{-1} \pi \left(q - \frac{\pi}{2} \right) \tag{4.17}$$

where

$$q = \left(\frac{\alpha_{\text{int}}}{v_1 v_2} \right)^{1/2} \ell.$$

This is the simplest backward wave oscillator generation regime. Real systems can be more complicated, especially in non-uniform media, and can have more than two interacting wave modes. But two necessary conditions always have to be met:

— the system must have two or more interacting wave modes; and
— the group velocities of the interacting pair of waves must be in opposite directions.

As we shall see, these conditions can be met in a CM for the distribution function of electrons exhibiting either a step or a delta-function (see Section 3.3). The BWO generation regime is impossible for a smooth distribution function of energetic electrons, when there is only the whistler mode, and the 'beam' mode is absent.

4.2 BWO regime for a charged particle distribution function with a step

Now we present the wave equation (2.132) with the resonant current on the right-hand side in the form of two interacting wave modes. For that we write the solution (2.116)–(2.117) in integral form, considering the wave amplitude $\mathcal{A}$ as some arbitrary function of (z, t). We have

$$v_{\|0} = v_\| + \frac{eN_\omega v_\perp}{mc\sqrt{2}} \operatorname{Re} \int_{-\infty}^{t} \mathcal{A}\{t', z - v_\|(t - t')\} \times \exp\left\{i\Delta_{\mathrm{w}}(t' - t) + i\Theta_{\mathrm{w}} - i\varphi\right\} dt' \tag{4.18a}$$

$$\equiv v_\| + a, \tag{4.18b}$$

$$v_{\perp 0} = v_\perp + b, \quad b = \frac{1 - N_\omega \beta_\|}{N\beta_\perp} a \tag{4.18c}$$

where

$$\Delta_{\mathrm{w}} = \omega_{\mathrm{w}} - \omega_B - k_{\mathrm{w}} v_\|, \quad \Theta_{\mathrm{w}} = \omega_{\mathrm{w}} t - k_{\mathrm{w}} z. \tag{4.19}$$

Here ω_{w} and k_{w} are connected by the cold plasma dispersion relation. In (4.18) we take into account that the amplitude $\mathcal{A}$ under the integral is written in the variables $\vec{v}$, z as

$$\mathcal{A}(t', z') = \mathcal{A}\{t', z - v_\|(t - t')\}. \tag{4.20}$$

Let us consider the resonant current (2.129)

$$J = e \int d^3\vec{v}\, v_\perp \exp\{i(\varphi - \Theta_{\mathrm{w}})\}\, \mathcal{F}(\vec{v})$$

for the distribution function with a step (3.34). Using (3.42), (3.44)–(3.45) and (4.18) we can write:

$$J = \frac{e^2 N_{\mathrm{w}}}{\sqrt{2}\, mc} \int_0^\infty v_\perp^3\, dv_\perp\, \mathcal{F}_0(-v_0, v_\perp) \int_0^{2\pi} \frac{d\varphi}{2\pi} \times \int_{-\infty}^{t} \mathcal{A}\{t', z + v_0(t - t')\}\, e^{i\Delta_{\mathrm{w}}(t - t')}\, dt', \tag{4.21}$$

where $\Delta_{\mathrm{w}} = \omega_{\mathrm{w}} - \omega_B + k_{\mathrm{w}} v_0 + k_{\mathrm{w}} a(\varphi)$ is a function of φ. This is a nonlinear integral equation connecting J and $\mathcal{A}$. We transform it to a differential equation in the linear approximation on $\mathcal{A}_0$ ($\Delta_{\mathrm{w}} \approx \omega_{\mathrm{w}} - \omega_B + k_{\mathrm{w}} v_0 \equiv \Delta_{\mathrm{w}0}$). Differentiating (4.21) with respect to z and t and taking the combination $\dfrac{\partial J}{\partial t} - v_0 \dfrac{\partial J}{\partial z}$ (see (4.2)), we obtain (for $v_0 > 0$)

$$\frac{\partial J}{\partial t} - v_0 \frac{\partial J}{\partial z} - i\Delta_{w0} J = \alpha_1 \mathcal{A}, \tag{4.22}$$

$$\frac{\partial \mathcal{A}}{\partial t} + v_g \frac{\partial \mathcal{A}}{\partial z} = \alpha_2 J, \tag{4.23}$$

where $\Delta_{w0} = \omega_w - \omega_B + k_w v_0$, and

$$\alpha_1 = \frac{e^2 N_w}{\sqrt{2}mc} \int_0^\infty v_\perp^3 \, dv_\perp \, \mathcal{F}_0(-v_0, v_\perp), \quad \alpha_2 = \frac{2\pi\omega\sqrt{2}}{\partial(N_w^2\omega^2)/\partial\omega}. \tag{4.24}$$

Putting $\mathcal{A}$, $J \sim \exp(i\Omega t - i æ z)$, we obtain the dispersion equation:

$$(\Omega - æ v_g)(\Omega + æ v_0 + \Delta_{w0}) = -\alpha_{int}, \tag{4.25}$$

where

$$\alpha_{int} = \frac{\pi e^2}{[\partial(N_w\omega)/\partial\omega]\, mc} \int_0^\infty v_\perp^3 \, dv_\perp \, \mathcal{F}_0(-v_0, v_\perp) > 0. \tag{4.26}$$

For the case $æ = 0$, (4.25) turns into (3.38). Now we see that Equation (4.25) is practically the same as (4.4), and that all the conditions required for the BWO generation regime are met. In particular, the group velocities of the whistler and the electron beam wave are oppositely directed (v_g and $-v_0$), and a CI takes place (minus sign in front of α_{int}). The optimal conditions for the absolute instability in the system (4.25) are realized when $\Delta_{w0} = 0$. The threshold and the growth rate γ_{BWO} are determined by the general relations (4.16)–(4.17), and in our case we have:

$$\left[\frac{\gamma_{st}\,\ell}{(v_g v_0)^{1/2}}\right]_{thr} = \frac{\pi}{2}, \tag{4.27}$$

$$\gamma_{BWO} = \frac{\pi}{\ell}\left(\frac{1}{v_g} + \frac{1}{v_0}\right)^{-1} \left(\frac{\gamma_{st}\,\ell}{\sqrt{v_g v_0}} - \frac{\pi}{2}\right) \tag{4.28}$$

where the growth rate on the step on the distribution function γ_{st} is determined by the relation (3.41), because $\alpha_{int} = \gamma_{st}^2$.

We shall use expressions (4.27) and (4.28) for the discussion of the generation mechanisms of discrete ELF/VLF emissions in the Earth's magnetosphere in Chapter 11.

4.3 BWO regime for a helical charged particle beam with small velocity spread

This case is well-known in laboratory CMs, where $N_w = 1$. The linear and nonlinear stages of the BWO regime in this case have been investigated both experimentally and theoretically with the help of some analytical models, and by computational modelling (Shevchik and Trubetskov, 1975; Ginzburg and Kuznetsov, 1981). These results are interesting for space CMs too, because they permit us to predict possible qualitative effects in space systems. At the same time such beams can occur in the Earth's magnetosphere.

The analytical description of the gyrating electron beam with small velocity dispersion is based on the distribution function in the form of a delta-function (3.26). Then, the BWO-generation regime is not described by (4.1)–(4.2), because there are three interacting wave modes (see Table 3.1). We now proceed similarly to Section 4.2 and present the wave equation (2.132) in the form of the system of interacting waves. For that we transform the current J on the right-hand side of (2.132), using (3.46). According to (3.46) we write J in the form:

$$J = -\frac{en_h}{2\pi}\int_0^{2\pi} e^{i\varphi - i\Theta_{\rm w}}\left\{2(b)_{v_\parallel=-v_0} + v_0\left(\frac{\partial a}{\partial v_\parallel}\right)_{v_\parallel=-v_0}\right\} d\varphi, \tag{4.29}$$

where a and b are described by the integral relations (4.18). The term $\left(\frac{\partial a}{\partial v_\parallel}\right)_{v_\parallel=-v_0}$ can be written as:

$$\left(\frac{\partial a}{\partial v_\parallel}\right)_{v_\parallel=-v_0,\, v_\perp=v_{\perp 0}} = k\,\frac{eN_{\rm w}v_{\perp 0}}{mc\sqrt{2}}\,\mathrm{Re}\, i\int_{-\infty}^{t}(t-t')\,\mathcal{A}\left\{t', z+v_0(t-t')\right\}$$
$$\times\ \exp\left\{i\Delta_{\rm w0}(t'-t) + i\Theta_{\rm w} - i\varphi\right\} dt'. \tag{4.30}$$

Remembering that

$$(b)_{v_\parallel=-v_0} \approx \frac{e\omega_B}{\omega m}\int_{-\infty}^{t}\mathcal{A}\{t', z+v_0(t-t')\}\exp\left\{i\Delta_{\rm w0}(t'-t) + i\Theta_{\rm w} - i\varphi\right\} dt' \tag{4.31}$$

and substituting (4.30)–(4.31) into (4.29), we obtain:

$$J = \frac{\omega_{\rm ph}^2}{4\pi\sqrt{2}}\int_{-\infty}^{t} dt'\left\{\frac{\omega_B}{\omega} + i\,(t-t')\,\frac{\beta_{\perp 0}^2\,\omega\,N_{\rm w}^2}{2}\right\}$$
$$\times\ \mathcal{A}\{t', z+v_0(t-t')\}\exp\left\{i\Delta_{\rm w0}(t'-t)\right\}. \tag{4.32}$$

For the combination $\frac{\partial J}{\partial t} - v_0\,\frac{\partial J}{\partial z}$ we have, from (4.32),

$$\frac{\partial J}{\partial t} - v_0\,\frac{\partial J}{\partial z} = -i\Delta_{\rm w0}\,J + \frac{\omega_B\,\omega_{\rm ph}^2}{\omega\,4\pi\sqrt{2}}\,\mathcal{A} + i\,\frac{\omega_{\rm ph}^2}{4\pi\sqrt{2}}\,\frac{i\beta_{\perp 0}\,\omega\,N_{\rm w}^2}{2}$$
$$\times\int_{-\infty}^{t}\mathcal{A}\{t', z+v_0(t-t')\}\exp\left\{i\Delta_{\rm w0}(t'-t)\right\}. \tag{4.33}$$

Differentiating (4.33) again and combining different terms, the final equation for J is found:

$$\left(\frac{\partial}{\partial t} - v_0\,\frac{\partial}{\partial z} + i\Delta_{\mathrm{w}0}\right)\left(\frac{\partial J}{\partial t} - v_0\,\frac{\partial J}{\partial z} + i\Delta_{\mathrm{w}0}\,J\right)$$
$$= \frac{\omega_B}{\omega}\,\frac{\omega_{\mathrm{ph}}^2}{4\pi\sqrt{2}}\left(\frac{\partial \mathcal{A}}{\partial t} - v_0\,\frac{\partial \mathcal{A}}{\partial z}\right)$$
$$+ i\,\frac{\omega_{\mathrm{ph}}^2}{4\pi\sqrt{2}}\left(\Delta_{\mathrm{w}0}\,\frac{\omega_B}{\omega} + \beta_\perp^2\,\omega\,N_{\mathrm{w}}^2\right)\mathcal{A}. \tag{4.34}$$

The equation for $\mathcal{A}$ is the same as in the previous section (4.23):

$$\frac{\partial \mathcal{A}}{\partial t} + v_{\mathrm{g}}\,\frac{\partial \mathcal{A}}{\partial z} = \alpha_2\,J. \tag{4.23}$$

Equations (4.34) and (4.23) describe the interaction of two beam waves $J_{1,2}$ and the whistler wave in a CM. If we put $\frac{\partial}{\partial t} = i\Omega$ and $\frac{\partial}{\partial z} = 0$, we obtain the dispersion equation (3.29) for an infinite system. For a solution of the boundary value problem, we must formulate the boundary conditions for the system (4.34) and (4.23). Two boundary conditions are the same as in the previous case and are described by the relation (4.7). The third boundary condition can be obtained using relations (4.18) and (4.33). Because the charged particle velocity is undisturbed at the entrance of the beam to the system $z = \ell$, we can put $v_\parallel(z = \ell) = v_{\parallel 0}$, and from (4.18) we obtain:

$$\mathrm{Re}\int_{-\infty}^{t} \mathcal{A}(t', \ell - v_\parallel t)\,e^{-i\Delta_{\mathrm{w}}(t'-t)}dt' = 0. \tag{4.35}$$

Taking into account (4.35) and $J\,|_{z=\ell} = 0$, $\mathcal{A}\,|_{z=\ell} = \mathcal{A}_{\mathrm{ext}}$, we find from (4.33) that

$$\left(\frac{\partial J}{\partial t} - v_0\,\frac{\partial J}{\partial z}\right)_{z=\ell} = \frac{\omega_B\,\omega_{\mathrm{ph}}^2}{\omega\,4\pi\sqrt{2}}\,\mathcal{A}_{\mathrm{ext}}. \tag{4.36}$$

For the weak electron beam with $\omega_{\mathrm{ph}}^2 \to 0$, we can put the right-hand side of (4.36) equal to zero. Thus, as in the previous section, we can formulate the boundary value problem for determining the threshold parameters of the CM and the growth rate in the BWO generation regime. Putting $\Delta_{\mathrm{w}0} = 0$ and $\frac{\partial}{\partial t} = i\Omega$ in (4.34), (4.23) and (4.36), we obtain:

$$\left(i\Omega - v_0\,\frac{\partial}{\partial z}\right)^2 J = \frac{\omega_{\mathrm{ph}}^2}{4\pi\sqrt{2}}\,\frac{\omega_B}{\omega}\left(i\Omega\mathcal{A} - v_0\,\frac{\partial \mathcal{A}}{\partial z}\right) + i\,\frac{\omega_{\mathrm{ph}}^2\,\beta_{\perp 0}^2\,\omega\,N_{\mathrm{w}}^2}{8\pi\sqrt{2}}\,\mathcal{A}, \tag{4.37}$$

$$i\Omega\mathcal{A} + v_{\mathrm{g}}\,\frac{\partial \mathcal{A}}{\partial z} = \alpha_2\,J. \tag{4.38}$$

The boundary conditions are (in the shifted coordinate system $z' = z - l$):

$$z' = 0, \quad J = 0, \quad \mathcal{A} = \mathcal{A}_{\text{ext}} \text{ and } \frac{\partial J}{\partial z} \simeq 0. \tag{4.39}$$

For a uniform medium we can present the solution of (4.37) to (4.39) in the form:

$$\mathcal{A} = \sum_{j=1}^{3} C_j \exp\{-i æ_j z\}, \quad J = \frac{i}{\alpha_2} \sum_{j=1}^{3} (\Omega - i æ_j v_g)\, C_j \exp\{-æ_j z\} \tag{4.40}$$

where the eigenvalues $æ_j$ are determined from the equation:

$$(\Omega + æ v_0)^2 (\Omega - æ\, v_g) = -\alpha_2 \frac{\omega_{\text{ph}}^2}{4\pi\sqrt{2}} \left[\frac{\omega_B}{\omega} (\Omega + æ v_0) + \frac{\beta_{\perp 0}^2 N_{\text{w}}^2}{2}\, \omega \right]. \tag{4.41}$$

For a weak beam we can omit the first term in brackets on the right-hand side (see Section 3.3). Putting $æ + \dfrac{\Omega}{v_0} = 2\pi (\ell)^{-1} \eta$, and using the negative frequency shift, $\Omega = -|\Omega|$ which is needed for instability according to (3.31), we transform Equation (4.41) into the form:

$$\eta^3 + 2b\eta^2 - g^3 = 0 \tag{4.42}$$

where

$$2b = |\Omega| \left(\frac{1}{v_0} + \frac{1}{v_{\text{g}}} \right) \pi^{-1} \ell, \tag{4.43}$$

$$g^3 = \left(\frac{\ell}{2\pi} \right)^3 \frac{\beta_{\perp 0}^2\, \omega}{v_0^2\, v_{\text{g}}} \left(1 - \frac{\omega}{\omega_B} \right) \omega_{\text{ph}}^2 = \frac{2\ell^3}{3\pi^3\sqrt{3}} \frac{\gamma_\delta^3}{v_0^2\, v_{\text{g}}}. \tag{4.44}$$

Here γ_δ is the growth rate (3.33) for the cyclotron beam–plasma instability.

The boundary conditions (4.39) can be written as follows:

$$\sum_{j=1}^{3} C_j = \mathcal{A}_{\text{ext}}, \tag{4.45}$$

$$\sum_{j=1}^{3} \eta_j C_j = 2b\mathcal{A}_{\text{ext}}, \tag{4.46}$$

$$\sum_{j=1}^{3} \eta_j^2 C_j = 4b^2 \mathcal{A}_{\text{ext}} \tag{4.47}$$

where η_j is the j-th root of Equation (4.42). From (4.45)–(4.47) we obtain:

$$C_1 = \frac{\mathcal{A}_{\text{ext}}}{\Delta}(\eta_3 - 2b)(\eta_2 - 2b)(\eta_3 - \eta_2), \tag{4.48}$$

$$C_2 = \frac{\mathcal{A}_{\text{ext}}}{\Delta}(\eta_3 - 2b)(\eta_1 - 2b)(\eta_1 - \eta_3), \tag{4.49}$$

$$C_3 = \frac{\mathcal{A}_{\text{ext}}}{\Delta}(\eta_2 - 2b)(\eta_1 - 2b)(\eta_2 - \eta_1). \tag{4.50}$$

Here $\Delta = (\eta_3 - \eta_2)(\eta_3 - \eta_1)(\eta_2 - \eta_1)$ is the determinant of the system (4.45)–(4.47).

As in Section 4.1 we determine the BWO generation threshold, putting the return transmission coefficient $T^{-1} = \mathcal{A}_{\text{ent}}/\mathcal{A}_{\text{ext}}$ to be zero, where $\mathcal{A}_{\text{ent}}$ is the amplitude at the entrance $z' = -\ell$ to the generator. Using (4.50) and $\mathcal{A}_{\text{ent}} = \sum_{j=1}^{3} C_j \exp\{i\,2\pi\eta_j\}$, we obtain:

$$T^{-1} = \sum_{j,k,l} (\eta_j - 2b)(\eta_k - 2b)(\eta_j - \eta_k) \exp\{i\,2\pi\eta_\ell\} = 0 \tag{4.51}$$

where $j \neq k \neq \ell$ are cyclic.

The real and imaginary parts of (4.51) are two equations for the parameters b and g at the threshold of backward wave oscillation. The characteristic equation (4.51) together with Equation (4.42) for the roots η_j is very similar to analogous equations in the theory of the laboratory gyro-BWO but with other values of b and g (Shevchik and Trubetskov, 1975; Ginzburg and Kuznetsov, 1981). Based on this analogy it is possible to find the minimal threshold value for BWO generation, which is equal to:

$$\left[\frac{\gamma_\delta \ell}{(v_0^2 v_g)^{1/3}}\right]_{\text{thr}} \approx 1.5\,\pi. \tag{4.52}$$

We leave more detailed investigation of (4.51) to the reader. Comparison of (4.27) and (4.52) reveals the common feature that the BWO threshold is determined by the corresponding growth rate of the instability for the infinite system.

Interesting features of the BWO regime, important for the interpretation of natural discrete ELF/VLF radio emissions, are found in the nonlinear stage of the BWO instability, which is discussed in Chapter 5.

4.4 Role of magnetic field inhomogeneity

According to the analysis of the BWO regime in a homogeneous magnetic field, the threshold parameters depend on the frequency deviation Ω from the resonant condition. For example, in the case of the step-like deformation, the absolute instability is, in principle, absent when (see (4.25))

$$|\Omega| \geq 2\alpha_{\text{int}}^{1/2}. \tag{4.53}$$

In an inhomogeneous (parabolic) magnetic field such a deviation from the resonance condition $\Delta = 0$ appears due to the dependence of Δ on z: $\Delta = \omega + kv_0 - \omega_{BL}\,(1 + z^2/a^2)$. Thus, this case is the subject of special consideration. The strict solution of this problem has not yet been obtained. However, it is possible to write a qualitative criterion, when the results of the 'uniform' consideration are applicable to the inhomogeneous magnetic field case. For that, the phase mismatch $\Delta\varphi = \int \frac{\Delta}{v_{\|}}\,dz$ must not exceed $\pi/2$ over the interaction length:

$$\Delta\varphi \equiv \int_0^{\ell} \left|\frac{\Delta}{v_{\|}}\right| dz \leq \frac{\pi}{2}\,. \tag{4.54}$$

Applying this to the Earth's equatorial plane where the magnetic field at a particular field line has its lowest value (subscript 'L'), this condition can be written as:

$$\frac{k_{\rm L}\,\ell^3}{3a^2} < \frac{\pi}{2}\,. \tag{4.55}$$

Here the length

$$\ell = \left(\frac{3\pi}{2k_{\rm L}a}\right)^{1/3} a \tag{4.56}$$

can be considered as the length of the magnetosphere BWO generator in the homogeneous approximation (Demekhov *et al.*, 2003).

4.5 Summary

This chapter has explained how a backward wave oscillator works, both physically and mathematically. Expressions have been defined for

- the minimum value of the generation length which has to be exceeded, and
- the growth rate of the backward wave oscillator (4.17).

For a step distribution, the growth rate is determined by (4.28). For a beam of gyrating energetic electrons, the minimum value of the generation length is found from (4.53). When the ambient magnetic field is inhomogeneous, as in the magnetosphere, the length of the BWO generator has been estimated (4.56).

Chapter 5

Nonlinear wave–particle interactions for a quasi-monochromatic wave

This chapter is devoted to a discussion of CMs with uniform magnetic fields. This is usually the case for laboratory CMs. The nonlinear theory of laboratory plasma devices has been developed over many years, and has received excellent experimental confirmation (Gaponov-Grekhov and Petelin, 1981). Space CMs differ from their laboratory analogues, even for the homogeneous magnetic field situation. The main difference concerns the smooth charged particle distribution function and various wave spectral features in space plasmas in comparison with the mono-energetic beams and high Q resonators, with waves in a narrow frequency band, in laboratory devices. However, many important nonlinear effects are similar in both laboratory and space CMs, and the experience of laboratory CM investigations can be relevant to importing their results to the nonlinear theory of CMs in space.

In particular, we are concerned with the nonlinear theory of the cyclotron generation of electromagnetic quasi-monochromatic waves by a beam of electrons, rotating about a homogeneous magnetic field. Pioneering investigations of this problem were performed by Yulpatov (1965) and Gaponov *et al.* (1967). The modern nonlinear theory of laboratory electron generators includes many features which are important for space CMs. The so-called backward wave oscillator (Kuznetsov and Trubetskov, 1977; Ginzburg and Kuznetsov, 1981), which seems to be very important for an explanation of a wide range of discrete ELF/VLF signals in the magnetosphere, especially the so-called chorus emissions, is discussed here and also in Chapter 11.

Nowadays the basic elements of the nonlinear theory of CMs can be specified; they provide the foundations on which, in the future, more sophisticated theories will be built. These elements include the nonlinear theory of gyroresonant wave–particle interactions in a wave field of given amplitude and nonlinear saturation effects, sideband generation phenomena, and feedback interactions in spatially limited systems. Dungey (1963), Brossier (1964), Roberts and Bushsbaum (1964), Laird and Knox (1965), Engel (1965), Bell (1965), Lutomirski and Sudan (1966), Sonnerup and Su (1967), Das (1968), and Shapiro and Shevchenko (1968) considered different aspects of the nonlinear theory of a monochromatic whistler-mode wave of given amplitude interacting with an electron. The nonlinear saturation effects and deformation of the distribution function of electrons under their gyroresonant interaction with a whistler-mode wave were considered by Erokhin and Masitov (1968), Ashour-Abdulla (1970), Bud'ko *et al.* (1971, 1972), Matsumoto and Kimura (1971), Palmadesso and Schmidt (1971) and Ossakov *et al.* (1972 a, b) following the classical papers by O'Neil (1965), Mazitov (1965) and Altshul and Karpman (1965). A detailed discussion of this problem can be found in the review by Galeev and Sagdeev (1973). Knox (1969) investigated the nonlinear evolution of a whistler wave packet under gyroresonant interaction with electrons.

The sideband instability, which plays an important role in the nonlinear dynamics of quasi-monochromatic signals, was first considered by Malmberg *et al.* (1968), Kruer *et al.* (1969), Shapiro and Shevchenko (1969), Goldman (1970) and Yagishita and Ichikawa (1970). As applied to whistler and Alfvén waves, Bud'ko *et al.* (1971, 1972), Brinca (1972) and Gokhberg *et al.* (1972) took the first steps.

5.1 The motion of charged particles in the field of a given wave. Phase space considerations

When considering nonlinear problems we continue the approach developed in previous chapters. This includes the solution of the equations of motion for a single charged particle and the application of Liouville's theorem to estimate collective effects. The system (2.114)–(2.116) of equations of motion will be used initially, with one important simplification. This relies on the fact that, in real CMs, the parameter (see Palmadesso and Schmidt (1971) for more details):

$$\delta = (h\omega_B^3/k^3\, v_\perp^3)^{1/2} \ll 1 \tag{5.1}$$

where $h = B_\sim/B$ is the magnetic field amplitude of the wave relative to the ambient (constant) magnetic field, with $B_\sim = N_\mathrm{w} A$ where A is the amplitude of the wave electric field. The inequality (5.1) permits us to omit the second term in Equation (2.116) for the phase ψ. This latter term gives the phase change $\Delta\psi \sim h\omega_B/kv_\perp$, while the phase change which is due to the change of $v_\parallel$ is proportional to $(h\omega_B/kv_\perp)^{1/2}$ (see below). Thus, the initial system of equations has the form:

$$\frac{dv_\perp}{dt} = -\omega_\mathrm{h} c\, \frac{1 - N_\mathrm{w}\beta}{N_\mathrm{w}} \cos\psi, \tag{5.2}$$

$$\frac{dv_\parallel}{dt} = -\omega_\mathrm{h} v_\perp \cos\psi, \tag{5.3}$$

$$\frac{d\psi}{dt} = \omega - \omega_B - kv_\parallel \equiv k(v_R - v_\parallel), \tag{5.4}$$

where the gyrofrequency in the wave magnetic field $\omega_h = \dfrac{eAN_w}{mc} = \dfrac{eB_\sim}{mc}$, and the cyclotron resonance velocity $v_R = \dfrac{\omega - \omega_B}{k}$. From (5.2)–(5.3), the motion integral follows:

$$v_\perp^2 + v_\parallel^2 - 2v_\parallel v_\phi = C_1 \tag{5.5}$$

where the phase velocity of the wave $v_\phi = c/N_w$.

Differentiating (5.4) with respect to time t, we find that

$$\frac{d^2\psi}{dt^2} = k\omega_h v_\perp \cos\psi, \tag{5.6}$$

where we have used (5.3) for $\dot{v}_\parallel$. The velocity $v_\perp$ is obtained from (5.5)

$$v_\perp = \left(C_1 + 2v_\parallel v_\phi - v_\parallel^2\right)^{1/2} \approx \left(C_1 + 2v_R v_\phi - v_R^2\right)^{1/2}, \tag{5.7}$$

where the last equality is fulfilled to an accuracy $\sim \delta$.

The motion of the charged particle is considered in the field of a given wave with $B_\sim = \text{const}$. In this case the second motion integral follows from (5.6):

$$(\dot{\psi})^2/2 - k\omega_h v_\perp \sin\psi = k\mathcal{H}, \tag{5.8}$$

where the function

$$\mathcal{H} = ku_\parallel^2/2 - \omega_h v_\perp \sin\psi \tag{5.9}$$

with (5.7) taken into account is the Hamiltonian of the system (5.2)–(5.4) with canonical variables

$$u_\parallel = v_R - v_\parallel \equiv \dot{\psi}/k \quad \text{and} \quad \psi. \tag{5.10}$$

This system of equations can thus be written as:

$$\dot{u}_\parallel = -\frac{\partial\mathcal{H}}{\partial\psi}, \quad \dot{\psi} = \frac{\partial\mathcal{H}}{\partial u_\parallel}. \tag{5.11}$$

To use tabulated functions we introduce the 'new' phase:

$$2\xi = \frac{\pi}{2} + \psi. \tag{5.12}$$

After that, (5.9) can be written in the form:

$$\frac{d\xi}{dt} = \frac{1}{æ\tau_0}\left(1 - æ^2 \sin^2\xi\right)^{1/2} \tag{5.13}$$

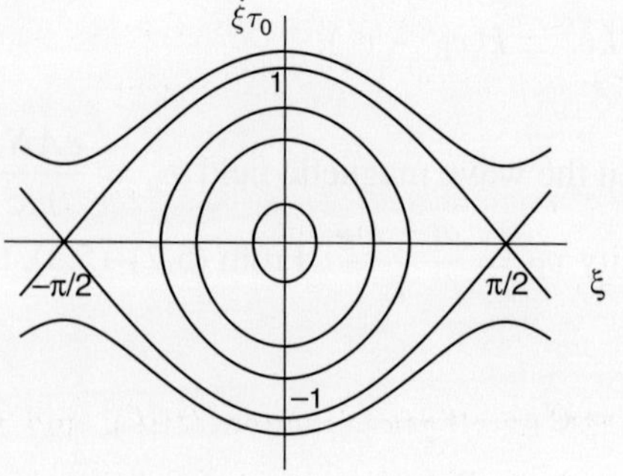

Figure 5.1 Phase plane representing trajectories of particles untrapped and trapped by the wave potential.

where the sign of æ corresponds to the sign of $d\xi/dt$,

$$æ^2 = \frac{2\omega_h v_\perp}{\mathcal{H} + \omega_h v_\perp}, \quad \tau_0 = \Omega_{tr}^{-1} = (k\omega_h v_\perp)^{-1/2}. \tag{5.14}$$

Ω_{tr} is the so-called ‘trapping’ frequency of electrons in the wave magnetic field.

The phase diagram of Equation (5.13) is shown in Fig. 5.1. There are particles trapped by the wave ($|æ| > 1$) and also untrapped particles ($|æ| < 1$). The trajectories of trapped particles (subscript ‘tr’) are closed and near the centre of rotation ($u_\parallel = 0$, $\xi = 0$); these are described by the relation ($|æ| \gg 1$):

$$\xi_{tr} \approx \frac{1}{æ} \sin æ(t - t_0) + \xi_0. \tag{5.15}$$

The trajectories of the untrapped particles (subscript ‘un’) are open along the ξ-axis. For $|æ| \ll 1$ we have, from (5.13):

$$\xi_{un} \simeq \frac{t}{æ\tau_0}\left(1 - \frac{æ^2}{4}\right) - \frac{æ^2}{8} \sin \frac{2t}{æ\tau_0} + \xi_0. \tag{5.16}$$

For the case of any arbitrary value of æ, $0 \leq |æ| < \infty$, the solution of (5.13) can be written in the form of the elliptic integral $F(æ, \xi)$:

$$\int_0^{\xi} \frac{d\xi'}{\sqrt{1 - æ^2 \sin^2 \xi'}} \equiv F(æ, \xi) = \frac{1}{æ\tau_0}(t - t_0). \tag{5.17}$$

As in the analyses of the CI using linear theory, in the non-uniform magnetic field case (Section 2.121) it is convenient to introduce into (5.17) a ‘new’ time ξ' instead of t'. When applying Liouville’s theorem we use the relation:

$$t_0 = t - æ\tau_0 F(æ, \xi). \tag{5.18}$$

Using the Jacobian elliptic functions as inverse functions of F (see the appendix), we can write the nonlinear solutions of (5.2)–(5.4) as follows:

$$v_{\|0} = v_{\|} + u_{\|} = v_{\|} + \frac{2\dot{\xi}}{k} = v_{\|} + \frac{2}{k æ \tau_0}\, dn\left(F(\xi, æ) - \frac{t}{æ\tau_0}, æ\right),$$

$$v_{\perp 0} = v_{\perp} + \frac{1 - N\beta_{\|}}{N} u_{\|}, \tag{5.19}$$

where $dn(u, æ)$ is the Jacobian elliptic function (JEF), and the modulus of æ lies between 0 and ∞. Usually, presentations for the JEF are given for values $0 \le |æ| \le 1$. So we shall use for $|æ| > 1$ the following relation:

$$dn\left(F(\xi, æ) - \frac{t}{æ\tau_0}, æ\right) = cn\left(æF - \frac{t}{\tau_0}, \frac{1}{æ}\right) \tag{5.20}$$

where cn is another JEF. The necessary information about JEFs, used in this and the following chapters, is given in the appendix to this chapter.

5.2 Deformation of the distribution function

Now, taking into account the inequalities

$$|u_{\|}|/|v_{\|}| \ll 1, \quad |\Delta v_{\perp}|/v_{\perp} \ll 1, \tag{5.21}$$

we can write the solution for the distribution function of the gyroresonant particles in the form:

$$\mathcal{F}(v_{\perp}, æ, \xi, t) = \mathcal{F}_0(v_{\perp} + \Delta v_{\perp}, v_{\|} + u_{\|}) \tag{5.22}$$
$$\approx \mathcal{F}_0(v_R, v_{\perp}) + \frac{2}{k æ \tau_0}\left\{\frac{\partial \mathcal{F}_0}{\partial v_{\|}} + \frac{1 - N\beta_{\|}}{N\beta_{\perp}}\frac{\partial \mathcal{F}_0}{\partial v_{\perp}}\right\}_{v_{\|}=v_R} dn\left(F - \frac{t}{æ\tau_0}, æ\right)$$

where a Taylor series expansion has been used for $\mathcal{F}_0$.

The distribution function (5.22) includes actually two dependencies. The first one on ξ is the phase modulation (bunching), which has the wave length scale and is important as the source of seed waves in the problem of triggered emissions (see Chapters 6 and 11). The dependence of $\mathcal{F}$ (5.22) on t is slower and is characterized by the nonlinear time τ_0. This dependence is important for the sideband instability (Section 5.4), which can serve as the amplifier of triggered emissions.

We proceed to discuss some properties of the distribution function (5.22). The JEFs, $dn\left(F - \frac{t}{æ\tau_0}, æ\right)$ for the untrapped particles and $cn\left(æF - \frac{t}{\tau_0}, \frac{1}{æ}\right)$ for the trapped particles, are double periodic functions of the first argument; moreover, their period depends on æ and τ_0. This can be seen from the Fourier cosine expansions of these functions, given in (A.5.100)–(A.5.101). Their main period is τ_0, and after several τ_0 these oscillations are very fast and small. They show that,

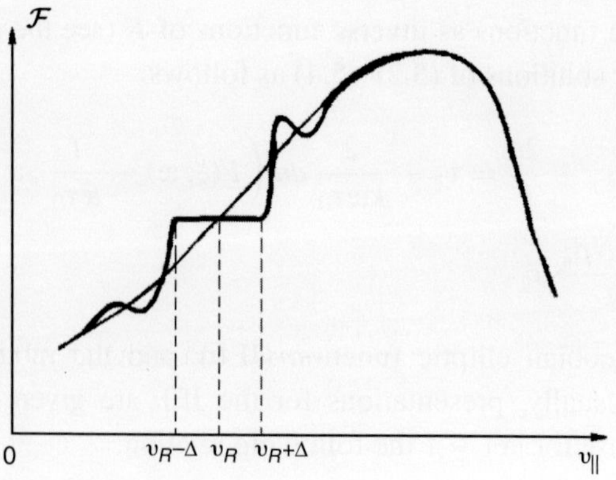

Figure 5.2 The ergodic distribution function $\mathcal{F}_E$ (thick line), $\Delta = (k\tau_0)^{-1}$.

as $t \to \infty$, the distribution function (5.22) goes to the ergodic distribution function $\mathcal{F}_E$, which can be found by the time averaging of (5.22). We take into account the fact that, according to (A.5.100) and (A.5.101), averaging over time gives:

$$\overline{dn(u, æ)} = \frac{\pi}{2K(æ)}, \quad \overline{cn(u, æ)} = 0, \tag{5.23}$$

where $K(æ)$ is the complete elliptic integral (A.5.86). Thus, we have a 'plateau' in the region of the trapped particles for which $|æ| > 1$. The final form of the ergodic function is:

$$\mathcal{F}_E - \mathcal{F}_0(v_R) = \begin{cases} \dfrac{\pi}{kæ\tau_0 K(æ)} \left(\dfrac{\partial \mathcal{F}_0}{\partial v_\|} + \dfrac{1 - N\beta_\|}{N\beta_\perp} \dfrac{\partial \mathcal{F}_0}{\partial v_\perp} \right)_{v_\| = v_R}, & |æ| \le 1, \\ 0, & |æ| > 1. \end{cases} \tag{5.24}$$

This function is shown in Fig. 5.2.

Problem 1
Find the distribution function $\mathcal{F}(t, \xi, v_\perp, æ)$, using the approximate solutions (5.15)–(5.16) for the charged particle trajectories.

Problem 2
Show that the deformation (5.22) of the distribution function conserves the total number of charged particles.

5.3 Nonlinear growth rate

To estimate the nonlinear growth rate γ_{N}, we use Equation (2.133), which can be written for a uniform medium in the form:

$$\frac{\partial A}{\partial t} = \frac{2\pi e}{N_{\mathrm{w}}^2} \left(1 - \frac{\omega}{\omega_B}\right) \int_0^\infty v_\perp^2 \, dv_\perp \int_{-\infty}^{\infty} \int_0^{2\pi} dv_\| \, d\varphi \, \mathcal{F}(\vec{v}) \cos\psi. \tag{5.25}$$

Here we must substitute the distribution function (5.22) and use the new variables F and æ instead of φ and $v_{\|}$. The Jacobian of this transformation is found in the appendix (A.5.106) and is given by:

$$dv_{\|}\, d\varphi = J\, dF\, d\text{æ}, \qquad J = \frac{2}{k\tau_0 \text{æ}^2}\,. \tag{5.26}$$

Taking into account that $\dfrac{1}{k\tau_0^2} = \dfrac{eB_\sim}{mc} v_\perp = \dfrac{eNA}{mc} v_\perp$, and using (5.22) and (5.26) we obtain:

$$\gamma_{\mathrm{N}} \equiv \frac{1}{A}\frac{\partial A}{\partial t} = \frac{8\pi e^2}{N_{\mathrm{w}} k}\left(1 - \frac{\omega}{\omega_B}\right)\int_0^\infty v_\perp^3\, dv_\perp \iint_S \frac{d\text{æ}\, dF}{\text{æ}^3}\, \mathrm{sn}\,(2F, \text{æ})$$
$$\times\; dn\left(F - \frac{t}{\text{æ}\tau_0}, \text{æ}\right)\mathcal{F}'_v, \tag{5.27}$$

where the integration region S is shown in Fig. 5.A (in the appendix), and

$$\mathcal{F}'_v = \left(\frac{\partial \mathcal{F}}{\partial v_\|} + \frac{1 - N_{\mathrm{w}}\beta_\|}{N_{\mathrm{w}}\beta_\perp}\frac{\partial \mathcal{F}}{\partial v_\perp}\right)_{v_\| = v_{\mathrm{R}}}. \tag{5.28}$$

For the further transformation of (5.27) we use the trigonometric series expansions (A.5.99) and (A.5.101) for *sn* and *dn*. After substitution of this expansion in (5.27) and integraton over F, we obtain (see Problem 5.3 for details):

$$\gamma(t) = \frac{2\gamma_{\mathrm{L}}}{\pi \int_0^\infty v_\perp^3 dv_\perp\, \mathcal{F}'_v}\int_0^\infty v_\perp^3\, dv_\perp \mathcal{F}'_v \int_0^\infty \frac{d\text{æ}}{\text{æ}}\sum_{n=0}^{\infty}\left(Q_{n1}(\text{æ})\sin\frac{n\pi t}{\text{æ}K(\text{æ})\tau_0}\right.$$
$$\left. +\; Q_{n2}(\text{æ})\sin\frac{(2n+1)\,\pi t}{2K(\text{æ})\,\tau_0}\right), \tag{5.29}$$

where γ_{L} is the linear growth rate, and the coefficients Q_{n1} and Q_{n2} are determined by the relations

$$Q_{n1} = \frac{64\pi^2 n}{\text{æ}^4 K^2(1 + q^{2n})(1 + q^{-2n})}\,, \tag{5.30}$$

$$Q_{n2} = \frac{32(2n+1)\pi^2 \text{æ}^2}{K^2(1 + q^{2n+1})(1 - q^{-2n-1})} \tag{5.31}$$

where the Jacobi parameter q is given by relations (A.5.92) and (A.5.94). The first term Q_{n1} describes the contribution of untrapped particles, and the second term Q_{n2} is due to trapped particles. The linear stage corresponds to the case $\dfrac{t}{\text{æ}\tau_0} < 1$, when only the term Q_{11} is important. Taking into account that according to (5.30) in this case $Q_{11} \simeq 1$ and $K(æ \to 0) \to \pi/2$, we find that:

$$\gamma(t) = \frac{2\gamma_L}{\pi}\int_0^\infty v_\perp^3\,dv_\perp\,\mathcal{F}_v'\int_0^1\frac{dæ}{æ}\sin\left(\frac{2t}{æ\tau_0}\right)\Big/\int_0^\infty v_\perp^3\,dv_\perp\,\mathcal{F}_v' \quad (5.32)$$

where γ_L is the linear growth rate (3.15). From (5.32), $\gamma(t) \longrightarrow \gamma_L$, because $\int_0^1 \frac{dæ}{æ}\sin\left(\frac{2t}{æ\tau_0}\right) = \int_{2t/\tau_0}^\infty \frac{\sin x}{x}\,dx \to \frac{\pi}{2}$ as $t \to 0$.

If $t \gg \tau_0$, the function under the integrals in (5.29) is rapidly oscillating, and $\gamma(t) \to 0$ when $t \to \infty$.

It is possible to obtain the total change of the wave amplitude according to the relation:

$$A(t) = A(0)\exp\left\{\int_0^t \gamma(t')\,dt'\right\}. \quad (5.33)$$

The dependence of $\int_0^t \gamma(t')$ of t' (5.33) on t is shown in Fig. 5.3 taken from Palmadesso and Schmidt (1971). Integrating (5.29), up to infinity, we find (see Problem 5.3) that:

$$\int_0^\infty \gamma\,dt \approx 2\gamma_L\frac{\int_0^\infty v_\perp^3\,dv_\perp\,\mathcal{F}_v'\tau_0(v_\perp)}{\int_0^\infty v_\perp^3\,dv_\perp\,\mathcal{F}_v'} \approx 2\gamma_L\tau_0^*(v_{\perp T}) \quad (5.34)$$

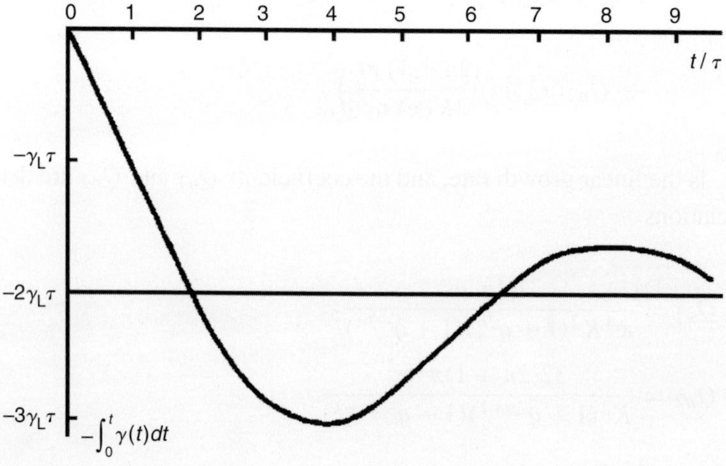

Figure 5.3 The total logarithmic damping rate of the wave amplitude vs. time (from Palmadesso and Schmidt, 1971). (Copyright American Institute of Physics, reproduced with permission.)

where $\tau_0^*(v_{\perp T})$ is the trapping period with the mean velocity $v_{\perp T}$. Therefore our approach for a given whistler wave is valid if

$$2\gamma_L \tau_0^* \ll 1. \tag{5.35}$$

Problem 3
Obtain the expressions (5.29)–(5.32) and (5.34), using the properties of JEF.

Problem 4
Analyse the nonlinear growth rate on the basis of the solutions (5.15)–(5.16).

5.4 Sideband instability

An important issue is the stability of the ergodic function relative to the excitation of other waves, having different frequencies and wave vectors. Some nonlinear phenomena observed in space plasmas are connected with this question. An example is the generation of new spectral forms via the cyclotron interaction of monochromatic whistler-mode waves with energetic electrons in the magnetosphere. Another major question, connected with this, is the transition to stochastic behaviour of the electrons and to noise-like whistler-mode signals.

We consider this problem, using the approach given in Chapter 3. Accordingly we have to find the perturbations to the motion integrals of the initial system under the action of the probe wave which, in this case, is the sideband wave (subscript 's'). Referring to Equations (5.5) and (5.8), these integrals are C_1 and $\mathcal{H}$ (or æ).

The physical basis for the instability of the plasma, which is in equilibrium with the given strong wave, is the deformation of $\mathcal{F}$ (Fig. 5.2) and the periodic motion of the resonant particles with the characteristic frequency $\Omega_{\rm tr}$ of the oscillations of particles trapped in the wave field. This frequency is determined by the expression (5.14). Thus, we analyse the behaviour of the waves which can be in resonance with this motion. This occurs if the frequency and wave vector of the probe wave is near the pump wave ω and k:

$$\omega_{\rm s} \sim \omega \pm \Omega_{\rm tr}, \qquad k_{\rm s} \approx k.$$

In this situation, phase effects are most important; these are described by the Hamiltonian $\mathcal{H}$. It is possible to estimate the change of $\mathcal{H}$ under the action of the probe wave, using an iterative procedure. The equations of motion for two waves can be written under the assumption (5.1) in the form:

$$\frac{dv_\perp}{dt} = -\omega_{\rm h}\, c\frac{1-N_{\rm w}\beta_\parallel}{N_{\rm w}}\cos\psi - \omega_{\rm hs}\, c\frac{1-N_{\rm w}\beta_\parallel}{N_{\rm w}}\cos(\psi+\delta\omega\,\Delta t-\delta k\,\Delta z),$$

$$\frac{dv_\parallel}{dt} = -\omega_{\rm h} v_\perp \cos\psi - \omega_{\rm hs} v_\perp \cos(\psi + \delta\omega\,\Delta t - \delta k\,\Delta z),$$

$$\frac{d\psi}{dt} = \omega - \omega_B - kv_\parallel \,, \tag{5.36}$$

where $\delta\omega = \omega - \omega_\mathrm{s}$, $\delta k = k - k_\mathrm{s}$, $\omega_\mathrm{h} = (e/mc)B_\sim$, $\Delta t = t - t_0$, $\Delta z = z - z_0$ and the subscript 's' refers to the sideband wave. From (5.36) we have

$$\frac{d^2\psi}{dt^2} + k\omega_\mathrm{h} v_\perp \cos\psi = -k\omega_\mathrm{hs} v_\perp \cos(\psi + \delta\omega\,\Delta t - \delta k\,\Delta z). \tag{5.37}$$

Using the definitions (5.8)–(5.12), we obtain from (5.37):

$$\frac{\partial\mathcal{H}}{\partial t} \approx -\dot{\psi}\omega_\mathrm{hs} v_\perp \cos(\psi + \delta\omega\,\Delta t - \delta k\,\Delta z). \tag{5.38}$$

In the linear approximation in B_s, the phase ψ is related to t by the relation (5.17), and

$$z \approx v_\mathrm{R}(t - t_0) + z_0 + \int^t v_\parallel^{(0)}\, dt' \tag{5.39}$$

where $v_\parallel^{(0)}$ is given by (5.19). Using the relations (5.17)–(5.19) and (5.39) and the notation in the appendix, we can write the solution of (5.38) in integral form:

$$\mathcal{H} - \mathcal{H}_0 = \Delta\mathcal{H} = 2\int_{F-\frac{t}{æ\tau_0}}^{F} \omega_\mathrm{hs} v_\perp dn\,F'\,dF' \sin\{2am\,F' + \delta\tilde{\omega}æ\tau_0(F' - F)$$
$$+\,\delta\omega t - \delta k z\}, \tag{5.40}$$

where we have substituted $\psi = 2\xi + \pi/2 = 2am\,F + \pi/2$, $\delta\tilde{\omega} = \delta\omega - \delta k v_{\mathrm{R}0}$, $t_0 = t - æ\tau_0 F$, $z_0 = z - v_\mathrm{R}æ\tau_0 F$,

$$\dot{\psi}(t) = \frac{2}{\tau_0 æ}\left(1 - æ^2\sin^2\xi\right)^{1/2} = \frac{2}{æ\tau_0}\,dn(F', æ), \quad dt' = æ\tau_0\,dF'.$$

Bearing in mind the small values of $\delta\omega$ and δk, we have omitted from the phase term the additional terms $\delta\omega(t - t_0) - \delta k(z - z_0)$, which are connected with the last term in (5.39). Now we take the Taylor series expansion of $\mathcal{F}_E$:

$$\mathcal{F}_E(æ_0) = \frac{\partial\mathcal{F}_E}{\partial æ}(æ_0 - æ) + \mathcal{F}_E(æ) \tag{5.41}$$

and bear in mind the relation which follows from (5.14):

$$æ - æ_0 = \Delta æ = \frac{\partial æ}{\partial\mathcal{H}}\Delta\mathcal{H} = -\frac{k_0\tau^2}{4}æ^3\,\Delta\mathcal{H}. \tag{5.42}$$

The sideband instability growth rate can be found from relation (5.25), where we have to substitute the ergodic function $\mathcal{F}_E$ and use the expressions (5.40) and (5.42).

For the sideband wave, the phase ψ_s is equal to:

$$\psi_s = \psi + \delta\omega t - \delta kz = 2am F - \frac{\pi}{2} + \delta\omega t - \delta kz. \tag{5.43}$$

Collecting (5.40)–(5.43) and using (5.25) with the new variables according to (5.26), we have

$$\frac{\partial A}{\partial t} = \frac{4\pi e^2 A}{mcN_w}\left(1 - \frac{\omega}{\omega_B}\right)\int_0^\infty v_\perp^3\, dv_\perp \int_{-1}^{1} d\text{æ} \int_{-K}^{K} \tau_0 \text{æ} \frac{\partial \mathcal{F}_E}{\partial \text{æ}}\, \Pi\, dF \tag{5.44}$$

where

$$\Pi = \frac{1}{2}\int_{F-\frac{t}{\text{æ}\tau_0}}^{F} dF'\, dn F' \cos\{2am F - 2am F' + \delta\tilde{\omega}\text{æ}\tau_0(F - F')\}. \tag{5.45}$$

The limit $t \to \infty$ has to be taken because the ergodic distribution function is considered; the appendix gives an explanation.

In (5.44) only the untrapped particles are taken into account because, for the trapped particles, $\partial\mathcal{F}_E/\partial\text{æ} = 0$. In (5.45), the following relation has been used:

$$\begin{aligned}
&\sin(2am F + \delta\omega t - \delta kz)\sin\{2am F' + \delta\tilde{\omega}\text{æ}\tau_0(F' - F) + \delta\omega t - \delta kz\}\\
&= \frac{1}{2}\big[\cos\{2am F - 2am F' + \delta\tilde{\omega}\text{æ}\tau_0(F - F')\}\\
&\quad + \cos\{am F + am F' - \delta\tilde{\omega}\text{æ}\tau_0(F - F') + 2\delta\omega t - 2\delta kz\}\big].
\end{aligned} \tag{5.46}$$

The rapidly oscillating second term on the right-hand side of (5.46) has been omitted. Transforming (5.45), using the new variable $\Lambda = F - F'$, the difference between two elliptical functions, and using the Fourier series expansion for $dn F'$ (5.101), we obtain:

$$\begin{aligned}
\Pi &= -\int_{-\frac{t}{\text{æ}\tau_0}}^{0} d\Lambda \frac{\pi}{4K}\left\{1 + 4\sum_{m=0}^{\infty}\frac{q^{m+1}}{1+q^{2(m+1)}}\cos\left[(m+1)\frac{\pi(F-\Lambda)}{K}\right]\right\}\\
&\quad\times \cos(2am\Lambda + \delta\tilde{\omega}\text{æ}\tau_0\Lambda)\\
&= -\int_{-\frac{t}{\text{æ}\tau_0}}^{0} d\Lambda \frac{\pi}{4K}\cos(2am\Lambda + \delta\tilde{\omega}\text{æ}\tau_0\Lambda).
\end{aligned} \tag{5.47}$$

After integration over F in (5.44) the second term in the curly brackets of (5.47) is zero. For further transformation of (5.47) we use the Fourier series expansion for

$\exp(i2am\Lambda)$ (A.5.97). Its substitution into (5.47) results in:

$$\Pi = + \sum_{n=-\infty}^{\infty} \alpha_n \frac{\pi}{2K} \frac{\sin\Big((n\pi/K\text{æ}\tau_0) + \delta\tilde{\omega}\Big)t}{(n\pi/K) + \delta\tilde{\omega}\text{æ}\tau_0}$$

$$\underset{t\to\infty}{\to} \pm \sum_{n=-\infty}^{\infty} \frac{\pi^2\alpha_n}{2K} \frac{\delta\Big(\delta\tilde{\omega} + (n\pi/K\text{æ}\tau_0)\Big)}{|\text{æ}\tau_0|} \tag{5.48}$$

where the sign + (−) refers to positive (negative) values of æ, and $\delta(x_n)$ is the Dirac delta-function, and α_n is determined by (A.5.98). We have used the asymptotic relation:

$$\frac{\sin x_n t}{x_n} \underset{t\to\infty}{\to} \pi\,\delta(x_n). \tag{5.49}$$

Substituting (5.48) into (5.44) and taking into account (A.5.98), we find that:

$$\int_{-1}^{1} d\text{æ} \int_{-K}^{K} dF\, \Pi \text{æ}\tau_0 \frac{\partial \mathcal{F}_E}{\partial \text{æ}} = \pm \sum_{n=-\infty}^{\infty} \frac{\alpha_n \text{æ}_n}{|n\text{æ}_n|} \operatorname{sign} \frac{\partial(1/\text{æ}K)}{\partial \text{æ}}$$

$$= \pi \sum_{n=1}^{\infty} \frac{\pi^2}{\text{æ}_n^2 K^2} \left(\frac{q^n}{1-q^{4n}} + \frac{q^{-n}}{1-q^{-4n}} \right) = \pi \sum_{n=1}^{\infty} \frac{\pi^2 q^n}{\text{æ}_n^2 K^2 (1+q^{2n})} \tag{5.50}$$

where q is determined by the relation (A.5.92) and

$$K(\text{æ}_n)\,\text{æ}_n = (\delta\tilde{\omega}\cdot\tau_0)^{-1}\pi n. \tag{5.51}$$

Finally, using (5.50) in (5.44) we can write:

$$\gamma = \frac{1}{A}\left|\frac{\partial A}{\partial t}\right| = \gamma_{\mathrm{L}} \int_0^{\infty} v_\perp^3\, dv_\perp\, \mathcal{F}'_v \sum_{n=1}^{\infty} \frac{4\pi^2 q^n}{\text{æ}_n^2\, K^2(1+q^{2n})} \left(\int_0^{\infty} v_\perp^3\, dv_\perp\, \mathcal{F}'_v \right)^{-1} \tag{5.52}$$

where γ_{L} is the linear growth rate.

If we consider a large frequency shift $\delta\omega\tau_0 \gg 1$, we find from (5.52) ($\text{æ}_n \to 0$, $K \to \pi/2$, $q_1 \to \frac{\text{æ}_1^2}{16}$, $q_{n>2} \to 0$) that $\gamma = \gamma_{\mathrm{L}}$. This result is clear from a physical point of view: for $\delta\omega\tau_0 \gg 1$ the disturbance of the distribution function is small, and γ is determined by the initial $\mathcal{F}_0$. We remove from the integral (5.52) the slowly changing sum $\sum_{n=1}^{\infty}$ to obtain the ratio of the sideband growth rate γ_{SB} to the linear

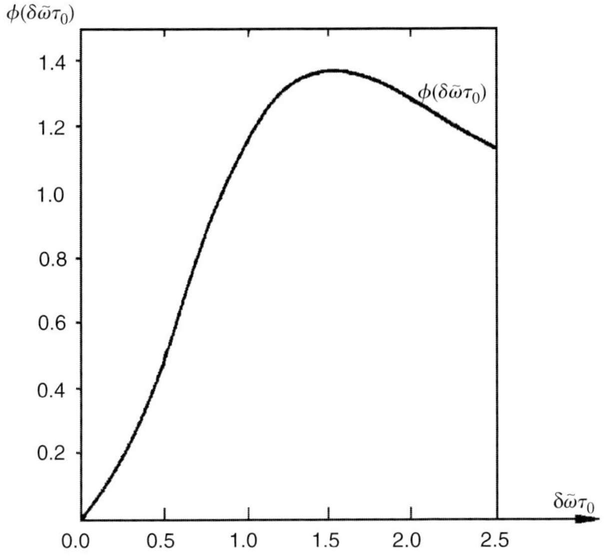

Figure 5.4 The frequency dependence of the sideband growth rate, γ_{SB}.

growth rate γ_L as (see, as well, Bud'ko *et al.*, 1972):

$$\frac{\gamma_{SB}}{\gamma_L} = \sum_{n=1}^{\infty} \frac{4\pi^2 q^n}{æ_n^2 K^2(1+q^{2n})}\bigg|_{æ_n = æ_n(v_{\perp T})}. \tag{5.53}$$

The dependence γ_{SB} on the frequency shift $\delta\tilde{\omega}\tau_0$ is shown in Fig. 5.4.

Two sideband waves are preferentially excited, with frequency shift $\delta\tilde{\omega}\tau_0 \approx \pm 1.5$ and growth rate $\gamma_{SB}^{\pm}/\gamma_L \approx 1.4$. The sideband instability is possible only for an unstable plasma, when the linear growth rate $\gamma_L > 0$. These results could be predicted qualitatively from Fig. 5.2, where the distribution function modification is shown under the action of a strong wave. Actually, the amplification (or damping) of waves is determined in the linear approach by the value and the sign of the derivative of the distribution function. So, taking into account the sign of this derivative in Fig. 5.2, we could qualitatively obtain the above results.

The sideband instability leads to some important consequences. First, it shows the direction of the possible temporal evolution of the wave frequency spectrum through discrete 'satellite' lines above and below the initial frequency. Secondly, it follows that the quasi-monochromatic wave generation regime is generally speaking a transient regime, which leads to more complicated and probably stochastic generation regimes over a rather wide range of frequencies.

It has to be borne in mind that a uniform infinite medium has been considered here. The situation can be drastically changed in the case of a non-uniform magnetic field or for the large, but space-limited, systems with positive feedback which are real CMs. These are considered in the final section of this chapter and in the next chapters.

Problem 5
Analyse the sideband instability, employing the approximate solutions (5.15)–(5.16).

5.5 Self-consistent effects for large (infinite) systems

In previous sections a wave of given amplitude has been considered. Now we shall estimate some effects in a self-consistent way. It is clear that for effects such as charged particles trapped by the wave, or untrapped by the wave, a transition to the ergodic distribution function and the sideband instability will be retained. Analytical studies using the self-consistent approach are very difficult, and numerical computations are used as a rule. The necessary computational experience has been acquired for electron CM devices in the laboratory, and we summarize this briefly in the next section.

Here we investigate the important question of the saturation wave amplitude, which is achieved during the process of CI development. We start by supposing that a quasi-monochromatic wave is generated and neglecting the sideband instability. According to the previous analysis (Section 5.2), after the stationary state is reached the ergodic distribution function (5.23) is formed; this depends on the wave amplitude only. Thus, we can use the principle of energy conservation to find this amplitude.

The application of the law of the conservation of energy in this case is explained by Fig. 5.2. We must calculate the change of the kinetic energy of the electrons when the initial distribution function $\mathcal{F}_0$ (thin line in Fig. 5.2) becomes the ergodic distribution function $\mathcal{F}_E$ (thick line in Fig. 5.2), and compare this with the wave energy density (see Galeev and Sagdeev, 1973):

$$\left(\int d^3\vec{v}\, W\mathcal{F}\right)_{t=0} - \left(\int d^3\vec{v}\, W\mathcal{F}\right)_{t\to\infty} = \int W_0\mathcal{F}_0\, d^3\vec{v} - \int W\mathcal{F}_E\, d^3\vec{v} = \mathcal{E}_{\mathrm{w}}. \tag{5.54}$$

Here $\mathcal{E}_w = \dfrac{B_\sim^2}{8\pi}\dfrac{1}{1-\omega/\omega_B}$ is the energy density of whistler waves (see Chapter 2), $B_\sim$ is the amplitude of the wave magnetic field, and the kinetic energy of a charged particle is $W = \dfrac{m}{2}(v_\perp^2 + v_\parallel^2)$. We note that the total charged particle number, under the deformation shown in Fig. 5.2, is conserved (see Problem 2). It is convenient to use variables (æ, F) instead of $v_\parallel$ and φ under the calculations of the integrals in (5.54) with integration limits which are determined by Fig. 5.A. In these variables the element of the phase space is equal to

$$d^3\vec{v} = 2v_\perp\, dv_\perp \frac{d\text{æ}\, dF}{k\,\tau_0\, \text{æ}^2}\,. \tag{5.55}$$

Here we have used the expression (A.5.106). Further we take into account a Taylor series expansion for $\mathcal{F}_0$ (5.22) and the motion integral (5.5). From (5.5) we obtain:

$$W_0 - W = mv_\phi u_\parallel. \tag{5.56}$$

The substitution of (5.22) and (5.56) into (5.54) gives us the following relation:

$$\int W_0 \mathcal{F}_0\, d^3\vec{v} - \int W \mathcal{F}_E\, d^3\vec{v} = \int d^3\vec{v}\, \mathcal{F}'_v m v_\phi u_\parallel^2 (t=0). \tag{5.57}$$

In (5.57) we have taken into account that only the trapped resonant electrons (æ $\geq$ 1) give a contribution to the energy conservation law (5.54), and the expression (5.24) is valid for the ergodic distribution function $\mathcal{F}_E$. Coming to new variables according (5.55) we obtain:

$$\begin{aligned}
&\int d^3\vec{v}\, \mathcal{F}'_v m v_\phi\, u_\parallel^2 (t=0) \\
&= \int_0^\infty u_\perp\, du_\perp\, (\mathcal{F}'_v m v_\phi) \frac{8}{(k\tau_0)^3} \\
&\quad \times 2 \int_1^\infty \frac{d\text{æ}}{\text{æ}^4} \int_0^{\frac{1}{\text{æ}} K\left(\frac{1}{\text{æ}}\right)} cn^2 \left(\text{æ}F, \frac{1}{\text{æ}}\right) dF,
\end{aligned} \tag{5.58}$$

where we have used (5.26), (5.19) and (5.20) for $t = 0$. With the help of Byrd and Friedman (1954) and making the change æ $\to$ æ$^{-1}$, we find that:

$$\begin{aligned}
&2 \int_1^\infty \frac{d\text{æ}}{\text{æ}^4} \int_0^{\frac{1}{\text{æ}} K\left(\frac{1}{\text{æ}}\right)} cn^2 \left(\text{æ}F, \frac{1}{\text{æ}}\right) dF \\
&= 2 \int_0^1 \text{æ}'\, d\text{æ}' \int_0^{K(\text{æ}')} cn^2(u, \text{æ}')\, du \\
&= 2 \int_0^1 \text{æ}'\, d\text{æ}' \left[E(\text{æ}') - \left(1 - \text{æ}'^2\right) K(\text{æ}') \right] = \frac{16}{9},
\end{aligned} \tag{5.59}$$

where E and K are the complete elliptic integrals (see (A.5.84)–(A.5.86)).

Substituting (5.58) and (5.59) into (5.54), we obtain:

$$\hat{\gamma}_0 \tau_0 = \left(\frac{3\pi}{16}\right)^2 \tag{5.60}$$

where the expression for the linear CI growth rate (3.15) has been used. The hat symbol above γ_0 means that it must be considered as an integral operator over $v_\perp$, because τ_0 is a function of $v_\perp$. Making rough estimates we can take $\tau_0 = (\omega_{\rm h} k v_\perp)^{-1/2}$ at the middle point $v_\perp \approx v_{\perp T}$. Thus, the relation (5.60) permits us to estimate the whistler wave amplitude for the final stage of CI development in the quasi-monochromatic regime.

Problem 6
Show that untrapped particles do not make any contribution to the energy conservation law (5.54).

5.6 Finite systems. BWO generation regime

Up to now we have discussed the nonlinear dynamics of the cyclotron instability for homogeneous and infinite systems. Both assumptions are violated in real conditions. We shall discuss the role of the inhomogeneity of magnetospheric parameters in Chapters 6 and 7. In this section we consider some new effects which take place in finite-in-space systems.

It is clear that the limited dimensions of a system start to play an important role if the 'trapping' period τ_0 (5.14) of the resonant electrons in the coordinate system moving with the group velocity of a wave packet is comparable with the wave packet duration T', which is equal in this coordinate system to $T' = \ell/|\vec{v}_{\rm g} - \vec{v}_{\rm R}|$, where ℓ is the length of the packet:

$$\tau_0 \equiv \Omega_{\rm tr}^{-1} \sim \ell/|\vec{v}_{\rm g} - \vec{v}_{\rm R}|. \tag{5.61}$$

In the case of the cyclotron interaction (CI) considered with $\omega < \omega_B$, $\vec{v}_{\rm g} \uparrow\downarrow \vec{z}_0 \uparrow\uparrow \vec{v}_{\rm R}$, and $|\vec{v}_{\rm g} - \vec{v}_{\rm R}| = |v_{\rm g}| + v_{\rm R}$. It is clear that when (5.61) is met the boundary conditions should be taken into account.

There is a rich experience in the investigation of such devices, for electron beam propagation in laboratory conditions. The most interesting for CMs in space is the so-called Backward Wave Oscillator (BWO), which occurs when the group velocity is opposite to the electron beam velocity. For magnetospheric conditions this generation regime manifests itself as the absolute cyclotron instability, which develops in a local region near the equatorial cross-section of a magnetic flux tube without any mirrors for waves at the ends. The linear stage of BWO generation was analysed in Chapter 4. Here we shall discuss the nonlinear theory of a BWO generator.

As we know from the analysis of Chapter 4, BWO generation takes place for well-organized electron beams with delta or step-like distributions over the field-aligned component of velocity. The nonlinear theory of BWO generation in the case of a delta-function distribution is well-developed as applied to laboratory devices (Ginzburg and Kuznetsov, 1981). For magnetospheric conditions a step-like deformation of the distribution over $v_\parallel$ is appropriate for well-organized electron beams (see Chapter 7), and we shall analyze the nonlinear stage of BWO generation for that case.

We take the initial system of equations in the form (2.133)–(2.134):

$$\frac{\partial A_m}{\partial t} + v_{gz}\frac{\partial A_m}{\partial z} = \alpha_2\, e \int d^3\vec{v}v_\perp \mathcal{F}\cos\psi \equiv \alpha_2 j_1,$$

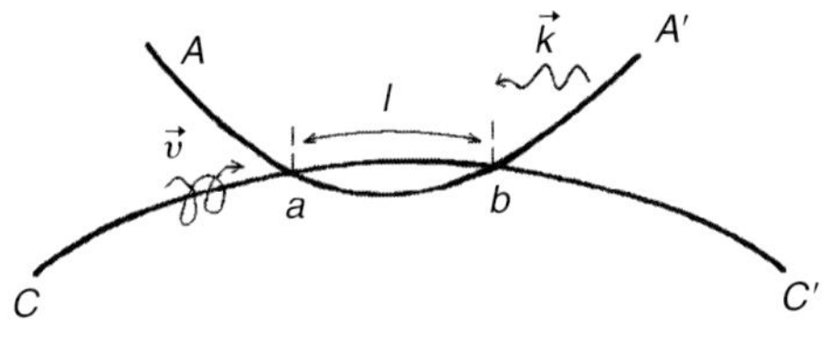

Figure 5.5 The BWO generator with two interaction points a and b, where the cyclotron resonance condition is fulfilled.

$$A_m \left(\frac{\partial \Delta\Theta}{\partial \theta} + v_{gz} \frac{\partial \Delta\Theta}{\partial z} \right) = \alpha_2 \, e \int d^3\vec{v} v_\perp \mathcal{F} \sin \psi = \alpha_2 j_2 \tag{5.62}$$

where $\alpha_2 = \dfrac{2\pi\,|v_g|}{cN_w}$, $\mathcal{A} = A_m \exp\{i\Delta\Theta\}$ is the slowly changing amplitude of a wave electric field, and $\psi = \varphi - \Theta$ is the relative phase between an electron and wave (see Section 5.4). Following Section 4.2 we can transform the expression for j_1 in the case of a step-like distribution function to the following form:

$$\begin{aligned} j_1 &= e \int_0^\infty v_\perp^2 \, dv_\perp \int_0^{2\pi} d\varphi \cos \psi \int_{a+v_0}^\infty dv_\parallel \, \mathcal{F}_0(v_\parallel, v_\perp) \\ &= -e \int_0^\infty v_\perp^2 \, dv_\perp \int_0^{2\pi} d\varphi \sin \psi \left(\frac{\partial a}{\partial \varphi} \right) \mathcal{F}_0(v_0 + a, v_\perp) \end{aligned} \tag{5.63}$$

where $\mathcal{F}_0$ is the smooth part of the distribution function, and a is determined by the formula (4.18); the wave packet propagates in the $(-z)$-direction, $v_{gz} < 0$, and the electron beam resonant velocity $v_\parallel \approx v_0 + a > 0$; v_0 is the modulus of the step velocity. The wave–electron interaction region is shown in Fig. 5.5.

It is convenient for space-limited systems to use the variables:

$$z = v_0 t + z_0 \text{ and } \zeta \equiv t_0 = t - z/v_0. \tag{5.64}$$

In these variables

$$a = \frac{1}{k} \frac{d\psi}{dz} = \int_0^z \frac{eN_w v_\perp}{mcv_0^2} A_m(z', \zeta) \cos \psi'(z', \zeta) \, dz' \tag{5.65}$$

and

$$\begin{aligned} j_1 = e \int_0^\infty v_\perp^2 \, dv_\perp \, \mathcal{F}_0(v_0 + a, v_\perp) \int_0^{2\pi} d\varphi \sin \psi(z, \varphi) \int_0^z \frac{eN_w v_\perp}{mcv_0} \\ \times A_m(z', \zeta) \sin \psi'(z, \varphi) \, dz'. \end{aligned} \tag{5.66}$$

In space plasmas as a rule, the change of the velocity components $v_\parallel$, $v_\perp$ is small, $|a| \ll v_0$, and the main nonlinear effect is dephasing (phase mismatch), so the nonlinearity is important only for the phase change. In this approach the current

j_1 (5.66) will be written in the form:

$$j_1 = en_{\text{st}} \int_0^{2\pi} d\psi_0 \int_0^z \frac{eN_{\text{w}} A_m(z', \zeta)}{2mc} [\cos(\psi - \psi') - \cos(\psi + \psi')] \, dz' \tag{5.67}$$

where

$$n_{\text{st}} = \frac{1}{v_0} \int_0^\infty v_\perp^3 \, dv_\perp \, \mathcal{F}_0(v_z = v_0, v_\perp) \tag{5.68}$$

is the effective density of the electrons, and a new variable ψ_0 has been introduced instead of φ, which corresponds to the initial relative phase (at $t = t_0$) and has a homogeneous distribution in the interval $[0, 2\pi]$. The wave equation (5.62) in the variables (z, ζ) has the form:

$$\left(\frac{1}{v_0} + \frac{1}{v_{\text{g}}} \right) \frac{\partial A_m}{\partial \zeta} - \frac{\partial A_m}{\partial z} = -\frac{2\pi}{N_{\text{w}} c} j_1(z, \zeta). \tag{5.69}$$

The relations (5.65), (5.67) and (5.69) together describe the nonlinear stage of BWO generation in a quasi-monochromatic regime. The analytical solution of this system of equations is impossible. We shall concern ourselves with the basics of this model, corresponding to the electron–wave interaction at two discrete points $z = 0;\ \ell$, at either ends of the interaction region (Fig. 5.5). In laboratory conditions this model corresponds to the interaction inside two resonators, localized at points $z = 0;\ \ell$. In the magnetosphere these two points can be connected with the condition of exact cyclotron resonance (CR) in the inhomogeneous magnetic field, placed symmetrically about the equator. To describe this model we should introduce the interaction coefficient ν in the form:

$$\nu = \nu_0 \left[\delta(z') + \delta(z' - \ell) \right] \tag{5.70}$$

one under the integral over z in the expression for a (5.65), and the second as the multiplier on the right side of (5.69) of the current j_1. After that the solution of (5.65) can be presented in the following form:

$$\psi(z, \zeta) = \begin{cases} \psi_0, & z = 0 - \varepsilon \quad (\varepsilon \to 0), \\ \alpha_{10} \, A_m(0, \zeta) \cos \psi_0 \cdot z + \psi_0, & 0 < z < \ell, \\ \alpha_{10}[A_m(0) \cos \psi_0 \cdot z & \\ \quad + A_m(l) \cos \psi(\ell) \, (z - \ell)] + \psi_0, & z \geq \ell \end{cases} \tag{5.71}$$

where the coefficient α_{10} is equal to:

$$\alpha_{10} = \frac{eN_{\text{w}} \overline{v}_\perp k \nu_0}{2mcv_0^2}, \tag{5.72}$$

and $\overline{v}_\perp$ is the characteristic value of $v_\perp$.

The normalizing length ℓ in (5.70)–(5.72) can be selected from the condition for the correct transition to the linear limit of BWO generation. Inside the interaction region $0 \le z \le \ell$ the arguments $[\psi(z,\zeta) \mp \psi'(z',\zeta)]$ in (5.67) can be written as

$$\psi(z) \mp \psi(z') = \alpha_{10} A_m(0,\zeta)\,(z \mp z')\cos\psi_0. \tag{5.73}$$

Using the relations (Gradstein and Ryzhik, 1971) for the Bessel functions $J_n(y)$

$$\int_0^{2\pi} \cos(y\cos\psi_0)\,d\psi_0 \equiv 2\int_0^{\pi} \cos(y\cos\psi_0) = \pi\, J_0(y),$$

$$\int_0^{2\pi} \cos(y\cos\psi_0 + 2\psi_0)\,d\psi_0 = \int_0^{2\pi} \cos(y\cos\psi_0)\cos 2\psi_0\,d\psi_0$$

$$= -\pi J_2(y)$$

and

$$J_0(y) + J_2(y) = \frac{2}{y}\,J_1(y)$$

we obtain, after integration of (5.67) over z' and ψ_0:

$$j_1(\zeta) = \left(\frac{e^2 n_{\mathrm{st}} N_{\mathrm{w}} v_0}{mc}\right) A_m(0,\zeta)\cdot\left(\frac{J_1(y)}{y}\right), \tag{5.74}$$

where $J_1(y)$ is the Bessel function of the first order, the argument y is equal to

$$y = \alpha_{10}\, A_m(0,\zeta)\,\ell. \tag{5.75}$$

Now the wave equation (5.69) can be written as

$$\left(\frac{1}{v_0} + \frac{1}{v_{\mathrm{g}}}\right)\frac{\partial A_m}{\partial \zeta} - \frac{\partial A_m}{\partial z} = \frac{\omega_{ps}^2 N_{\mathrm{w}} v_0^2\, A_m(\zeta,0)}{c^2}\left(\frac{J_1(y)}{y}\right)[\delta(z) + \delta(z-\ell)] \tag{5.76}$$

where $\omega_{ps}^2 = 4\pi e^2 n_{\mathrm{s}}/m$, the square of the plasma frequency of energetic electrons s.

Integrating (5.75) over z in the limits $-\varepsilon + \ell < z < \ell + \varepsilon$ ($\varepsilon \to 0$) and taking into account the boundary condition $A_m(\ell + \varepsilon) = 0$, we obtain:

$$-\int_{\ell-\varepsilon}^{\ell+\varepsilon} \frac{\partial A_m}{\partial z'}\,dz' = A_m(\ell,\zeta) = \frac{\omega_{ps}^2 v_0^2}{c^2}\left(\frac{J_1(y)}{y}\right) A_m(0,\zeta). \tag{5.77}$$

Inside the interaction region $\varepsilon < z < \ell - \varepsilon$, Equation (5.69) gives

$$\left(\frac{1}{v_0} + \frac{1}{v_{\mathrm{g}}}\right)\frac{\partial A_m}{\partial \zeta} - \frac{\partial A_m}{\partial z} = 0. \tag{5.78}$$

We just obtain from (5.78):

$$A_m(\ell, \zeta) = A_m(0, \zeta + T), \tag{5.79}$$

where

$$T = \ell\left(\frac{1}{v_0} + \frac{1}{v_g}\right). \tag{5.80}$$

Using (5.79), we find the final equation, describing the nonlinear dynamics of BWO generation, as:

$$A_m(0, \zeta + T) = 2\lambda\left(\frac{J_1(y)}{y}\right) A_m(0, \zeta) \tag{5.81}$$

where

$$\lambda = \frac{\omega_{ps}^2 v_0^2}{2c^2}. \tag{5.82}$$

Putting $y \ll 1$, we obtain from (5.80) the equation ($y = 2\sqrt{2} \cdot p, z = 0$)

$$p(t + T) = \lambda\, p(t)\,[1 - p^2(t)] \tag{5.83}$$

which is well known in the theory of laboratory BWO devices in the case of a δ-distribution over velocity (a helical beam (Ginzburg and Kuznetsov, 1981)). BWO generation starts when $\lambda > 1$. With an increase of λ the equation (5.83) reveals different regimes of wave generation which are shown in Fig. 5.6, from Ginzburg and Kuznetsov (1981). The Lamerey diagram (Arnol'd, 1978), shown in Fig. 5.6, serves to plot the solution of Equation (5.83); $f(p)$ is the function on the right side of (5.83). We can see from the right-hand side of Fig. 5.6 that the generation regimes are changing from oscillations with a constant amplitude ($1 < \lambda < 2$) to oscillations with periodic amplitude modulation ($\lambda > 2$) and to oscillations with stochastic modulation, when $\lambda > 2.303$. A more detailed analysis of the model (5.83) will be performed in Chapter 11 where the generation of chorus is discussed.

5.7 Summary

The main nonlinear effect under the resonant wave–particle interaction is the trapping of the resonant electrons by the wave. Electrons oscillate in the potential well of the wave with the period τ_0, which is determined by the relation (5.14).

These oscillations lead to phase mixing, which changes the wave damping (growth) rate. On time scales $t \gg \tau_0$, the distribution function goes to the so-called ergodic function, which is determined by (5.24), and the damping (growth) rate goes to zero. In an unstable plasma this ergodic state is unstable in relation to the excitation

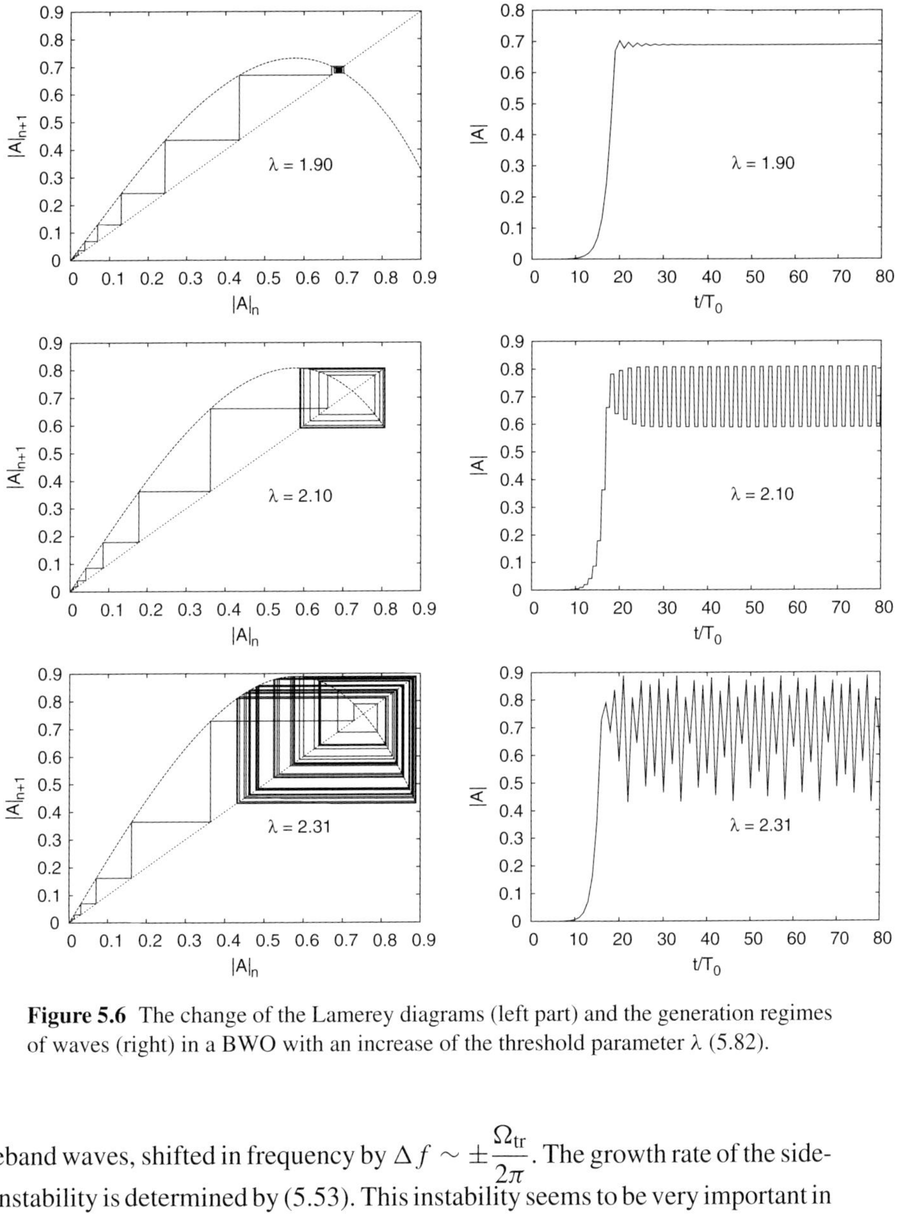

Figure 5.6 The change of the Lamerey diagrams (left part) and the generation regimes of waves (right) in a BWO with an increase of the threshold parameter λ (5.82).

of sideband waves, shifted in frequency by $\Delta f \sim \pm\frac{\Omega_{tr}}{2\pi}$. The growth rate of the sideband instability is determined by (5.53). This instability seems to be very important in determining the general dynamics of the interaction of quasi-monochromatic waves with an unstable plasma. Analytical considerations are possible if the wave amplitude is almost constant for times $\Delta t \gg \Omega_{tr}^{-1}$. The growth of quasi-monochromatic waves in the unstable plasma, which manifest themselves as the instability develops, is not described analytically, but it is possible to find the wave amplitude and the distribution function for the final stage of the instability when $t \rightarrow \infty$ (see the relation (5.60)).

New effects appear in a spatially limited unstable plasma. The most interesting phenomenon is the backward wave oscillator generator of waves, which operates in a small region of the magnetic flux tube without any reflections of waves from boundaries of the unstable region. This is a generator of discrete signals (see Fig. 5.6), and can serve as a real mechanism for the generation of discrete electromagnetic emissions in the Earth's magnetosphere.

Appendix The properties of the elliptic integrals and Jacobi elliptic functions

1. The main definitions.

The normal elliptic integral of the first kind:

$$\int_0^{\varphi} \frac{d\xi}{\sqrt{1 - æ^2 \sin^2 \xi}} = F(\varphi, æ). \tag{A.5.84}$$

The normal elliptic integral of the second kind:

$$\int_0^{\varphi} d\xi \sqrt{1 - æ^2 \sin^2 \xi} = E(\varphi, æ). \tag{A.5.85}$$

Complete elliptic integrals:

$$F\left(\frac{\pi}{2}, æ\right) \equiv K(æ), \quad E\left(\frac{\pi}{2}, æ\right) \equiv E(æ), \quad K(0) = E(0) = \frac{\pi}{2}. \tag{A.5.86}$$

Jacobian elliptic functions (JEFs) are inversions of elliptical integrals. If $F(\varphi, k) = u$, inverse functions are defined by the relations:

$$\varphi = am(u, æ) \equiv amu, \quad y_1 \equiv \sin\varphi = sn(u, æ) \equiv snu \tag{A.5.87}$$

where amu is the amplitude u, and snu is the sine of the amplitude u. Correspondingly,

$$cn(u, æ) = \sqrt{1 - y_1^2} = \cos\varphi \tag{A.5.88}$$

and

$$dn(u, æ) \equiv dnu = \sqrt{1 - æ^2 y_1^2} = \sqrt{1 - æ^2 \sin^2\varphi}. \tag{A.5.89}$$

The functions (A.5.87)–(A.5.89) are Jacobian elliptic functions.

2. Developments of elliptical integrals.

$$K(æ) = \frac{\pi}{2}\left(1 + \frac{1}{4}æ^2 + \frac{9}{64}æ^4 + \frac{25}{256}æ^6 + \ldots\right) \tag{A.5.90}$$

$$K(æ) = \frac{\pi}{2}\left[1 + 4\sum_{m=1}^{\infty} \frac{q^m}{1 + q^{2m}}\right]_{\text{nome}}, \tag{A.5.91}$$

where q is Jacobi's nome (Borwein and Borwein, 1987)

$$q = \exp\left\{-\frac{\pi K'}{K}\right\}, \tag{A.5.92}$$

$$K' = K\left(\sqrt{1-æ^2}\right) \equiv K(æ'), \tag{A.5.93}$$

where $æ'$ is the complementary modulus

$$q = \frac{æ^2}{16}\left[1 + \frac{æ^2}{2} + \frac{21æ}{64} + \frac{31æ}{128} + \dots\right], \quad æ^2 < 1, \tag{A.5.94}$$

$$E(æ) = \frac{\pi}{2}\left[1 - \frac{æ^2}{4} - \frac{3}{64}æ^4 - \frac{5}{256}æ^6 - \dots\right]. \tag{A.5.95}$$

3. Fourier series of Jacobian elliptic functions.

$$amu = \frac{\pi u}{2K} + 2\sum_{m=0}^{\infty}\frac{q^{m+1}}{(m+1)\left[1+q^{2(m+1)}\right]}\sin\left[(m+1)\frac{\pi u}{K}\right] \tag{A.5.96}$$

$$\exp\{2iam(u, æ)\} = \sum_{n=-\infty}^{\infty} 2\alpha_n \exp\left\{\frac{in\pi u}{K(æ)}\right\} \tag{A.5.97}$$

where

$$\alpha_n = \left[\frac{\pi}{æ\, K(æ)}\right]\frac{nq^n}{1-q^{4n}}, \qquad n \neq 0, \tag{A.5.98}$$

$$\alpha_0 = \frac{1}{2} - \frac{1}{æ^2}\left[1 - \frac{E(æ)}{K(æ)}\right]$$

$$snu = \frac{2\pi}{æK}\sum_{m=0}^{\infty}\frac{q^{m+1}}{1-q^{2m+1}}\sin\left[(2m+1)\frac{\pi u}{2K}\right] \tag{A.5.99}$$

$$cnu = \frac{2\pi}{æK}\sum_{m=0}^{\infty}\frac{q^{m+1/2}}{1+q^{2m+1}}\cos\left[(2m+1)\frac{\pi u}{2K}\right] \tag{A.5.100}$$

$$dnu = \frac{\pi}{2K} + \frac{2\pi}{K}\sum_{m=0}^{\infty}\frac{q^{m+1}}{1+q^{2(m+1)}}\cos\left[(m+1)\frac{\pi u}{K}\right]. \tag{A.5.101}$$

The useful relations

$$\frac{E(æ)}{\pi} - \frac{\pi}{4K(æ)} = \frac{2\pi}{K}\sum_{n=1}^{\infty}\frac{1}{(1+q^{2n})(1+q^{-2n})} \tag{A.5.102}$$

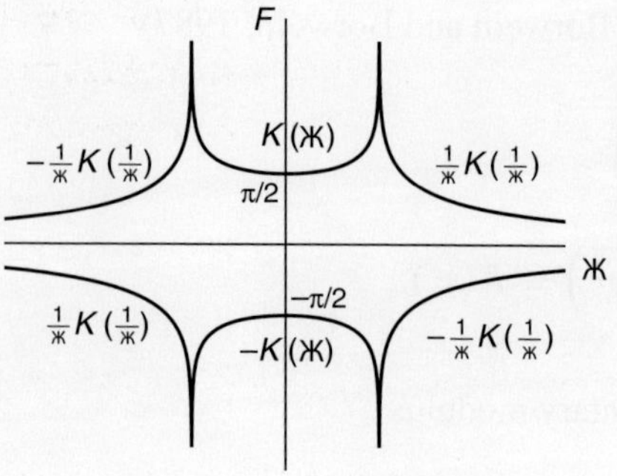

Figure 5.A Diagram showing the velocity space in new variables æ and F.

$$\frac{æ}{\pi}\Big[E(æ) + (æ^2 - 1)\,K(æ)\Big] = \frac{2\pi æ}{K}\sum_{n=0}^{\infty}\frac{1}{(1+q^{2n+1})(1+q^{-2n-1})}. \tag{A.5.103}$$

4. The transformation of the Jacobian J of the variables $(v_\parallel, \varphi)$ to (æ, F).

$$J = \begin{vmatrix} \dfrac{\partial v_\parallel}{\partial F} & \dfrac{\partial v_\parallel}{\partial æ} \\ \dfrac{\partial \varphi}{\partial F} & \dfrac{\partial \varphi}{\partial æ} \end{vmatrix}. \tag{A.5.104}$$

The relation between $(v_\parallel, \varphi)$ and (æ, F) is determined by the relations which follow from (5.9), (5.13) and (5.14)

$$\varphi = am(u, æ), \qquad v_\parallel = \frac{2}{k\tau_0}\frac{1}{æ}\sqrt{1 - æ^2\sin^2\varphi} = \frac{2}{k\tau_0}\frac{dn(æ, u)}{æ}, \qquad F \equiv u. \tag{A.5.105}$$

From (A.5.104)–(A.5.105) we have:

$$J = \frac{2}{k\tau_0}\begin{vmatrix} \dfrac{1}{æ}\dfrac{\partial dn(u, æ)}{\partial u} & \dfrac{\partial}{\partial æ}\left(\dfrac{1}{æ}\,dnu\right) \\ \dfrac{\partial am(u, æ)}{\partial u} & \dfrac{\partial am(u, æ)}{\partial æ} \end{vmatrix} = \frac{2}{k\tau_0 æ^2}, \tag{A.5.106}$$

where the derivatives are taken from the *Handbook of Elliptic Integrals* (Byrd and Friedman, 1954). Using (A.5.84), (A.5.86) and (A.5.105), we find new integration limits in the variables (æ, F). These limits are shown in Fig. 5.A.

Chapter 6

Nonlinear interaction of quasi-monochromatic whistler-mode waves with gyroresonant electrons in an inhomogeneous plasma

The development of the nonlinear theory of gyroresonant quasi-monochromatic wave–particle interactions in an inhomogeneous plasma was stimulated by experiments on the injection of quasi-monochromatic VLF radio waves into the magnetosphere, which led to the discovery of a broad class of triggered VLF emissions. We shall discuss this phenomenon in Chapter 11. The aim of this chapter is to give the foundations of the theory, putting off the applications to later chapters.

The case of cyclotron resonant wave–particle interactions in an inhomogeneous plasma is much more complicated than the case for a homogeneous plasma considered in the previous chapter. It is connected with the mismatch of the cyclotron resonance condition, which occurs due to the change of the electron gyrofrequency along the inhomogeneous magnetic field, and to the change of the magnitude of the wave vector. Helliwell (1967) made a fundamental contribution to the mechanism generating whistler-mode waves in an inhomogeneous magnetic field when he suggested the idea of second-order cyclotron resonance (CR). Then the CR mismatch is compensated by the change with time of the frequency of a generated wave packet. The first papers on a quantitative consideration of the nonlinear CR interaction of electrons with a quasi-monochromatic whistler wave in an inhomogeneous magnetic field were published between 1971 and 1974 (Dysthe, 1971, Nunn, 1971, 1973, 1974, Sudan and Ott, 1971, Helliwell and Crystal, 1973, and Brinca 1973). A systematic approach to the problem of gyroresonant wave–particle interactions in an inhomogeneous plasma was developed by Karpman and Shklyar (1972, 1974), Istomin *et al.* (1973, 1975) and Karpman *et al.* (1974 a, b; 1975 a b). Important contributions to this problem were made by Roux and Pellat (1978), Matsumoto (1979), Vomvoridis and Denavit (1979, 1980), Matsumoto and Omura (1981), Vomvoridis *et al.* (1984, 1986), Winglee (1985), Sagdeev *et al.* (1985), Molvig *et al.* (1988), Sa and Helliwell (1988), Omura and Matsumoto (1989) and Carlson *et al.* (1990), and Sa (1990). A comprehensive review of work carried out during this period was published by Omura *et al.* (1991).

Further investigations up to the present have been devoted to a quantitative analysis of electron acceleration by a whistler-mode wave packet, with either a constant or changing (in space and time) frequency (Hobara *et al.*, 2000, Trakhtengerts *et al.*, 2003a); phase bunching effects have been taken into account (Trakhtengerts *et al.*, 2003b). Computer simulations of particular cases have dealt with triggered VLF emissions (Nunn and Smith, 1996, Smith and Nunn, 1998) and VLF chorus and discrete emissions (Nunn *et al.*, 1997). An original approach, based on a resonance-averaged Hamiltonian, is being developed by Albert (2000).

It should be borne in mind that the theory for arbitrary inhomogeneous plasma parameters as applied to gyroresonant wave–particle interactions is very complicated and has no general solution. In space plasmas the situation is simplified thanks to the very small changes of both plasma parameters in the medium and magnetic field strength when compared with the wavelength. This permits us to use the approach of geometrical optics, which was used in the previous chapter. We further use the drift theory for the description of charged particle motions in an inhomogeneous magnetic field, which was given in Chapter 2.

6.1 Particle trajectories in a wave field of constant frequency and given amplitude

As stated in Chapter 1 whistler waves can propagate in the Earth's magnetosphere in ducted and unducted modes. We consider the first case, when a whistler wave packet is trapped inside a magnetic field aligned column of enhanced plasma density, which serves as a waveguide.

The equations of an electron's motion in an inhomogeneous magnetic field $\vec{B}$, in the presence of a whistler wave whose wave vector $\vec{k}$ is antiparallel to $\vec{B}$, were discussed in Chapter 2 (see (2.121)–(2.123)). They can be written in their simplest form if we use as variables the kinetic energy W and the first adiabatic invariant $J_\perp$ of the electron, and if we take the coordinate z along the magnetic field line instead of time t:

$$W = \frac{m}{2}(v_\parallel^2 + v_\perp^2) \qquad J_\perp = \frac{mv_\perp^2}{2B}$$
$$v_\parallel = \sqrt{\frac{2}{m}(W - J_\perp B)} \tag{6.1}$$

where, as earlier, $v_\parallel$ and $v_\perp$ are the electron's velocity components along and across the geomagnetic field, $B = |\vec{B}|$, and m is the electron mass. The relation between

z and t for a test electron in terms of their initial values z_0 and t_0 is

$$t - t_0 = \int_{z_0}^{z} \frac{dz'}{\sqrt{\frac{2}{m}\left(W - J_\perp B(z')\right)}} \tag{6.2}$$

where z' is the variable of integration along the field line.

The equations of a test electron's motion in these variables are written in the form (2.124)–(2.126):

$$\frac{dW}{dz} = -e\left(\frac{J_\perp B}{W - J_\perp B}\right)^{1/2} \mathrm{Re}\left(A \exp(i\psi)\right) \tag{6.3}$$

$$B\frac{dJ_\perp}{dz} = -e\left(\frac{J_\perp B}{W - J_\perp B}\right)^{1/2} (1 - N\beta_\parallel)\,\mathrm{Re}\left(A \exp(i\psi)\right) \tag{6.4}$$

$$\frac{d\psi}{dz} = \left(\frac{2}{m}(W - J_\perp B)\right)^{-1/2} (\omega - \omega_B - kv_\parallel) \tag{6.5}$$

where $\psi = \Theta - \varphi$ is the phase difference between the wave electric field and the electron's perpendicular velocity. Here, by definition, $\frac{\partial \Theta}{\partial t} = \omega$, $\frac{\partial \Theta}{\partial z} = -k$, and $\frac{\partial \varphi}{\partial t} = \omega_B$, A is the amplitude of the whistler-mode wave electric field, $N = |k|c/\omega$ is the whistler-mode refractive index, $\beta_\parallel = v_\parallel/c$, and c is the velocity of light in free space. In (6.5) we have omitted, following (5.1), the term which is proportional to the wave amplitude A and which gives a small change to the solution of (6.3)–(6.5). We consider an electron moving in the $(+z)$-direction, which interacts via cyclotron resonance

$$\omega - \omega_B - kv_\parallel \approx 0 \tag{6.6}$$

with a whistler-mode wave with $k < 0$ (i.e. propagating in the $(-z)$-direction) as shown in Fig 6.1. We assume that the whistler wave amplitude is given and changes only slowly along the trajectory of the energetic electrons. For the gyroresonant electrons, when the equality (6.6) is satisfied, the system of equations (6.3)–(6.5) has an integral of motion, which can be written as

$$W - \frac{mc}{e}\omega J_\perp = \text{const.} \tag{6.7}$$

This can be demonstrated by differentiation of (6.7) with respect to z, and using (6.3) and (6.4). This integral exists if the wave frequency ω is constant.

For the case of a small-amplitude whistler $B_\sim \ll B$, where $B_\sim$ is the wave magnetic field amplitude, and a slowly varying geomagnetic field, B, we use the adiabatic approach to solve the system of equations (6.3)–(6.5) in the following way. The right-hand side of (6.5) is presented in the form of a Taylor series expansion

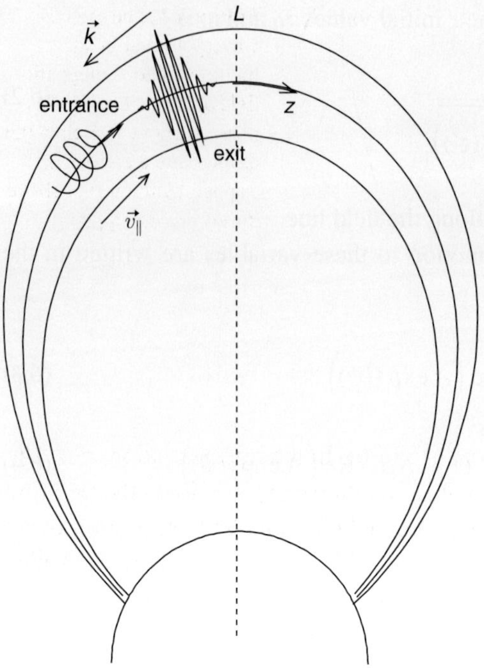

Figure 6.1 Scheme of cyclotron interaction of a whistler wave and RB electrons.

over z near the point of the exact resonance (6.6) (see below). Differentiating both parts of Equation (6.5) with respect to z, and using (6.3)–(6.4), we obtain:

$$\frac{d^2\psi}{dz^2} - \ell_{\mathrm{tr}}^{-2}\cos\psi + \alpha_B = 0 \tag{6.8}$$

where the trapping length ℓ_{tr} is equal to

$$\ell_{\mathrm{tr}} = \frac{v_{\mathrm{R}}}{\Omega_{\mathrm{tr}}}, \tag{6.9}$$

$$\Omega_{\mathrm{tr}} = (|k| v_{\perp} \omega_{B_\sim})^{1/2} \tag{6.10}$$

is the frequency of electron oscillations in the potential well of the wave (see Fig 6.2), and $\omega_{B_\sim} = (e/mc)B_\sim$ is the electron gyrofrequency in the wave magnetic field. The gyroresonance velocity for the electrons is

$$v_{\mathrm{R}} = \frac{\omega - \omega_B}{k} > 0. \tag{6.11}$$

The inhomogeneity factor α_B is

$$\alpha_B = k\,\frac{d}{dz}\left(\frac{v_{\mathrm{R}}}{\sqrt{\dfrac{2}{m}\,(W - J_{\perp}B)}}\right)_{z_{\mathrm{R}}} \tag{6.12}$$

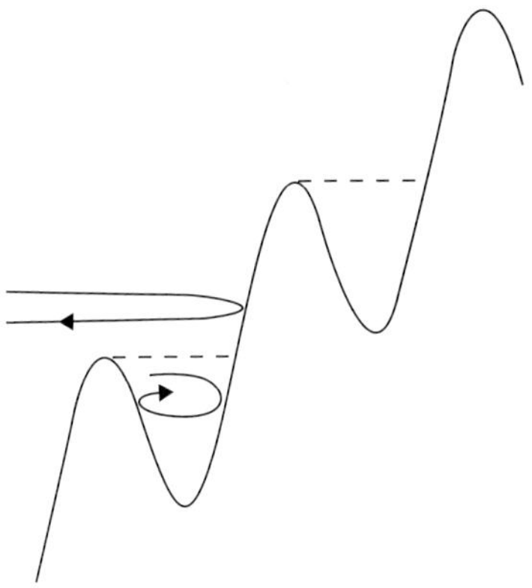

Figure 6.2 The effective wave potential in an inhomogeneous magnetic field as a function of the phase variable ζ. Below the dashed line electrons are trapped by the wave; above it they are untrapped.

where z_{R} is the point of exact gyroresonance (6.6). In the differentiation in (6.12), W and $J_\perp$ are assumed to be independent of z. Equation (6.8) is valid for the electrons with $v_\parallel$ close to the resonance velocity v_{R}, and includes only first order terms of the small parameters $\varepsilon_1 = B_\sim/B$ and $\varepsilon_2 = \ell_{\mathrm{tr}}/\ell$; here, ℓ is the characteristic scale length of B and k changes. To have the solution of (6.8) in a standard form we introduce a new variable ξ, which is connected to ψ by the relation

$$2\xi = \psi - 3\pi/2. \tag{6.13}$$

For ℓ_{tr} and α_B constant over z, Equation (6.12) has the energy integral

$$æ = (\dot{\xi}^2\ell_{\mathrm{tr}}^2 + \sin^2\xi + 2\alpha_B\ell_{\mathrm{tr}}^2\xi)^{-1/2} = \text{const}. \tag{6.14}$$

where $\dot{\xi} = d\xi/dz$. Values of $|æ| \geq 1$ correspond to electrons trapped by the wave, while $|æ| \leq 1$ corresponds to untrapped particles. It follows from (6.14) that there is a trapping threshold, which is determined by the condition that the effective potential $U_{\mathrm{eff}} = \sin^2\xi + 2\alpha_B\ell_{\mathrm{tr}}^2\xi$ has a minimum (see Fig. 6.2), i.e.

$$|2\alpha_B\ell_{\mathrm{tr}}^2| \leq 1. \tag{6.15}$$

Equation (6.8) is strictly true for the case of a linearly changing magnetic field. For real magnetic field parameters α_B and ℓ_{tr} are slowly changing functions of z. In this case $æ(z)$ is not constant, but can be found using an adiabatic approach. For our problem this approach is formulated as follows. In the case $\alpha_B = 0$ and $\ell_{\mathrm{tr}} = \text{const.}$, the solution of (6.8) is purely periodic and coincides with the solution (5.17), if t is changed to z/v_{R}. The spatial period of this solution is

$$T_{\mathrm{sp}} = \begin{cases} 2\ell_{\mathrm{tr}}\, æ\, K(æ), & \text{for } æ < 1 \\ 4\ell_{\mathrm{tr}}\, K(1/æ) & \text{for } æ > 1 \end{cases} \tag{6.16}$$

where K is the complete elliptic integral of the first type (see (5.A.86)). We take this solution as the zero-order approximation to the solution, when $\alpha_B \neq 0$ and ℓ_{tr} is a slowly changing function of z. According to the adiabatic approach, this is possible if two inequalities are fulfilled for Equation (6.8):

$$\alpha_B \ell_{\rm tr}^2 \ll 1 \text{ and } \ell_{\rm tr}/\ell \ll 1. \tag{6.17}$$

Following the usual adiabatic approach, we can obtain an equation for slowly varying æ. For that it is necessary to differentiate (6.14) with respect to z and to average the right-hand side of the relation obtained over z, using the zero-order solution. We shall have:

$$\frac{d}{dz}\left(\frac{1}{æ^2}\right) = \overline{\dot{\xi}^2}\,\frac{d\ell_{\rm tr}^2}{dz} + 2\alpha_B \ell_{\rm tr}^2 \overline{\dot{\xi}} \tag{6.18}$$

where the bar above $\dot{\xi}^2$ and $\dot{\xi}$ means averaging over z. To find $\overline{\dot{\xi}}$ and $\overline{\dot{\xi}^2}$, we use the zero-order solution of (6.8) (see (5.19)) in the form:

$$\dot{\xi} = \frac{1}{æ\ell_{\rm tr}} \times \begin{cases} dn(u, æ) & æ < 1 \\ cn(v, æ^{-1}) & æ > 1 \end{cases} \tag{6.19}$$

where dn and cn are Jacobian elliptic functions (JEF, see Chapter 5); the arguments u and v are equal to:

$$u = F(\xi, æ) - \frac{z}{æ\ell_{\rm tr}}, \qquad v = æF - \frac{z}{\ell_{\rm tr}} \tag{6.20}$$

where F is the elliptic integral (5.17), and $dn(u, æ)$ and $cn(v, æ^{-1})$ are periodic functions with the period given by (6.16). According to the properties of JEF (see the Appendix to Chapter 5),

$$\overline{\dot{\xi}} \sim \overline{dn(u, æ)} = \frac{\pi}{2K(æ)}, \quad æ < 1, \text{ and } \overline{cn(v, æ^{-1})} = 0, \quad æ > 1. \tag{6.21}$$

The second condition in (6.21) means that the averaged value of the field-aligned velocity component for trapped electrons coincides with the resonance velocity:

$$\overline{v}_{\parallel} = v_{\rm R} \equiv \frac{\omega - \omega_B}{k}. \tag{6.22}$$

To find $\overline{\dot{\xi}^2}$ we take into account (6.19) and obtain for æ < 1:

$$æ^2 \ell_{\rm tr}^2 \overline{\dot{\xi}^2} = \frac{1}{T_{\rm sp}} \int_0^{T_{\rm sp}} dn^2 u \, du = \frac{\pi^2}{4K^2} + \frac{4\pi^2}{K^2 T_{\rm sp}} \int_0^{T_{\rm sp}} du \sum_{m=1}^{\infty} \frac{q^{2m}}{\left[1+q^{2m}\right]^2} \cos^2\left[(m+1)\frac{\pi u}{K}\right]$$

$$= \frac{\pi^2}{K^2}\left[\frac{1}{4} + 2\sum_{m=0}^{\infty}\frac{1}{(q^{2m}+1)(q^{-2m}+1)}\right]$$

$$= \frac{\pi^2}{K^2}\left[\frac{1}{4} + K\left(\frac{E(æ)}{\pi^2} - \frac{1}{4K}\right)\right] = \frac{E(æ)}{K(æ)}, \quad æ < 1 \tag{6.23}$$

where we have used the Fourier series for JEF (A.5.101) and the relation (A.5.102); $E(æ)$ is the complete elliptic integral of the second kind.

In the case $æ > 1$ the relations (A.5.100) and (A.5.103) should be used. We obtain:

$$æ^2\ell_{\rm tr}^2\overline{\dot{\xi}^2} = \frac{1}{T_{\rm sp}}\int_0^{T_{\rm sp}} cn^2u\,du = \frac{2\pi^2}{æ'^2K^2}\sum_{m=0}^{\infty}\frac{1}{(1+q^{2m+1})(1+q^{-2m-1})}$$

$$= \frac{1}{Kæ'^2}\left[E(æ') + (æ'^2-1)\,K(æ')\right]$$

$$= \frac{æ^2}{K(1/æ)}\left[E\left(\frac{1}{æ}\right) - \left(1-\frac{1}{æ^2}\right)K\left(\frac{1}{æ}\right)\right], \quad æ' = \frac{1}{æ}. \tag{6.24}$$

Substituting the relations (6.23) and (6.24) into Equation (6.18), we finally have:

$$\frac{d}{dz}\left[\frac{E(æ)}{æ\ell_{\rm tr}}\right] = \frac{\pi\alpha_B}{2}, \quad æ < 1 \text{ (for untrapped particles)} \tag{6.25}$$

and

$$\frac{d}{dz}\left[\frac{E\left(\frac{1}{æ}\right) - (1-æ^{-2})\,K\left(\frac{1}{æ}\right)}{\ell_{\rm tr}}\right] = 0, \quad æ > 1 \text{ (for trapped particles).} \tag{6.26}$$

These relations (6.25) and (6.26) are very important because they permit us to determine the condition for the stable trapping of resonant particles. The sufficient condition for that is, clearly:

$$\frac{dæ^2}{dz} > 0 \quad (æ^2 > 1). \tag{6.27}$$

Equation (6.27) means that a trapped particle moves to the bottom of the wave potential well. Using a Taylor series expansion of $E\left(\frac{1}{æ}\right)$ and $K\left(\frac{1}{æ}\right)$ for $æ > 1$ (relations (A.5.90) and (A.5.95)), it is possible with good accuracy to find the solution (6.26)

$$\frac{æ^2}{æ_0^2} \approx \frac{\ell_{\rm tr0}}{\ell_{\rm tr}}, \quad æ_0 > 1. \tag{6.28}$$

Here the subscript zero means the initial values of $\ell_{\rm tr}$ and æ, when an electron starts from the point z_0. According to (6.28), the inequality (6.27) is fulfilled if

$$\frac{d\ell_{\rm tr}}{dz} < 0 \tag{6.29}$$

i.e. if the trapping length decreases with increasing distance from the initial point z_0.

The relations (6.7) and (6.22), together with the conditions (6.15) and (6.29) for stable trapping, permit us to consider the acceleration of electrons by a whistler-mode wave in an inhomogeneous magnetic field. They also enable us to study the formation of an electron beam as it leaves the wave packet, shown as 'exit' in Fig. 6.1.

6.2 Electron acceleration by a whistler-mode wave packet of constant frequency

It follows from the analysis of the previous section that electron acceleration (i.e. net growth of energy W) by a whistler-mode packet in an inhomogeneous magnetic field can be considerable in the case of a large change of the resonance velocity along the length of the wave packet. However, the accelerated electron must be kept trapped in the potential well of the wave during the entire interaction time. The necessary and sufficient conditions for such trapping are given by the relations (6.15) and (6.29).

It is important to realize that changes of the electron's velocity components field-aligned and perpendicular to the magnetic field have opposite signs. It is easy to see that if we rewrite the integral of motion (6.7) in the following form, taking (6.22) into account:

$$\Delta W_\perp \left(1 - \frac{\omega}{\omega_B}\right) = W_{\perp\,\rm ent}\left(\frac{\omega}{\omega_B} - \frac{\omega}{\omega_{B\,\rm ent}}\right) - \Delta W_\parallel \tag{6.30}$$

$$\Delta W_\parallel \equiv W_\parallel - W_{\parallel\,\rm ent} = W_{\rm R} - W_{\rm R\,ent}, \quad \text{with } W_{\rm R} = \frac{m}{2}\, v_{\rm R}^2$$

where $\Delta W_\perp$ and $\Delta W_\parallel$ are the increases of the components of energy of the electrons across and along the magnetic field, respectively; the subscript 'ent' means the position of the forward edge (i.e. entrance) of a wave packet (see Fig. 6.1).

The resonance velocity $v_{\rm R}$ for whistler-mode waves of constant frequency decreases in the direction of decreasing magnetic field value, i.e. towards the equatorial plane. As is seen from (6.30) this decrease is accompanied by the growth of $\Delta W_\perp$. Moreover, $\Delta W_\perp > |\Delta W_\parallel|$, which means that net electron acceleration takes place when a whistler packet propagates away from the magnetic equatorial plane.

We now analyse the especially important situation of the geomagnetic field and energetic electrons trapped in the Earth's radiation belts. There are actual sources of quasi-monochromatic whistler-mode signals in the Earth's magnetosphere, which emanate from ground-based VLF radio transmitters used for communicating with submarines.

For further calculations we need particular expressions for the k, v_{R} and v_{g}; these are obtained using (6.15), (6.29) and (6.30). The resonance velocity v_{R} is equal to:

$$|v_{\mathrm{R}}| = \frac{\omega_B - \omega}{|k|} = \frac{\omega_B}{p}\left(1 - \frac{\omega}{\omega_B}\right)^{3/2} \tag{6.31}$$

where we have used a valid approximation for the expression (2.92) for the whistler-mode wavenumber k in the form:

$$k = \frac{N_\omega \omega}{c} = p\left(1 - \frac{\omega}{\omega_B}\right)^{-1/2} \tag{6.32}$$

with $p = \left(\dfrac{\omega_{\mathrm{p}}^2\,\omega}{c^2\omega_B}\right)^{1/2}$; the electron plasma frequency is $\omega_{\mathrm{p}} = \left(\dfrac{4\pi e^2 n_{\mathrm{c}}}{m}\right)^{1/2}$, and n_{c} is the cold plasma density. Further, we assume that p does not depend on z, i.e. for a particular geomagnetic flux tube the cold plasma density is proportional to the geomagnetic field strength. The group velocity of whistler-mode waves can be written as

$$v_{\mathrm{g}} = \frac{2\omega}{\omega_B}\, v_{\mathrm{R}} = \frac{2\omega}{p}\left(1 - \frac{\omega}{\omega_B}\right)^{3/2}. \tag{6.33}$$

Now we analyse the trapping conditions (6.15) and (6.29). The magnetic inhomogeneity factor α_B (6.12) can be written with (6.31) being taken into account as follows:

$$\alpha_B = \frac{k}{2}\left(\frac{2\omega_B + \omega}{\omega_B - \omega} + \frac{v_\perp^2}{v_{\mathrm{R}}^2}\right)\frac{d\ln\omega_B}{dz}. \tag{6.34}$$

Substitution of (6.34) into (6.15) gives an inequality which permits us to determine the region in the electrons' perpendicular velocity component $v_\perp$ space where the electrons are trapped by a wave of given amplitude $B_\sim$:

$$v_\perp^2 + v_{\mathrm{R}}^2\left(\frac{2\omega_B + \omega}{\omega_B - \omega}\right) - \frac{\omega_{B\sim} v_\perp}{|d\ln\omega_B/dz|} \le 0. \tag{6.35}$$

The solution of equation (6.35) is

$$v_{\perp 1,2}(z) \le \frac{\omega_{B\sim}}{2|d\ln\omega_B/dz|} \mp \sqrt{\frac{\omega_{B\sim}^2}{4\,|d\ln\omega_B/dz|^2} - \frac{\omega_B^2}{p^2}\left(1 - \frac{\omega}{\omega_B}\right)^2\left(2 + \frac{\omega}{\omega_B}\right)}. \tag{6.36}$$

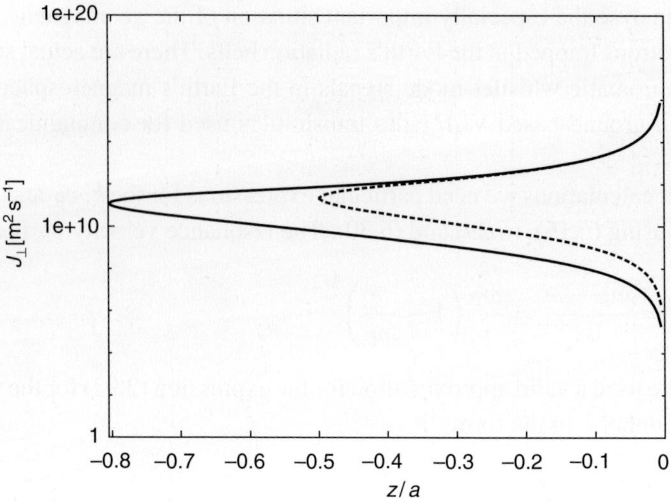

Figure 6.3 Typical examples of the trapping region (right side of the curves) as a function of the normalized distance from the equator along the $L = 4$ field line and the adiabatic invariant, for the wave amplitude $B_\sim = 60$ pT and two different cold plasma densities ($n_c = 30\,\text{cm}^{-3}$ (dashed line) and $n_c = 100\,\text{cm}^{-3}$ (solid line)).

Trapping is possible only within the interval

$$v_{\perp 1}(z) \leq v_\perp \leq v_{\perp 2}(z) \tag{6.37}$$

shown as a function of z in Fig. 6.3.

When the inhomogeneity of the plasma parameters is strong, trapping is absent (Fig 6.4). The limiting condition for trapping is determined by the expression under the square root in (6.36) being zero:

$$\left(\omega_{B\sim} \Big/ 2\left|\frac{d\omega_B}{dz}\right|\right) \geq \frac{1}{p}\left(1 - \frac{\omega}{\omega_B}\right)\left(2 + \frac{\omega}{\omega_B}\right)^{1/2}. \tag{6.38}$$

In (6.38) $\omega_{B\sim}$ is also a function of z, because the whistler wave mode amplitude $B_\sim$ is changing along the magnetic flux tube. In the case of ducted propagation this is in accordance with the law of the conservation of wave energy flux:

$$\sigma_A v_g \frac{B_\sim^2}{8\pi} = \text{const.} = \sigma_{AL} v_{gL} \frac{B_{\sim L}^2}{8\pi}, \tag{6.39}$$

where σ_A is proportional to $1/B$, the cross-sectional area of a magnetic flux tube, and v_g is the group velocity. Taking (6.33) into account we find that:

$$B_\sim = \frac{\omega_B}{\omega_{BL}}\left(\frac{v_{RL}}{v_R}\right)^{1/2} B_{\sim L}. \tag{6.40}$$

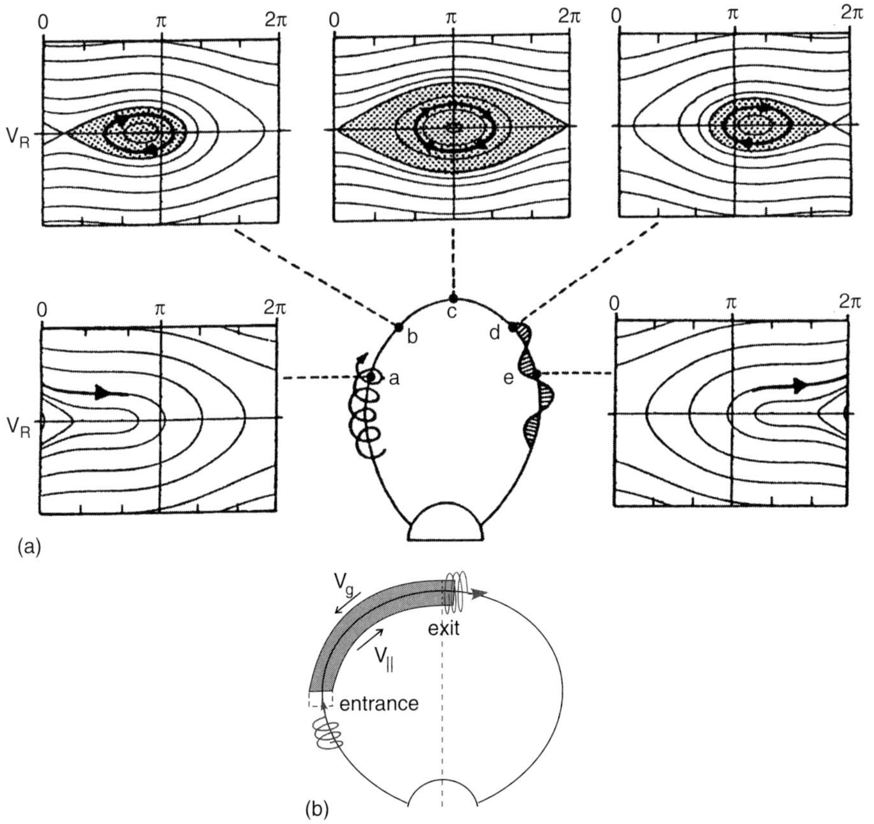

Figure 6.4 (a) Different shapes of $v_\parallel - \xi$ diagrams at different locations along a geomagnetic field (Omura *et al.*, 1991). The horizontal and vertical axes correspond to ξ and $v_\parallel$, respectively. Panels at (a), (b), (c), (d) and (e) correspond to the cases with $S = -1$; -0.5; 0.0; 0.5 and 1.0, respectively; (b) The optimal scheme for electron acceleration by a monochromatic wave packet.

When the amplitude of the whistler-mode signal is small, trapping takes place only near the equator, where a parabolic approximation for the geomagnetic field is valid (see (2.8)):

$$\omega_B = \omega_{\mathrm{BL}}(1 + s^2). \tag{6.41}$$

Here the dimensionless variable $s = z/a$, with $a = R_0 L\sqrt{2}/3$, where R_0 is the Earth's radius, $|s| < 1$. Taking into account (6.31), (6.40) and (6.41), we find from (6.38) (for $s_* < 1$):

$$s_* = \frac{a}{4}\frac{B_{\sim\mathrm{L}}}{B_{\mathrm{L}}}\,pQ^{-1}, \quad Q = \left(1 - \frac{\omega}{\omega_{\mathrm{BL}}}\right)\left(2 + \frac{\omega}{\omega_{\mathrm{BL}}}\right)^{1/2}. \tag{6.42}$$

The acceleration of energetic electrons takes place on the segment of the magnetic field line for which $0 \le |s| \le s_*$ (Fig. 6.4), when the front edge of the constant frequency whistler-mode wave packet moves in the direction of increasing geomagnetic field. In this case the resonance velocity of a trapped electron is less as it leaves (exits from) the trailing edge of the wave packet than its initial value as it enters the wave packet; that corresponds to the net growth of electron energy in accordance with (6.30).

To have a quantitative evaluation of the electron acceleration effects, we consider the relations (6.30) in more detail. Taking (6.31) into account, we find that:

$$\Delta W_{\parallel} = \frac{m}{2p^2}\left[\omega_{B\,\mathrm{ext}}^2\left(1-\frac{\omega}{\omega_{B\,\mathrm{ext}}}\right)^3 - \omega_{B\,\mathrm{ent}}^2\left(1-\frac{\omega}{\omega_{B\,\mathrm{ent}}}\right)^3\right] < 0, \quad (6.43)$$

$$\Delta W_{\perp} = \frac{W_{\perp\,\mathrm{ent}}\left(\dfrac{\omega}{\omega_{B\,\mathrm{ext}}}\right)\left(1-\dfrac{\omega_{B\,\mathrm{ext}}}{\omega_{B\,\mathrm{ent}}}\right)}{1-\omega/\omega_{B\,\mathrm{ext}}} - \frac{\Delta W_{\parallel}}{1-\omega/\omega_{B\,\mathrm{ext}}}. \quad (6.44)$$

So

$$\Delta W = \frac{\omega}{\omega_{B\,\mathrm{ent}}}\left(1-\frac{\omega_{B\,\mathrm{ext}}}{\omega_{B\,\mathrm{ent}}}\right)\left[\frac{W_{\perp\,\mathrm{ent}}}{1-\omega/\omega_{B\,\mathrm{ext}}} + \frac{m\omega_{B\,\mathrm{ent}}^2}{2p^2} \times \frac{\omega_{B\,\mathrm{ext}}+\omega_{B\,\mathrm{ent}}-2\omega}{\omega_{B\,\mathrm{ent}}}\right] \quad (6.45)$$

where the subscript 'ent' (entrance) means the coordinate of the forward front of the wave packet at the moment the electron crosses it, and the subscript 'ext' (exit) corresponds to the position of the trailing edge when the electron escapes from the wave packet. These coordinates are connected by the relation:

$$\tau = \int_{z_{\mathrm{ent}}}^{z_{\mathrm{ext}}}\left(\frac{1}{v_{\mathrm{R}}}+\frac{1}{v_{\mathrm{g}}}\right)dz \quad (6.46)$$

where τ is the duration of the wave packet.

The acceleration will be greatest when $z_{\mathrm{ext}} \lesssim z_* = s_* a$ and $z_{\mathrm{ent}} = 0$ as shown in Fig. 6.4. This means that, for a wave of a given amplitude, there is an optimum pulse duration

$$\tau_{\mathrm{opt}} = \int_{-z_*}^{0}\left(\frac{1}{v_{\mathrm{R}}}+\frac{1}{v_{\mathrm{g}}}\right)dz \quad (6.47)$$

for the acceleration of the electrons to be a maximum. This acceleration will take place only if the sufficient condition for trapping (6.29) is met.

We find from (6.9), with (6.10), (6.31), (6.32) and (6.40) being taken into account, that:

$$\frac{d\ell_{\mathrm{tr}}}{dz} \sim \frac{d}{dz}\left[\omega_B\left(1-\frac{\omega}{\omega_B}\right)^{17/8}\right] = \left(1-\frac{\omega}{\omega_B}\right)^{9/8}\left(1+\frac{9\omega_B}{8\omega}\right)\frac{d\omega_B}{dz}. \tag{6.48}$$

Thus, for electron acceleration by the constant frequency whistler-mode wave, when ω_B is decreasing along the electron trajectory ($d\omega_B/dz < 0$), this derivative is negative. That also corresponds to the condition for stable trapping.

6.3 Acceleration of electrons by a whistler wave packet with changing frequency

The cyclotron interaction of an energetic electron with a whistler of changing frequency $\omega(z, t)$ is described by the same set of equations (6.3)–(6.5). Now, $\omega(z, t)$ and $k(\omega(z, t), z)$ are known functions of z and t. As a result, an additional term appears in the expression for the inhomogeneity factor α_B given by (6.12). The expression given in (6.7) is not an integral of the motion in this case. Thus, it is necessary to generalize the results, obtained above, for the case of a changing frequency $\omega(z, t)$.

The change of frequency of a wave packet along its ray path is given by the standard equation of geometrical optics in a medium whose properties do not change with time:

$$\frac{\partial\omega}{\partial t} - v_{\mathrm{g}}\frac{\partial\omega}{\partial z} = 0. \tag{6.49}$$

Here v_{g} is the group velocity; the fact that $\vec{k}$ is parallel to $\vec{v}_{\mathrm{g}}$ and antiparallel to $\vec{v}_{\parallel}$ and the z-direction is taken into account (see Fig. 6.1). Thus, ω is a function of the argument $t + \int^z \frac{dz'}{v_{\mathrm{g}}}$. In accordance with relation (6.2), we should take $\omega(z, t)$ for the frequency experienced by the fast electron at z, t in the form

$$\omega(z,t) = \omega\left(t + \int^z \frac{dz'}{v_{\mathrm{g}}}\right)$$

$$= \omega\left(t_0 + \int_{z_0}^{z}\frac{dz'}{\sqrt{\frac{2}{m}\left(W - J_\perp B(z')\right)}} + \int^z \frac{dz'}{v_{\mathrm{g}}}\right). \tag{6.50}$$

The new inhomogeneity factor α_N now has two terms:

$$\alpha_N = \frac{d}{dz}\left[k\left(\frac{v_R}{v_\|} - 1\right)\right]_{z\text{R}} = \alpha_B + \alpha_{\Delta\omega}. \tag{6.51}$$

Here

$$\alpha_B = \frac{k}{2}\left(\frac{2\omega_B + \omega}{\omega_B - \omega} + \frac{v_\perp^2}{v_\text{R}^2}\right)\frac{d\ln\omega_B}{dz} \tag{6.52}$$

is the value of α_B for the case of $\omega = \text{const.}$ defined by (6.34), and the new term $\alpha_{\Delta\omega}$ results from differentiation over ω which, in our case, is itself a function of z, as determined by the expression (6.50):

$$\alpha_{\Delta\omega} = -\frac{k}{2}\frac{2\omega_B + \omega}{\omega_B - \omega}\frac{d\ln\omega}{dz}. \tag{6.53}$$

To obtain (6.51)–(6.53) we took into account that for whistler waves v_R and k are determined by the relations (6.31) and (6.32).

Trapping of electrons by the wave takes place under the same condition (6.15), but with α_B replaced by α_N. The trapping is stable if the condition (6.29) is obeyed, where the expression (6.50) for $\omega\big(z, t(z)\big)$ should be taken into account. In the regime of stable trapping, the field-aligned velocity component $v_\|$ of the electron when it exits from the whistler packet is determined by the relation (6.31).

The system of equations (6.3)–(6.5) enables us to find the change of the first adiabatic invariant $J_\perp$. From Equations (6.3)–(6.4) we have, similarly to (6.7):

$$\frac{dJ_\perp}{dz} = \frac{e}{mc\,\omega(z,t)}\frac{dW}{dz}. \tag{6.54}$$

Putting the relation $W = J_\perp B + \dfrac{mv_\|^2}{2} \approx J_\perp B + \dfrac{mv_\text{R}^2}{2}$ into (6.54), after some algebraic manipulations we obtain

$$\frac{dJ_\perp}{dz} + \frac{d\omega_B/dz}{\omega_B - \omega}J_\perp + \frac{e}{2c(\omega_B - \omega)}\frac{dv_\text{R}^2}{dz} = 0 \tag{6.55}$$

where

$$\frac{dv_\text{R}^2}{dz} = \frac{(\omega_B - \omega)^2}{\omega_B\omega p^2}\left[\left(2 + \frac{\omega}{\omega_B}\right)\frac{d\omega_B}{dz} - \left(2 + \frac{\omega_B}{\omega}\right)\frac{d\omega}{dz}\right]. \tag{6.56}$$

The solution of the differential equation (6.55) for $J_\perp$ has the form

$$J_{\perp\,\mathrm{ext}} = J_{\perp\,\mathrm{ent}} - \frac{e}{2p^2c}\int_{z_\mathrm{ent}}^{z_\mathrm{ext}} dz\left[\left(\frac{2\omega_B}{\omega} - 1 - \frac{\omega}{\omega_B}\right)\frac{d\ln\omega_B}{dz} + \left(\frac{\omega_B}{\omega} + 1 - \frac{2\omega}{\omega_B}\right)\frac{d\ln\omega}{dz}\right]\exp\left\{\int_{\omega_B(z_\mathrm{ent})}^{\omega_B(z_\mathrm{ext})} (\omega'_B - \omega)^{-1}\, d\omega'_B\right\} \tag{6.57}$$

where z_ent and z_ext are the coordinates of the forward and trailing edges of the wave packet at the times of electron entry and exit from the packet, respectively, and $J_{\perp\,\mathrm{ent}}$ is the value of the magnetic moment $J_\perp$ when the test electron enters the wave packet. In the case of a constant wave frequency, Equation (6.57) is consistent with the results of the previous section.

When the rate of change of wave frequency is higher than that of ω_B (for example, near the equator, where $\omega_B \simeq$ const.), the change of magnetic moment and hence the electron acceleration is determined by the following expression:

$$J_{\perp\,\mathrm{ext}} - J_{\perp\,\mathrm{ent}} = \frac{e}{2p^2c}\left[\frac{\omega_B}{\omega_\mathrm{ent}} - \frac{\omega_B}{\omega_\mathrm{ext}} + \ln\frac{\omega_\mathrm{ext}}{\omega_\mathrm{ent}} - 2\,\frac{\omega_\mathrm{ext} - \omega_\mathrm{ent}}{\omega_B}\right]. \tag{6.58}$$

Here ω_ent and ω_ext refer to the wave frequency at the point of electron entrance and exit, respectively. Electron acceleration occurs if $\omega_\mathrm{ext} > \omega_\mathrm{ent}$.

6.4 Acceleration of electrons by a whistler generated by a lightning discharge

Now we consider an important case (Trakhtengerts *et al.*, 2003a) when a whistler wave packet with changing frequency is generated by a lightning discharge in which all frequencies are emitted instantaneously. As is well known, Equation (6.49) expresses the propagation and dispersion of the whistler wave $\omega(z, t)$ everywhere, and gives rise to the nose whistler phenomenon. Let the lightning generate an electromagnetic signal at the point $z = l/2$ (ground level of a particular magnetic field line) with the frequency

$$\omega = \omega_0 + \delta \cdot t, \quad 0 \le t \le \Delta \sim \delta^{-1}. \tag{6.59}$$

The parameter δ has a very large value for real lightning discharges (i.e. the causative atmospheric is approximated here by an extremely steep ramp in frequency). Integrating (6.49), we obtain the frequency $\omega(z, t)$ at any arbitrary point z:

$$\omega = \omega_0 + \delta \cdot t + \delta \int_{\ell/2}^{z} \frac{dz'}{v_\mathrm{g}(z', \omega)} \tag{6.60}$$

or, using (6.2),

$$\omega = \omega_0 + \delta \cdot t_0 + \delta \int_{z_0}^{z} \frac{dz'}{\sqrt{\frac{2}{m}\Big(W - J_\perp B(z')\Big)}} + \delta \int_{\ell/2}^{z} \frac{dz'}{v_g(z', \omega)}. \tag{6.61}$$

From this we find:

$$\frac{d\omega}{dz} = \left(\frac{1}{v_R} + \frac{1}{v_g}\right) \Bigg/ \left(\int_{\ell/2}^{z} \frac{dz'}{v_g^2} \frac{\partial v_g}{\partial \omega} + \delta^{-1}\right). \tag{6.62}$$

For a whistler wave, we define q as

$$q \equiv \int_{z}^{\ell/2} \frac{dz'}{v_g^2} \frac{\partial v_g}{\partial \omega} = \frac{p}{4} \int_{z}^{\ell/2} \frac{y^{3/2}(y-4)}{\omega^{3/2}(y-1)^{5/2}}\, dz' \approx \frac{T_g}{\omega} \frac{\omega_{BL} - 4\omega}{\omega_{BL} - \omega} \tag{6.63}$$

where $y = \omega_B/\omega$ is the magnetoionic variable and $T_g(\omega) = 2\int_{-\ell/2}^{\ell/2} dz/v_g(\omega)$ is whistler two-hop travel time. The last equality in (6.63) is valid near the equator ($z \approx 0$); as usual, the subscript 'L' refers to values taken at the equatorial plane. Substituting (6.62) and (6.63) into (6.53), and assuming that $\delta \to \infty$ so that $\delta^{-1} \to 0$, we obtain

$$\alpha_{\Delta\omega} = \left(1 + \frac{\omega_B}{2\omega}\right)^2 \left(v_R^2\, q\right)^{-1} \approx \left(1 + \frac{\omega_{BL}}{2\omega}\right)^2 \frac{8\omega}{v_{RL}^2 T_g} \frac{\omega_{BL} - \omega}{\omega_{BL} - 4\omega}. \tag{6.64}$$

To study electron acceleration due to the whistler near the equator ($\omega_B \gtrsim \omega_{BL}$), we use the expression (6.15) and write the condition for trapping (6.29) in the form

$$\left|2\alpha_{\Delta\omega}\ell_{tr}^2\right| = \frac{8 v_{RL}\, v_{gL}}{v_\perp\, \ell\, \omega_{B\sim L}} \left(1 + \frac{\omega_{BL}}{2\omega}\right)^2 \left|4 - \frac{\omega_{BL}}{\omega}\right|^{-1} \le 1. \tag{6.65}$$

Considering the case $\vec{k} \parallel \vec{B}$ we actually assume that, as a whistler wave propagates in a duct, the wave amplitude $B_\sim$ changes along this duct in accordance with the law for the conservation of energy flux (6.39). Neglecting absorption, this law relates the wave amplitude at the entrance to the duct in the ionosphere (I) to that in the equatorial plane (L):

$$B_{\sim L} = \left(\frac{\omega_{BL}}{\omega_{BI}} \frac{v_{R0}}{v_{RL}}\right)^{1/2} B_{\sim I} \approx \left(\frac{\omega_{BL}}{\omega_{BI}}\right)^{1/2} \left(1 - \frac{\omega}{\omega_{BL}}\right)^{-3/4} B_{\sim I}. \tag{6.66}$$

Substituting the expressions for v_{R0}, v_{RL}, and $B_{\sim L}$, and putting $\ell \simeq 1.38\, L R_E$ for a dipolar geomagnetic field line into (6.65), we obtain the trapping condition in the form

$$\frac{4\omega_{BL}^2 c^2 L^{1/2}}{1.38\, \omega_{pL}^2 v_\perp R_E\, \omega_{B\sim I}}\, \varphi(\omega) \le 1. \tag{6.67}$$

Here

$$\varphi(\omega) = \left(1 - \frac{\omega}{\omega_{\mathrm{BL}}}\right)^{15/4} \left(1 + \frac{\omega_{\mathrm{BL}}}{2\omega}\right)^2 \left(4 - \frac{\omega_{\mathrm{BL}}}{\omega}\right)^{-1}, \tag{6.68}$$

L is McIlwain's parameter for the flux tube, and R_E is the Earth's radius. The dependence of the left-hand side of (6.67) on L is determined by the multiplier $\omega_{\mathrm{BL}}^2 L^{1/2}/\omega_{\mathrm{pL}}^2$. If it is also assumed that on different geomagnetic flux tubes the electron density in the plasmasphere is proportional to B, this multiplier becomes $L^{-5/2}$.

The relations (6.58), (6.65), and (6.67) permit us to estimate the electron acceleration by a whistler for the conditions occurring in the Earth's magnetosphere. As follows from (6.57) and (6.58), acceleration near the equatorial plane takes place if $d\omega/dz > 0$, or $\omega(z_{\mathrm{ext}}) > \omega(z_{\mathrm{ent}})$. Therefore, electron acceleration in this case is effective only for the higher-frequency components of a whistler, above the nose frequency ω_{N}, where

$$\omega_{\mathrm{N}} \approx 0.25\,\omega_{\mathrm{BL}} \tag{6.69}$$

in our approximation (see Fig 6.5). From the inequality (6.29) it is easy to show that the condition $d\omega/dz > 0$ is also necessary for stable particle trapping. Trapping occurs when the condition (6.65) is met. For that it is necessary to move from the nose frequency to higher frequencies.

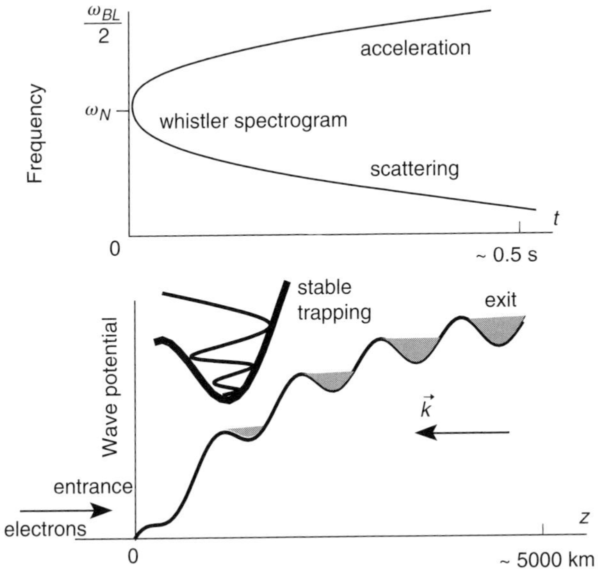

Figure 6.5 Variation of the effective potential of the whistler wave above its nose frequency ω_{N}, with distance z from the equator (where $z = 0$); the shaded regions show where gyroresonant electrons are phase trapped by the wave. In reality, there are far more oscillations than shown, because the wavelength in the medium is much less than $\sim$ 1000 km (Trakhtengerts *et al.*, 2003a). (Copyright American Geophysical Union, reproduced with permission.)

6.5 The distribution function of an electron beam after interacting with a whistler-mode wave packet of constant frequency

To estimate the contribution of electrons, accelerated by a whistler-mode wave packet, to the radiation belt population, and to investigate possible effects of secondary wave generation, we need to know the velocity distribution function of this electron beam. This function includes temporal, spatial and phase modulation along the magnetic field line (so-called phase bunching effects), as well as distribution over energy components along and across the magnetic field according to the relations (6.43)–(6.45). Phase bunching effects occur over a short distance after electrons have escaped from the trailing edge of the wave packet; they gradually disappear due to velocity dispersion. But their role can be very significant for the generation of secondary waves (i.e. triggered VLF emissions). These effects will be analysed later.

Now we shall consider the distribution function of escaping electrons over velocity (energy) components at different distances from the wave packet's trailing edge (Fig. 6.1), neglecting phase bunching effects.

The analytical solution of the problem formulated is rather difficult because of the complicated motion of electrons in the inhomogeneous geomagnetic field. To illustrate the main qualitative effects we shall consider the simplified model of electron ejection from the boundary $z = z_0$ (in a 1D approximation). Here we assume that the wave packet, whose trailing edge coincides with this boundary, moves to the left (in the negative z-direction) and the electrons ejected from the wave packet move in a homogeneous magnetic field (in the positive z-direction). This case is described by the following kinetic equation:

$$\frac{\partial \mathcal{F}}{\partial t} + v_{\|} \frac{\partial \mathcal{F}}{\partial z} = 0 \tag{6.70}$$

where t is time, z and $v_{\|}$ are the coordinate and electron velocity components along the magnetic field, respectively, $\mathcal{F}$ is the distribution function, and the integral $n = \int_{-\infty}^{\infty} \mathcal{F}\, dv_{\|}$ determines the total density of electrons. In such a formulation we can remove the boundary motion (with constant velocity) by transforming to a coordinate system moving with the boundary. We have two constants of motion from (6.70):

$$z - v_{\|} t = c_1, \quad v_{\|} = c_2. \tag{6.71}$$

The boundary condition is formulated as follows:

$$z = z_0 : \mathcal{F}\Big|_{z=z_0} = \mathcal{F}_0(v, t) = n_0 \delta\left(v_{\|} - v_0(t)\right) \mathrm{He}(t) \tag{6.72}$$

where $\mathrm{He}(t)$ is the Heaviside unit function, and so $\mathrm{He}(t) = 0$ for $t < 0$ and $\mathrm{He}(t) = 1$ for $t > 0$. The velocity of escaping electrons at the point $z = z_0$ is assumed to grow linearly with time, i.e. the electron acceleration α is a constant:

$$v_0(t) = v_{in} + \alpha t. \tag{6.73}$$

Here, the initial velocity v_{in} includes the velocity of the moving coordinate system.

In the real situation, the velocity of electrons at the moving boundary (the trailing edge of the whistler wave packet), coinciding with the local resonance velocity of those electrons, increases in association with the movement of this boundary to the left of the geomagnetic equator so as to increase the resonance velocities whereas, in a homogeneous magnetic field, the velocity component along the field line for each electron does not change in time and space after ejection from this boundary. Using (6.72) and (6.73), the solution of (6.70) at an arbitrary point and time is:

$$\mathcal{F} = n_0 \delta \left[v_\parallel - v_{in} - \frac{\alpha(z_0 - z + v_\parallel t)}{v_\parallel} \right] \cdot \mathrm{He}(z_0 + v_\parallel t - z). \tag{6.74}$$

Thus, two beams can occur at a given point z and instant of time t. Their field-aligned velocities and their densities are given by:

$$v_{1,2} = \frac{v_{in} + \alpha t}{2} \pm \sqrt{\frac{(v_{in} + \alpha t)^2}{4} - \alpha(z - z_0)} \tag{6.75}$$

$$n_{1,2} = \frac{n_0 \, v_{1,2}}{|v_1 - v_2|} \mathrm{He}(z_0 + v_{1,2} t - z) \tag{6.76}$$

where

$$v_1 - v_2 = 2 \left[\frac{(v_{in} + \alpha t)^2}{4} - \alpha(z - z_0) \right]^{1/2}. \tag{6.77}$$

The total density of the energetic electrons in the beam is

$$n_b = n_1 + n_2. \tag{6.78}$$

Numerical results are shown in Figs. 6.6 and 6.7. At small times $t < v_{in}/\alpha$ there is one beam, and its velocity v_1 grows from $z = z_0$. At $t_* = v_{in}/\alpha$ (and $z_* - z_0 = v_{in}^2/\alpha$), two beams appear, with velocities $v_{1,2}$ (Equation 6.75). The existence of two beams means that two groups of electrons that escape from the wave packet at different times with different velocities reach the same point in space at the same instant of time. During the process the beams are compressed, and their densities grow according to Equation (6.76), as shown in Fig. 6.7. After $t = t_*$ the density at the front $z_{max} = z_0 + (v_{in} + \alpha t)^2/4\alpha$ goes to infinity. The total number of electrons remains finite since the front region with infinite particle density has zero width in space. Of course, some small initial velocity dispersion removes the singularity, but the compression effect remains. If the dispersion of parallel velocities of escaping

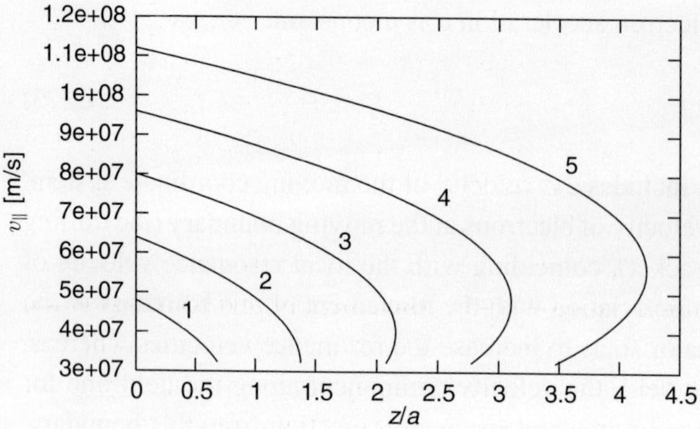

Figure 6.6 Snapshots of the electron beam parallel velocity $v_{\parallel}$ as a function of normalized distance z/a derived from a simple analytical consideration (6.75) ($a = 1.16 \cdot 10^7$ m, $v_{\text{in}} = 3.2 \cdot 10^7$ m s^{-1}, $\alpha = 2.0 v_{\text{in}}$, $z_0 = 0.0$ and curve labels $i = 1$–5 correspond to $t_i = 0.25i$ s; thus t_4 corresponds to 1 s).

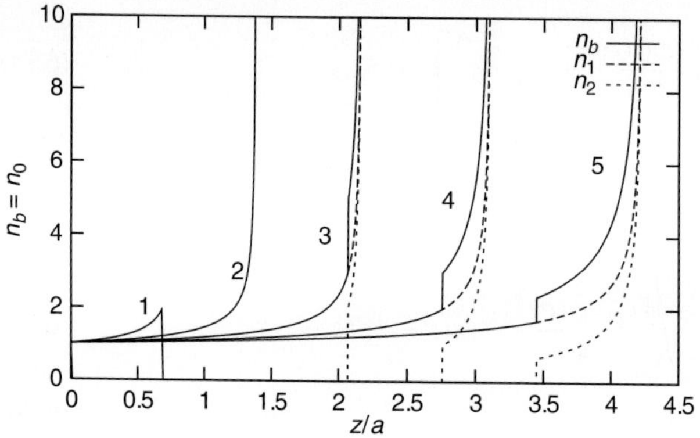

Figure 6.7 Snapshots of the beam densities; n_1, n_2 and n_b are defined in Equations (6.76) and (6.78), under the same conditions as in Fig 6.6.

electrons is δ_v, then the maximum beam density is

$$n_{b\,\text{max}} \sim \frac{n_0\, v_0^{1/2}}{\delta_v^{1/2}}. \tag{6.79}$$

In particular, if

$$\mathcal{F}_0(v, t) = \frac{n_0}{\delta_v \pi^{1/2}} \exp\left\{-\left(\frac{v - v_0(t)}{\delta_v}\right)^2\right\} \tag{6.80}$$

where $v_0(t)$ changes according to (6.73), then

$$n_{b\,\text{max}} = \frac{n_0}{2^{3/2}\,\pi^{1/2}} \Gamma\left(\frac{1}{4}\right)\left(\frac{v_{0*}}{\delta_v}\right)^{1/2} \tag{6.81}$$

where $v_{0*} = (v_{\text{in}} + \alpha t)/2 = [\alpha(z - z_0)]^{1/2}$ is the velocity of electrons at the front, and $\Gamma(x)$ is the gamma function of the argument x.

All the features which are evident in Fig. 6.6 take place in the more complicated case of an inhomogeneous magnetic field. This constitutes a specific acceleration mechanism for electrons interacting with a whistler wave packet. The relation (6.55) and (6.62) to (6.65) allow for these specific features and permit us to find the distribution function of an electron beam interacting with a whistler wave at an arbitrary point along the magnetic flux tube. For that the initial distribution function of the electrons has to be given at the entrance of the wave packet, and Liouville's theorem should be used. This states that the distribution function is conserved along the trajectories in phase space. In our particular case this problem is solved in two stages. First we find the distribution function at the exit of the wave packet as a function of $J_\perp$, v and t using the relations (6.43)–(6.45). Then we can find the distribution function from Liouville's theorem using Equation (6.2) at the arbitrary point (z, t), taking into account the conservation of $J_\perp$ and v and changing t_{ext} (considering the electrons which escaped from the wave packet at different times). For that we have to use the relation

$$z_{\text{ext}} = z_{\text{ext}\,0} - \int_0^{t_{\text{ext}}} v_{\text{g}}\, dt \tag{6.82}$$

connecting the position of the end of the wave packet at an arbitrary time t_{ext} with its position $z_{\text{ext}\,0}$ at the starting time $t = 0$ (the wave packet moves in the $(-z)$-direction). For the escaping phase, the parallel velocity at any point s and time t is given by:

$$v_\parallel = \sqrt{(W_{\text{ext}} - J_{\perp\,\text{ext}}\, B)\,\frac{2}{m}}\,. \tag{6.83}$$

In further calculations we shall assume the initial (escaping) distribution function to be a delta-function over $v_\parallel$: $F = F_0(J_{\perp\,\text{ext}}, v_{\text{ext}}) \cdot \delta(v_\parallel - v_{R\,\text{ext}})$. So, we have the similar problem as dealt with for Equations (6.70)–(6.72), with the integral of motion c_1 being

$$\int \frac{ds}{\sqrt{(W_{\text{ext}} - J_{\text{ext}}\, B)\,\frac{2}{m}}} - t = c_1. \tag{6.84}$$

The application of these results to the Earth's magnetosphere is considered in Chapter 12 in greater detail.

6.6 Phase bunching of the electron beam. Antenna effect

It was shown in Sections 5.2 and 5.3 that the distribution function of electrons in the field of a whistler-mode wave moves towards the so-called ergodic state, when the cyclotron growth (or damping) rate becomes equal to zero. However, the situation is completely different in an inhomogeneous plasma, when newly resonant electrons become involved in the cyclotron interaction due to change of the resonance velocity with position along the field line. Another important feature is the acceleration of electrons trapped in the potential well of the wave, considered in the four previous sections of this chapter. These electrons take energy from a pump wave and can dominate the untrapped electrons. Unlike this the compensation effects for a pump whistler amplitude can take place, when wave damping due to electron acceleration is compensated by wave amplification by new gyroresonant particles.

It is supposed that the initial distribution function of gyroresonant electrons is unstable, as happens in the case of the Earth's radiation belts. If the plasma inhomogeneity is very weak (the inequalities (6.17) are fulfilled, very close to the equator), the distribution function of trapped electrons can be considered at every instant of time as a quasi-ergodic one described by the relation (5.22). This function has a 'plateau' over $v_\parallel$ values, but is phase-modulated. At the exit from the trailing edge of a whistler wave packet this beam creates a current which serves as an active antenna and excites whistler-mode waves at other frequencies; the frequency range excited depends on the plasma (density and magnetic field) inhomogeneity. These seed waves play a very important role in triggered VLF emissions, determining an initial level of waves which can be amplified further by the beam, and so account for the wave amplitudes which are observed.

Our aim is now to estimate this antenna effect quantitatively. We suggest that, at the exit from the pump wave packet, trapped electrons can be approximately described by the ergodic distribution function. We consider that untrapped particles do not make any significant contribution to the antenna effect. In this case, the distribution function $\mathcal{F}_0$ can be written as

$$\mathcal{F}_0(\xi_*, \dot{\xi}_*, z = z_*) = \begin{cases} \mathcal{F}_{\text{erg}} = C_{\text{erg}}(v_{\perp *}), & \ell_{\text{tr}}^2 \dot{\xi}_*^2 + \sin^2 \xi_* \le 1; \\ 0, & \ell_{\text{tr}}^2 \dot{\xi}_*^2 + \sin^2 \xi_* > 1, \end{cases} \tag{6.85}$$

where we use the variables (see (6.8) and (6.13))

$$\begin{aligned} 2\xi &= \Theta_w + \varphi - 3\pi/2 = \int_{z_0}^{z} k_0 \, dz - \omega_0 t + \varphi - 3\pi/2 \\ 2\dot{\xi} &= k_0 \, (v_\parallel - v_{R_0}) \end{aligned} \tag{6.86}$$

for the trapped particles (ξ has a value between $-\pi/2$, and $\pi/2$), the subscript '$*$' refers to the escape point of the electrons from the primary wave packet, $v_\parallel$ is the velocity component parallel to the external magnetic field $\vec{B}_0$, φ is the phase of electron cyclotron rotation ($\tan \varphi = v_y/v_x$, with x and y being orthogonal axes transverse to $\vec{B}_0$), $d\varphi/dt = \omega_B$ is the local gyrofrequency, k_0 and ω_0 are, respectively, the wavenumber and frequency of the primary whistler wave packet ($\vec{k}_0 \parallel \vec{B}_0$), and ℓ_{tr} is determined by the relation (6.9); C_{erg} does not depend on ξ and $\dot{\xi}$ due to the phase mixing of trapped electrons. Note that we have omitted the subscript '$*$' in Equation (6.86), so that this definition can also be used for points other than z_* where necessary.

Choosing the distribution function of the form (6.85), we have implicitly assumed the acceleration in the primary wave packet to be sufficiently large compared with the mean energy of the electron distribution. For smaller accelerations, we should overestimate the height of the bump on the distribution function and, hence, the resulting secondary wave amplitude. As shown by Trakhtengerts *et al.* (2003), this overestimate can be corrected in a fairly simple way to take into account the actual height of the bump on the distribution function.

The resonant current related to the distribution function (6.85) is determined by the relation

$$\vec{j}_0 = e\int \mathcal{F}_0 \vec{v}_\perp \, d^3\vec{v} = 4e\int v_\perp^2 C_{\rm erg} \iint\limits_{\Omega} (\vec{x}_0 \cos\varphi + \vec{y}_0 \sin\varphi)\, \frac{d\xi\, d\dot{\xi}}{k_0}\, dv_\perp, \tag{6.87}$$

where the integration over ξ and $\dot{\xi}$ is performed over the phase space domain Ω in which the distribution function (6.85) is non-zero, and e is the charge on an electron.

We now show that the work done by the current $\vec{j}_0$ on the pump wave is equal to zero. The electric and magnetic field components of a right-hand circularly polarized whistler-mode wave are written as

$$B_{\sim 0x} = B_{\sim 0} \sin\Theta_0, \quad B_{\sim 0y} = B_{\sim 0}\cos\Theta_0, \tag{6.88}$$

$$E_{\sim 0x} = N^{-1} B_{\sim 0y}, \text{ and } E_{\sim 0y} = -N^{-1} B_{\sim 0x}, \tag{6.89}$$

where $\Theta_0 = \int_{z_0}^{z} k_0(z')\, dz' - \omega_0 t + \psi_0$ is the phase of the pump wave, and N is the refractive index of the whistler waves. The work done is:

$$\left(\vec{j}_0 \vec{E}_{\sim 0}\right) = \frac{4e\, E_{\sim 0}}{k_0} \int v_\perp^2 C_{\rm erg} \iint_{\Omega} \sin 2\xi \, d\xi\, d\dot{\xi}\, dv_\perp = 0. \tag{6.90}$$

This means that the pump wave is saturated, i.e. there is no net exchange of energy between the wave and the gyroresonant electrons.

After their escape from the pump wave packet ($z < z_{*0}$, see Fig. 6.4), the electrons move in an inhomogeneous magnetic field according to the integrals of motion

$$v^2 = v_\parallel^2 + J_\perp \omega_B(z) = \text{const. and } J_\perp = v_\perp^2/\omega_B(z) = \text{const.} \tag{6.91}$$

where $J_\perp$ is the first adiabatic invariant. Therefore, the distribution function at an arbitrary point z behind the pump wave packet is found according to Liouville's theorem, i.e. using the conservation of a phase-space volume along the charged particles' trajectories. To zeroth order in the secondary wave amplitude, the distribution function is determined by (6.85) where the variables $v_{\perp *}$, ξ_*, and $\dot{\xi}_*$ are determined by (6.86) and reference to the escape point z_* should be expressed in terms of the particle phase-space coordinates at the current point z. This is done using the following obvious relations:

$$\varphi - \varphi_* = \int_{z_*}^{z} \frac{\omega_B}{v_\parallel}\, dz' \tag{6.92}$$

and

$$t - t_* = \int_{z_*}^{z} \frac{dz'}{v_\parallel}. \tag{6.93}$$

We note that the escape point z_* is not the same as the current position z_t of the trailing edge of the pump wave packet, since this wave packet moves with its group velocity while the electrons which escaped at time t_* move in the opposite direction to the point z. Clearly,

$$\int_{z_*}^{z} \frac{dz'}{v_\parallel} = \int_{z_*}^{z_t} \frac{dz'}{v_{g0}}. \tag{6.94}$$

Using the above relations, it is easy to show that

$$2\xi = 2\xi_* + \int_{z_*}^{z} k_0\, dz' - \omega_0(t - t_*) + \varphi - \varphi_* = 2\xi_* - \int_{z}^{z_*} \frac{\Delta_0}{v_\parallel}\, dz' \tag{6.95}$$

where $\Delta_0 = \omega_B - \omega_0 + k_0 v_\parallel$ is the frequency mismatch from the local cyclotron resonance.

The beam of electrons excites a secondary wave, since it forms an active travelling wave antenna; the current is given by (6.87). On the other hand, it can amplify the secondary wave due to coherent processes. The latter effect requires a self-consistent approach, and this will be considered in Chapter 7. Below we consider the antenna effect which determines the starting wave amplitude for amplification.

If we know the distribution function $\mathcal{F}$ of the energetic electrons, we can calculate the resonant current $\vec{j}_0$. Then the magnetic field of the secondary wave $B_{\sim s}$ can be determined from the equation

$$\frac{\partial B_{\sim s}}{\partial t} + v_g \frac{\partial B_{\sim s}}{\partial z} = -\frac{2\pi}{c} v_g\, j_{E_{\sim s}} \tag{6.96}$$

where $j_{E_{\sim s}}$ is the projection of $\vec{j}_0$ to the direction of electric field of the secondary wave $E_{\sim s}$. It is easily shown that

$$j_{E_{\sim s}} = -\frac{4e}{k} \int v_\perp^2\, \mathcal{F} \sin(2u)\, du\, d\dot{u}\, dv_\perp \tag{6.97}$$

where u and $\dot{u}$ are similar variables to ξ and $\dot{\xi}$ (see (6.87)) but determined with respect to the secondary wave having frequency ω_s and wave vector k_s.

Since under magnetospheric conditions the scale length of the antenna due to escaping electrons is small compared with the inhomogeneity scale of the external magnetic field, we can use the approximation

$$d\xi\, d\dot{\xi} \simeq d\xi_*\, d\dot{\xi}_*. \tag{6.98}$$

Therefore, the resonant current producing the secondary wave $j_{E_{\sim s}}$ can be written in the form

$$j_{E_{\sim s}} \simeq -\frac{4e}{k_0(z_*)} \int v_\perp^2 \, C_{\text{erg}} \int_{-\pi/2}^{\pi/2} \int_{-\cos(\xi_*)/\ell_{\text{tr}}}^{\cos(\xi_*)/\ell_{\text{tr}}} \sin\left(\Theta - \Theta_0 + 2\xi_* + \int_{z_*}^{z} \frac{\Delta_0 \, dz'}{v_\parallel}\right) d\dot{\xi}_* \, d\xi_* \, dv_\perp \tag{6.99}$$

where

$$\Theta - \Theta_0 = \int_{z_0}^{z} (k - k_0) \, dz' - (\omega - \omega_0)t \tag{6.100}$$

(the integration limits in (6.99) are set according to (6.85)).

Taking into account that these processes are taking place near the equatorial plane, we can approximate the value $æ = (\omega_B - \omega + kv_\parallel)/v_\parallel \equiv \Delta/v_\parallel$ as

$$æ \simeq æ_{\text{L}} + k_{0\text{L}} \frac{z^2}{a^2} \left(\frac{\omega_{\text{BL}} + \omega}{\omega_{\text{BL}} + \omega} + \frac{v_{\perp\text{L}}^2}{v_{\parallel\text{L}}^2} \right) = æ_{\text{L}} + \ell^{-3} z^2 = æ_* + \ell^{-3}\left(z^2 - z_*^2\right) \tag{6.101}$$

(here the subscript 'L' refers to the equatorial plane exactly). Under this approximation, the integral in the phase part of Equation (6.99) is evaluated as

$$\int_{z_*}^{z} æ(z') \, dz' \simeq æ_*(z - z_*) + \ell^{-3}(z - z_*)^2 \left[z + \frac{2}{3}(z_* - z) \right]. \tag{6.102}$$

Moreover, by definition, $æ_* = \dot{\xi}_*/v_*$, where $v_* = v_{\text{R}0}(z_*)$. Therefore, under the approximation (6.101), integration over $\dot{\xi}_*$ in (6.99) can be performed, and after some simple transformations we obtain

$$j_{E_{\sim s}} \simeq -\frac{16ev_*}{k_0(z_*)\ell_{\text{tr}}} \int v_\perp^2 C_{\text{erg}} \sin(\phi) \, \mathcal{H}(x) \, dv_\perp \tag{6.103}$$

where

$$\phi = \Theta - \Theta_0 - \ell^{-3}(z - z_*)^2 \left[z + \frac{2}{3}(z_* - z) \right] \tag{6.104}$$

$$\mathcal{H}(x) = \int_0^{\pi/2} \sin^2 u \cos u \cos(x \cos u) \, du \tag{6.105}$$

and

$$x \simeq \frac{z_* - z}{\ell_{\text{tr}}} \tag{6.106}$$

is the normalized distance between the current point z and the corresponding injection point z_*. We recall that the relation between z and z_* is given by (6.94).

The solution of (6.96) can be written using new variables following waves and particles, respectively,

$$d\eta = \frac{dz}{v_{\rm g}} - dt \quad \text{and} \quad d\zeta = dz + |v_\parallel|\, dt. \tag{6.107}$$

In these variables, (6.96) becomes

$$v_\Sigma \frac{\partial B_{\sim \rm s}}{\partial \zeta} = -\frac{2\pi}{c}\, v_{\rm g}\, j_{E_{\sim \rm s}} \tag{6.108}$$

with $v_\Sigma = v_{\rm g} + |v_\parallel|$, whose solution is written as

$$B_{\sim \rm s} = -\frac{2\pi}{c} \int_{\zeta_0}^{\zeta} \frac{v_{\rm g}}{v_\Sigma}\, j_{E_{\sim \rm s}}\, d\zeta'. \tag{6.109}$$

Assuming that the frequency shift $\omega - \omega_0$ is small and expanding the wave vector k and the group velocity $v_{\rm g}$ near ω_0, we obtain

$$\Theta - \Theta_0 \simeq \frac{1}{2}(\omega - \omega_0)\left[\int_{z_*}^{z} \left(\frac{1}{v_{\rm g0}} - \frac{1}{v_\parallel}\right) dz' + \eta\right]. \tag{6.110}$$

Integration over ζ' in (6.109) is performed at $\eta = \text{const}$. Transforming (6.94), we obtain the necessary relation between z and x:

$$\left.\frac{\partial x}{\partial z}\right|_{\eta = \text{const.}} = \frac{v_{\rm g0} - v_{\rm g}}{v_{\rm g}} \frac{v_*}{v_{\Sigma 0}} \tag{6.111}$$

where $v_{\Sigma 0} = v_{\rm g0} + |v_*|$.

Finally, we obtain an expression for the amplitude of the magnetic field driven by the beam antenna current, $B_{\sim \rm ant}$, as

$$B_{\sim \rm ant} \approx b_* \frac{v_* \tau}{a\delta} \int_{x_0}^{x} \int_{v_{\perp 1}}^{v_{\perp 2}} \tilde{C}_{\rm erg} \frac{v_\perp^2\, dv_\perp}{v_0^3} \mathcal{H}(\xi) \sin[\phi(\xi) - \phi(x_0)]\, d\xi \tag{6.112}$$

where b_* is a factor independent of the pump wave amplitude and of the secondary wave frequency shift:

$$b_* \approx \frac{16}{\pi^{1/2}}\, e\, n_{\rm h} \frac{v_*}{c}\, a \frac{\omega_B}{4\omega_0 - \omega_B} \frac{v_{\Sigma 0}^2}{v_*^2}. \tag{6.113}$$

As discussed earlier, a is the characteristic scale of the geomagnetic field inhomogeneity, which is determined from (6.41).

6.7 Summary

In an inhomogeneous plasma (such as occurs naturally in the Earth's magnetosphere up to 1 or 2 Earth radii away from the magnetic equatorial plane), the cyclotron resonance condition, the wavenumber (or refractive index) and the cyclotron frequency all vary appreciably. The modified condition for the stable trapping of electrons in a whistler-mode wave packet is given by (6.27)–(6.29).

In this interaction, energetic electrons can be accelerated significantly. Expressions are derived for the energy increment of electrons for a constant frequency whistler-mode wave (6.45). For a natural whistler whose frequency changes markedly with time, components above the nose frequency accelerate electrons; Section 6.4 presents original work on this topic.

Figure 6.6 presents the spatio-temporal structure of an electron beam along the geomagnetic field after interacting with a constant frequency whistler-mode signal. The gyroresonant beam density grows by up to an order of magnitude (Fig. 6.7). Electrons which are phase bunched by the interaction radiate secondary waves, so acting as a travelling wave antenna. An expression is derived for the amplitude of the wave magnetic field driven by the beam antenna current (6.112).

Chapter 7

Wavelet amplification in an inhomogeneous plasma

In Chapter 6 the case of a whistler-mode wave with a given amplitude was considered. However, many processes which influence cyclotron maser behaviour demand a self-consistent approach, which takes into account the feedback effects of gyroresonant electrons on the wave field.

The first step in this approach is an analysis of the linear amplification (or damping) of a whistler-mode wave by electrons with different velocity distributions. This problem was tackled in Chapters 3 and 4 for a monochromatic whistler-mode wave. There it was shown that the cyclotron amplification for a broad velocity distribution function ($|\Delta v_\parallel|/v_\parallel \sim 1$) is the same for homogeneous and weakly inhomogeneous plasmas (compare Sections 3.2 and 3.4). The situation with amplification by well-organized electron beams (such as a step-like deformation or a delta-function in $v_\parallel$-velocity space) is more complicated (see Sections 3.3 and 3.5) and strongly differs from the cases of homogeneous and weakly inhomogeneous plasmas. This difference is as follows. First, the hydrodynamic type of instability of a well-organized beam in a homogeneous plasma is replaced by the kinetic-type instability in an inhomogeneous plasma. Secondly, the amplification in inhomogeneous plasmas strongly depends on the spatial gradients of the plasma parameters, specifically the plasma density and magnetic field strength. The amplification of a monochromatic wave is large very close to the equatorial plane of a magnetic flux tube. It decreases sharply (and even changes sign for a delta-function distribution) beyond this interval. The physical cause for this is the cyclotron resonance mismatch which destroys the amplification when a wave packet propagates a significant distance along the magnetic field line. On the other hand it is qualitatively clear that this mismatch can be at least partly compensated by a suitable change of the wave frequency in space and/or time. Helliwell (1967) was the first who applied this idea to explain triggered VLF emissions. Some aspects of this problem were discussed by Dysthe (1971), Sudan and Ott (1971),

Nunn (1974), Karpman *et al.* (1974a), Roux and Pellat (1978), Vomvoridis and Denavit (1979; 1980), Vomvoridis *et al.* (1982), Sagdeev *et al.* (1985), Molvig *et al.* (1988), Carlson *et al.* (1990), and Omura *et al.* (1991).

Following Trakhtengerts *et al.* (1999; 2003a) and Demekhov *et al.* (2000), we develop an approach which permits us to investigate both issues, first when the hydrodynamic-type instability is replaced by the kinetic instability and, second, how the frequency of the wavelet should vary in both space and time to have maximum wavelet amplification.

7.1 Generation of wavelet by a beam with a step-like distribution. Quantitative formulation of the problem

At first we consider a special type of well-organized electron beam represented by a step-like distribution function over the velocity component $v_\parallel$ along the magnetic field:

$$\mathcal{F} = n_\mathrm{b}\,\mathrm{He}(v_{*i} - v_{\parallel *})\,\mathcal{F}(v_{\parallel *})\,\Phi(J_\perp,\, z_*) \tag{7.1}$$

where $\mathrm{He}(x)$ is the Heaviside unit function,

$$\mathrm{He}(x) = 1, \quad x \geq 0$$
$$\mathrm{He}(x) = 0, \quad x < 0,$$

$\mathcal{F}(v_{\parallel *})$ is the smooth part of the distribution over parallel velocities, and the function $\Phi(J_\perp,\, z)$ characterizes the distribution over the first adiabatic invariant $J_\perp = mv_\perp^2/2B$ (where $v_\perp$ is the modulus of the velocity across the magnetic field, and B is the local magnetic field). Here v_{*i} is the step velocity at an injection point z_*. Further, it is assumed that Φ has a maximum at $J_\perp = J_*$ with a small dispersion, and can be approximated as

$$\Phi(J_\perp,\, z) = \delta\,[J_\perp - J_{\perp 0}(z, t)]. \tag{7.2}$$

The distribution function is normalized to the local beam density n_b:

$$2\pi \int \mathcal{F} v_\perp\, dv_\perp\, dv_\parallel = n_\mathrm{b}.$$

A particular form of the smooth part $\mathcal{F}(v_{\parallel *})$ of the distribution over $v_{\parallel *}$ is not critical since we consider only interactions with waves resonant with electrons at the step velocity $v_*(z, t)$.

The function (7.1) was considered in Chapter 3 for analyses of the cyclotron amplification of a monochromatic wave using the standard approach. But this approach is not applicable for treating wavelet amplification with waves of changing frequency. A strict consideration of the cyclotron instability for a well-organized beam was given in Chapter 4, but only for a homogeneous plasma. Here we generalize this approach to the case of an inhomogeneous plasma and for variable beams.

The parallel velocity $v_{\parallel *}$ in (7.1) corresponds to some point z_* where the local gyrofrequency is ω_{B_*}, and can be expressed as

$$v_{\parallel *} = \left[\frac{2}{m} (W - J_\perp B_*) \right]^{1/2} \tag{7.3}$$

where W is the electron beam energy. The point z_* can be considered as the injection point of the beam. The location of this point is allowed to vary slowly with time, as is the characteristic step velocity v_{*i} at this point. We assume that, after the injection, expansion of the electron beam occurs along the magnetic field line ($z \geq z_*$), and that this is described by the equation

$$\frac{\partial \mathcal{F}}{\partial t} + v_\parallel (J_\perp, W, z) \frac{\partial \mathcal{F}}{\partial z} = 0 \tag{7.4}$$

where $v_\parallel = \left\{ \frac{2}{m} \left[W - J_\perp \, B(z) \right] \right\}^{1/2}$; W and $J_\perp$ are integrals of the motion which are conserved along the magnetic field line. The distribution function (7.1) is unstable with respect to the generation of whistler-mode waves at the cyclotron resonance,

$$\omega(z, t) - k_\parallel v_* = \omega_B(z) \tag{7.5}$$

where ω and k are the frequency and wavenumber of the waves, and v_* is the parallel velocity of the step in the velocity distribution function considered at the point (z, t). Earlier we considered whistler-mode waves with $\omega < \omega_B$ and $\vec{k}$ directed along the magnetic field. In this case the wave packet resonant with the energetic electron beam moves against the beam, $k_\parallel \, v_* = -kv_*$, $v_* > 0$. From (7.5), a general problem for the hydrodynamic instability is evident: the magnetic field inhomogeneity destroys the cyclotron resonance but this mismatch can be partly compensated for by suitable dependencies of ω and v_* on z and t. A quantitative investigation of this problem is the subject of our consideration here.

Actually we deal with two subsystems – one is a cold, weakly inhomogeneous plasma, and the other is a weakly inhomogeneous and non-stationary beam of energetic electrons with small velocity spread (see (7.1) and (7.2)), whose density n_b is much less than the cold plasma density n_c. We adopt a description of the generation and propagation of electromagnetic waves in such a system based upon the geometrical optics approximation. In this approximation, the whistler-mode wave electric field amplitude can be presented in the form:

$$E_\sim(z, t) = A(z, t) \exp[i \, \Theta(z, t)] \tag{7.6}$$

where $A(z, t)$ is a slowly varying amplitude. The frequency ω and wave vector k are defined as temporal and spatial derivatives of the phase ψ:

$$\omega(z, t) = -\frac{\partial \Theta}{\partial t}, \qquad k(z, t) = \frac{\partial \Theta}{\partial z}. \tag{7.7}$$

In the case of a weak beam ($n_b/n_c \ll 1$), it is convenient to express the phase Θ as a sum

$$\psi = \Theta_w(z, t) + \Delta\varphi(z, t) \tag{7.8}$$

where Θ_w is determined by the cold plasma, and $\Delta\varphi(z, t)$ is due to the beam contribution. Thus

$$\omega = \omega_0(z, t) + \Omega(z, t), \qquad k = k_0(z, t) + æ(z, t) \tag{7.9}$$

where $\Omega(z, t)$ and $æ(z, t)$ are due to the beam particles and ω_0 and k_0 are related by the dispersion equation for the cold plasma,

$$\mathcal{D}(k_0, \omega_0, z) = 0, \qquad \omega_0 = \omega_0(k_0, z). \tag{7.10}$$

The group velocity is determined by the relation $v_g = \partial\omega_0/\partial k_0$, and the frequency ω_0 obeys the equation

$$\frac{\partial \omega_0}{\partial t} + v_g \frac{\partial \omega_0}{\partial z} = 0. \tag{7.11}$$

Further we omit the subscripts using the notation (ω, k) instead of (ω_0, k_0). With these definitions, the equation for the slowly varying complex amplitude $\mathcal{A} = A \exp(i\varphi)$ can be written in the form:

$$\frac{\partial \mathcal{A}}{\partial t} + v_g \frac{\partial \mathcal{A}}{\partial z} = \alpha_1 J \tag{7.12}$$

where we assume, for simplicity, that $v_g = \mathrm{const}(z)$,

$$\alpha_1 = \frac{2\pi\omega}{\partial N_w^2\, \omega^2/\partial\omega}, \qquad N_w = \frac{kc}{\omega}$$

and the current J (4.1) is determined by the beam. Referring to Section 3.4 for details, this is given as

$$\begin{aligned} J = {} & \frac{2\pi e^2}{m^2} \int \frac{J_\perp^{1/2} B^{3/2}\, dW\, dJ_\perp}{(W - J_\perp B)^{1/2}} \left(\frac{\partial \mathcal{F}_0}{\partial W} + \frac{e}{mc\omega} \frac{\partial \mathcal{F}_0}{\partial J_\perp} \right) \\ & \times \int_{z_0}^{z} dz' \left[\frac{J_\perp B(z')}{W - J_\perp B(z')} \right]^{1/2} B_\sim(z', \tilde{t}) \\ & \times \exp\left[i\Theta_w(z', \tilde{t}) - i\Theta_w(z, t) - \int_{z'}^{z} \frac{\omega_B(z'')}{v_\parallel(z'')} dz'' \right] \end{aligned} \tag{7.13}$$

where $\tilde{t} = t - \displaystyle\int_{z'}^{z} \frac{dz''}{v_\parallel(z'')}$.

Now we take into account the specific form of the energetic electron distribution function described by (7.1) and (7.2). After substituting these into (7.13), we obtain for the current

$$J = \frac{e^2 J_0^{1/2} n_{b*}}{m^2 v_{*i} v_* B_*} B^{3/2} \left(\frac{\omega_{B_*}}{\omega} - 1 \right) \int_{z0}^{z} dz' \frac{\left[J_0 B(z') \right]^{1/2}}{\tilde{v}_*(z'', z, t)} B_\sim(z', \tilde{t}_*)$$
$$\times \exp \left[i\Theta_{\rm w}(z', \tilde{t}_*) - i\Theta_{\rm w}(z, t) - \int_{z'}^{z} \frac{\omega_B(z'')}{\tilde{v}_*} dz'' \right] \tag{7.14}$$

where $\tilde{t}_* = t \;-\; \int_{z'}^{z} \frac{dz''}{\tilde{v}_*(z'', z, t)}$, $\tilde{v}_*(z'', z, t) = \sqrt{\frac{2}{m}} \left[W_*(z, t) - \Delta B(z'') J_0 \right]$, $\Delta B = B(z'') - B_*$, and $W_*(z, t)$ satisfies the equation

$$\frac{\partial W_*}{\partial t} + v_* \frac{\partial W_*}{\partial z} = 0 \tag{7.15}$$

which follows from (7.3). $W_*(z_*, t) = m v_{*i}^2/2$ is the parallel energy which the step electrons had at the injection point, and

$$v_* = \left\{ \frac{2}{m} \left[W_*(z, t) - \Delta B(z) J_0 \right] \right\}^{1/2} \tag{7.16}$$

is the instantaneous beam parallel velocity of the step at location (z, t).

It is convenient to rewrite the integral relation (7.14) in the form of a differential equation; for that, we take the combination $\frac{\partial (J B^{-3/2})}{\partial t} + v_* \frac{\partial (J B^{-3/2})}{\partial z}$. From (7.14) and (7.15) we obtain:

$$\frac{\partial (J B^{-3/2})}{\partial t} + v_* \frac{\partial (J B^{-3/2})}{\partial z} = \alpha_2 \mathcal{A} + i \Delta \cdot J B^{-3/2} \tag{7.17}$$

where

$$\alpha_2 = \frac{e^2 n_{b0}}{m^2} \frac{J_0 B^{1/2}}{v_{*i} v_* B_*} \left(\frac{\omega_{B_*}}{\omega} - 1 \right), \qquad \Delta = \omega + k v_* - \omega_B. \tag{7.18}$$

Equation (7.17) together with (7.12) describe the linear stage of the evolution of the cyclotron instability (CI) in an inhomogeneous magnetic field with an arbitrary initial dependence of ω (or k) on z and t. Taking into account that $\vec{v}_*$ is parallel to $\vec{z}$ and antiparallel to $\vec{v}_{\rm g}$ for the cyclotron resonance interaction, we obtain

$$\frac{\partial \mathcal{A}}{\partial t} - v_{\rm g} \frac{\partial \mathcal{A}}{\partial z} = \alpha_1 B^{3/2} \mathcal{J} \tag{7.19}$$

$$\frac{\partial \mathcal{J}}{\partial t} + v_* \frac{\partial \mathcal{J}}{\partial z} = \alpha_2 \mathcal{A} + i \Delta \cdot \mathcal{J} \tag{7.20}$$

where $\mathcal{J} = J B^{-3/2}$, and v_* obeys Equation (7.15).

7.2 General formula for amplification

In this section we investigate the solution of the set of equations (7.19) and (7.20). For that, we transform this set using new variables following the wave and the beam

$$\eta = t + z/v_g, \qquad \xi = \int_{z_*}^{z} \frac{dz'}{\tilde{v}_*[z', W_*(z,t)]} - t. \tag{7.21}$$

In these variables we have, instead of (7.19) and (7.20),

$$\left(1 + \frac{v_g}{v_*}\right)\frac{\partial \mathcal{A}}{\partial \xi} = -\alpha_1\left(1 + Q\frac{\partial W_*}{\partial \xi}\right)\mathcal{J}B^{3/2} \tag{7.22}$$

$$\left(1 + \frac{v_*}{v_g}\right)\frac{\partial \mathcal{J}}{\partial \eta} = \alpha_2\mathcal{A} + i\Delta \cdot \mathcal{J} \tag{7.23}$$

where

$$Q(\xi,\eta) = \int_{z_*}^{z(\xi,\eta)} \frac{dz'}{\tilde{v}_*^2}\frac{\partial \tilde{v}_*}{\partial W_*},$$

and W_* is the function of ξ only, according to (7.15).

Equations (7.22) and (7.23) can be rewritten as one second-order equation:

$$\frac{\partial^2 \mathcal{A}}{\partial \xi\, \partial \eta} - \left(i\frac{\Delta \cdot v_g}{v_\Sigma} - \frac{\partial \ln f_1}{\partial \eta}\right)\frac{\partial \mathcal{A}}{\partial \xi} + \alpha_{\text{eff}}^2\mathcal{A} = 0 \tag{7.24}$$

where

$$f_1 = \frac{v_\Sigma}{v_*\alpha_1(1 + Q\,\partial W_*/\partial \xi)}, \qquad v_\Sigma = v_* + v_g,$$

and

$$\alpha_{\text{eff}}^2 = \alpha_1\alpha_2\left(1 + Q\frac{\partial W_*}{\partial \xi}\right)B^{3/2}\frac{v_g v_*}{v_\Sigma^2} > 0. \tag{7.25}$$

In (7.24) we neglect the small term $\partial \ln f_1/\partial \eta$.

Equation (7.24) describes completely the linear stage of the cyclotron instability of a non-stationary beam in an inhomogeneous magnetic field, as well as some non-trivial generation regimes such as the backward wave oscillator (BWO) regime. It is convenient to rewrite Equation (7.24) in the form of an integro-differential equation:

$$\frac{\partial \mathcal{A}}{\partial \xi} = -\int_{\eta_0(\xi)}^{\eta} \alpha_{\text{eff}}^2\mathcal{A}(\eta',\xi)\exp\left\{i\int_{\eta'}^{\eta}\Phi(\xi,\eta'')\,d\eta''\right\}d\eta' \tag{7.26}$$

where

$$\Phi(\xi,\eta'') = \frac{v_g}{v_\Sigma}\Delta(\xi,\eta''), \tag{7.27}$$

and we have used the boundary (initial) condition:

$$t(\xi, \eta) = 0: \quad \mathcal{J}(t = 0) \sim \frac{\partial \mathcal{A}}{\partial \xi} = 0\,. \tag{7.28}$$

In (7.26), $\eta_0(\xi)$ is the solution of $t(\xi, \eta) = 0$, which gives us, from (7.21), the following relation:

$$\xi = \int_{z_*}^{\eta_0} \frac{dz'}{v_*\big(z', W_*(\xi)\big)}\,. \tag{7.29}$$

It is evident from (7.26) that the kinetic stage of the instability, when the amplification is proportional to the concentration of energetic electrons ($\sim \alpha_{\rm eff}^2$), takes place for a rather weak amplification, when $\mathcal{A}(\eta', \xi)$ can be taken out of the integral (7.26). We introduce the total logarithmic amplification (including the phase correction φ) as

$$\ln\left(\frac{\mathcal{A}}{\mathcal{A}_0}\right) = G \equiv \Gamma + i\varphi \tag{7.30}$$

where $\mathcal{A}_0$ is the initial wave amplitude. Hence, from (7.26), we obtain for the kinetic stage of the instability an expression for the amplification Γ in the form:

$$\Gamma(\eta, \xi) = -\alpha_{\rm eff}^2 \int_{\xi_0(\eta)}^{\xi} \int_{\eta_0(\xi')}^{\eta} \cos\left[\int_{\eta'}^{\eta} \Phi(\eta'', \xi')\, d\eta''\right] d\eta'\, d\xi' \tag{7.31}$$

where $\xi_0(\eta)$ is the inverse function to $\eta_0(\xi)$ which is determined from (7.29).

In the case of a stationary beam, which corresponds formally to $t = 0$ in (7.21) and $W_* = $ const., the expression (7.31) coincides with the formula used in Chapter 3 (see (3.68)).

7.3 Effects of second-order cyclotron resonance

The formula (7.31) serves as the basis for an analysis of the wave amplification with second-order cyclotron resonance effects and the non-stationary behaviour of the beam being taken into account. We can transform relation (7.31) to algebraic form if we recall that the phase Φ contains the large parameter

$$p_0 \equiv \frac{a\omega_{BL}}{v_*} \gg 1 \tag{7.32}$$

where a and v_* are the characteristic scales of the magnetic field inhomogeneity and the beam velocity, respectively. For example, in the Earth's magnetosphere this parameter for whistler-mode waves is $p_0 \sim 10^3$–10^4. Thus we may use the stationary phase method to calculate the integral in (7.31). The stationary point for the integral over η' is determined from the equation

$$\frac{\partial \psi}{\partial \eta'} = -\Phi(\eta', \xi') = 0, \qquad \eta'_{\mathrm{st}} = \eta'_{\mathrm{st}}(\xi'). \tag{7.33}$$

It can be assumed, without loss of generality, that Φ is represented in algebraic form as

$$\Phi = p_0 \prod_{i=1}^{n} \{\xi' - \xi_i[\eta, \omega(\eta)]\}\, \zeta \tag{7.34}$$

where p_0 is determined by (7.32), ξ_i are the real roots of the equation $\Phi = 0$ (7.33), and the symbol $\prod$ means the product; the coefficient ζ is determined by a particular dependence of ω_B, k and v_* on z and t. For simplicity, we take $n = 2$; this is sufficient to analyse all the principal features of the problem. Applying the stationary phase method twice to the integration over η' and ξ' in (7.31), with (7.33) and (7.34) being taken into account, we obtain:

$$\Gamma = \frac{2\pi\alpha_{\mathrm{eff}}^2}{p_0\, |\xi_2 - \xi_1|\, \zeta}. \tag{7.35}$$

It is clear that the maximum amplification Γ_m is achieved when there are multiple roots in relation (7.33):

$$\xi_i[\eta, \omega(\eta)] = \xi_j[\eta, \omega(\eta)] = \xi_0, \qquad i \neq j. \tag{7.36}$$

In this case the simple stationary phase method is not valid, and it is necessary to apply the modified stationary phase (MSP) method. On the other hand, the dependence $\omega(\eta)$ in (7.34) is arbitrary, and we can choose this dependence from the condition of multiple roots (7.36). It is actually the quantitative formulation of the second-order cyclotron resonance effects.

We now suppose that Equation (7.36) can be satisfied for a suitable choice for the dependence of ω on η, the variable, following the wave packet according to (7.21). In this case, Φ in (7.34) can be presented in the form

$$\Phi = p_0\, \zeta(\xi', \eta')\, [\xi' - \xi_0(\eta')]^2. \tag{7.37}$$

Then the point $\xi' = \xi_0(\eta)$ provides the main contribution to the integral (7.31), and we can neglect the contribution from the ordinary roots of Equation (7.34), with $\Phi = 0$. According to (7.37),

$$\left(\frac{\partial \Phi}{\partial \eta}\right)_{\mathrm{st}} \simeq \left\{-2(p_0\zeta)\, [\xi - \xi_0(\eta)]\, \frac{\partial \xi_0}{\partial \eta}\right\}_{\mathrm{st}} \to 0 \tag{7.38}$$

$$\left(\frac{\partial^2 \Phi}{\partial \eta^2}\right)_{\mathrm{st}} = 2(p_0\zeta)_{\mathrm{st}} \left(\frac{\partial \xi_0}{\partial \eta}\right)_{\mathrm{st}}^2 \neq 0.$$

Here the subscript 'st' denotes stationary. Applying the MSP method to the internal integration over η' in (7.31), we find that

$$\Gamma_m \simeq \frac{2\Gamma_0(1/3)\cos(\pi/6)}{3^{2/3}} \int_{\xi_0(\eta)}^{\xi} d\xi' \, \alpha_{\text{eff}}^2 \left[p_0 \zeta \left(\frac{\partial \xi_0}{\partial \eta} \right)^2 \right]_{\text{st}}^{-1/3}$$
$$\times \cos \left\{ \int_{\eta_{\text{st}}(\xi')}^{\eta} p_0 \zeta [\xi' - \xi_0(\eta'')]^2 \, d\eta'' \right\} \tag{7.39}$$

where $\Gamma_0(1/3)$ is the gamma function, and the stationary value η_{st} is found from the equation

$$\xi_0(\eta_{\text{st}}) = \xi'. \tag{7.40}$$

The MSP method can also be used to evaluate the integral (7.39). To find the stationary point ξ_{st}, the second and third derivatives of the phase in (7.39) should be calculated. They are equal to:

$$\frac{\partial \psi}{\partial \xi'} = \frac{\partial}{\partial \xi'} \left\{ \int_{\eta_{\text{st}}}^{\eta} p_0 \zeta \, [\xi' - \xi_0(\eta'')]^2 \, d\eta'' \right\} = 2p_0 \int_{\eta_{\text{st}}}^{\eta} \zeta (\xi' - \xi_0) \, d\eta''$$
$$\frac{\partial^2 \psi}{\partial \xi'^2} = 2p_0 \int_{\eta_{\text{st}}}^{\eta} \zeta \, d\eta'', \qquad \frac{\partial^3 \psi}{\partial \xi'^3} = -2p_0 \zeta \frac{\partial \eta_{\text{st}}}{\partial \xi_0}. \tag{7.41}$$

The stationary point ξ'_{st} is determined from the following equations:

$$\frac{\partial \psi}{\partial \xi'} = \frac{\partial^2 \psi}{\partial \xi'^2} = 0, \qquad \eta = \eta_{\text{st}}(\xi'_{\text{st}}). \tag{7.42}$$

If we recall that, according to (7.33), the equation $\partial \psi / \partial \xi' \equiv \Phi = 0$ is actually the cyclotron resonance condition ($\Phi \propto \omega - \omega_B - k v_\parallel$), then we can rewrite the second-order cyclotron resonance condition (7.42) in the form:

$$\omega - \omega_B - k v_\parallel = 0, \qquad \frac{\partial}{\partial \xi} (\omega - \omega_B - k v_\parallel) = 0. \tag{7.43}$$

Finally, we have for the amplification Γ_m:

$$\Gamma_m = \Gamma_0^2(1/3) \left[\alpha_{\text{eff}}^2 \left(3 p_0^2 \zeta^2 \frac{\partial \xi_0}{\partial \eta} \right)^{-1/3} \right]_{\text{st}}. \tag{7.44}$$

In comparison with the ordinary CI amplification Γ (7.35), the second-order cyclotron resonance effects give an additional gain (one order of magnitude, 20 dB), such that

$$\Gamma_m / \Gamma \sim p_0^{1/3} \gg 1. \tag{7.45}$$

Thus the relations (7.36) (or (7.43)) and (7.44) determine the spatio-temporal characteristics of the dynamical spectrum of the generated wave packet and the corresponding amplification Γ_m.

7.4 The case of a beam with a delta-function distribution

The basic formulation of the problem in the case of a delta-function is the same as in the previous section. We consider a quasi-monochromatic whistler-mode wave propagating along the ambient magnetic field but in the $(-z)$-direction. The wave frequency is allowed to vary in space and time, and the final goal is to determine the maximum amplification of such a wavelet, and also the corresponding spatio-temporal variation of its frequency. We assume that the distribution function of energetic electrons has a δ-like dependence on the parallel velocity component $v_\parallel$:

$$\mathcal{F} = n_\mathrm{b}\, \delta(v_* - v_{\parallel *})\, \mathcal{F}(v_{\parallel *})\, \Phi(J_\perp, z) \tag{7.46}$$

where the same definitions have been used, and the same suggestions taken for $\mathcal{F}$, Φ and $v_*(z, t)$ as for the step-like distribution function (see Section 7.3).

The basic equation for the wave amplitude written in the same form (7.12), with the current being given by (7.13), but putting the distribution function (7.46) into (7.13), gives another expression for the resonant current $\mathcal{J}$. This is similar to the current (4.32) in the case of a homogeneous plasma, but changing the integration over t to integration over z. Differentiating $\mathcal{J}(z, t)$ with respect to z and t, and making the appropriate combinations such as $\left(\dfrac{\partial}{\partial t} + v_* \dfrac{\partial}{\partial z}\right)$, we obtain the following system of differential equations for the wave electric field amplitude $\mathcal{A}$ and the current $\mathcal{J}$:

$$\frac{\partial \mathcal{A}}{\partial t} - v_\mathrm{g} \frac{\partial \mathcal{A}}{\partial z} = \alpha_1 \mathcal{J} \tag{7.47}$$

$$\left(\frac{\partial}{\partial t} + v_* \frac{\partial}{\partial z} + i\Delta\right)^2 \mathcal{J} = \alpha_2 \left(i \frac{k^2 v_\perp^2}{\omega_{B_*}} + \frac{\partial}{\partial t} + v_* \frac{\partial}{\partial z} + i\Delta\right) \mathcal{A} \tag{7.48}$$

where $\Delta = \omega + k v_* - \omega_B$, α_1 is determined by the same expression (6.12) as for the step, and

$$\alpha_2 = \frac{e^2 n_\mathrm{b}}{m\sqrt{2}} \left(\frac{\omega_{B_*}}{\omega} - 1\right). \tag{7.49}$$

Actually the system of equations (7.47) and (7.48) is the same as (4.23) and (4.34), but now Δ and v_* are functions of z and t. Unlike the case with a step-like distribution function, the system of equations (7.47) and (7.48) is of third-order. We already met this situation in Chapter 4 for the analysis of BWO regimes in a homogeneous plasma.

To find the amplification we introduce new variables (see (7.21)):

$$d\eta = v_g^{-1}\, dz + dt, \qquad d\xi = v_*^{-1}\, dz - dt. \tag{7.50}$$

With these variables (7.47) and (7.48) have the form

$$\frac{v_\Sigma}{v_*}\frac{\partial \mathcal{A}}{\partial \xi} = \alpha_1 \mathcal{J} \tag{7.51}$$

$$\left(\frac{v_\Sigma}{v_g}\frac{\partial}{\partial \eta} - i\Delta\right)^2 \mathcal{J} = i\alpha_2 \mathcal{A} - \alpha_2 \frac{\delta}{v_\Sigma}\left(\frac{v_\Sigma}{v_g}\frac{\partial}{\partial \eta} - i\Delta\right)\mathcal{A} \tag{7.52}$$

where

$$v_\Sigma = v_g + v_* \text{ and } \delta = \frac{\omega_B v_\Sigma}{(k^2 v_\perp^2)}.$$

Equations (7.51) and (7.52) can be written as a single third-order equation:

$$\left(\frac{\partial}{\partial \eta} - i\Phi\right)^2 \frac{\partial \mathcal{A}}{\partial \xi} = \alpha_0^2\left[i - \delta\left(\frac{\partial}{\partial \eta} - i\Phi\right)\right]\mathcal{A} \tag{7.53}$$

where $\Phi = \dfrac{\Delta v_g}{v_\Sigma}$ and

$$\alpha_0^2 = \alpha_1\alpha_2\left(\frac{v_g v_*}{v_\Sigma^2 \delta}\right). \tag{7.54}$$

It is worth noting that $\alpha_0^2 \simeq \alpha_{\text{eff}}^2 \delta^{-1}$, where α_{eff}^2 is determined by (7.25). Taking the boundary condition (7.28), we can write the formal solution of (7.53) in the form:

$$\ln\left(\frac{\mathcal{A}}{\mathcal{A}_0}\right) \equiv \Gamma + i\varphi \simeq i\int_{\xi_0(\eta)}^{\xi} \mathcal{A}^{-1}(\xi', \eta)\int_{\eta_0(\xi')}^{\eta} \alpha_0^2\, \mathcal{A}(\xi', \eta')\,(\eta - \eta' + i\delta)$$
$$\times \exp\left\{i\int_{\eta'}^{\eta} \Phi(\xi', \eta'')\, d\eta''\right\} d\eta'\, d\xi' \tag{7.55}$$

where, as in the previous section, $\xi_0(\eta)$ is the inverse function to $\eta_0(\xi)$, which is determined from the relation (7.29).

We can see from (7.55) that the kinetic stage of the cyclotron instability takes place when the amplification is relatively weak, and the amplitude $\mathcal{A}(\xi', \eta')$ can be taken out of the internal integral (7.55) at the point $\eta' \simeq \eta$. After that the dependence of the right-hand side of (7.55) on the amplitude $\mathcal{A}$ disappears, and we can find the amplification Γ in a similar way as for the step-like distribution, using the MSP

Table 7.1 Whistler-mode wave gain for step-like and beam-like electron distributions compared with that for a smooth distribution in the cases of the first- and second-order CR.

	First-order CR	Second-order CR
$\Gamma_{\text{m beam}}/\Gamma$	1	p_0
$\Gamma_{\text{m step}}/\Gamma$	1	$p_0^{1/3}$

method. Referring to Demekhov *et al.* (2000) for details, we write the maximum value of Γ:

$$\Gamma_m \simeq \frac{4\pi^2 v_* \alpha_0^2 \ell_{\text{st}}^3}{3v_g \Gamma_0(2/3)\Gamma_0(1/3)} \sim p_0\Gamma \tag{7.56}$$

where $\Gamma_0(x)$ is the gamma function, $\ell_{\text{st}}^3 \simeq \dfrac{1}{3p_0}$, and Γ is the amplification for first-order cyclotron resonance (see (7.35)).

Table 7.1 summarizes the wavelet amplification for the different distributions: smooth, step and delta-function.

7.5 The example of a linearly varying ambient magnetic field

As an example illustrating the application of the formulae obtained in the previous sections, we consider the case of the linearly varying magnetic field. We suppose the gyrofrequency ω_B to vary as

$$\omega_B = -\omega_{BL}\, s, \quad s < 0 \tag{7.57}$$

where $s = z/a$, z is the coordinate along the magnetic field line, and a is the characteristic scale length. To simplify the calculations we assume that the energetic electron velocity is changed mainly due to acceleration by the wave field, and that the magnetic field inhomogeneity can be neglected in Equation (7.21), so that $\xi = z/v_* - t$. For convenience we change the variables to

$$\tilde{\eta} = s + u_g\tau, \qquad \tilde{\xi} = s - u_*\tau \tag{7.58}$$

where s, τ, and u_g, u_* are dimensionless variables,

$$s = z/a, \quad u_g = v_g/v_0, \quad u_* = v_*/v_0, \quad \tau = v_0 t/a. \tag{7.59}$$

Further, we omit the tilde for brevity. The dependence of v_* on ξ is assumed to have the following form:

$$v_* = v_0(1 - \alpha\xi). \tag{7.60}$$

From (7.58) and (7.60) it follows that

$$\xi = \frac{s - \tau}{1 - \alpha\tau}, \quad s = (\xi u_g + \eta u_*)u_\Sigma^{-1}, \quad \tau = u_\Sigma^{-1}(\eta - \xi) \tag{7.61}$$

where $u_\Sigma = u_* + u_g$. Now the phase $\psi \equiv \int_{\eta'}^{\eta} \Phi\, d\eta''$ can be expressed in the dimensionless variables as

$$\psi = \int_{\eta'}^{\eta} p_0 \left(\frac{k u_* v_0 + \omega}{\omega_{BL}\, u_\Sigma} + \frac{\xi u_g + \eta'' u_*}{u_\Sigma^2} \right) d\eta'' \tag{7.62}$$

where the large parameter $p_0 = a\omega_{BL}/v_0$; k and ω are the wave vector and the frequency of the whistler wave, respectively. We shall consider the case $\omega \ll \omega_B$ and $\vec{k} \parallel \vec{B}$ when the following approximate relations can be used without any significant error:

$$k = \frac{\omega_p}{c} \frac{\omega^{1/2}}{\omega_B^{1/2}}, \qquad v_g \simeq \frac{2\omega}{k}, \qquad k = \frac{\omega_p^2 v_g}{2\omega_B\, c^2}, \qquad \omega = \frac{\omega_p^2 v_g^2}{4\omega_B\, c^2} \tag{7.63}$$

where ω_p is the electron plasma frequency, and we assume that the ratio $\omega_p^2/\omega_B = \text{const}(z)$. After that, Equation (7.62) can be written in the form:

$$\psi \equiv \int_{\eta'}^{\eta} \Phi\, d\eta'' = \int_{\eta'}^{\eta} \frac{p_0}{q u_\Sigma^2}\, u_g\, (u_* - u_1)(u_* - u_2)\, d\eta'' \tag{7.64}$$

where $q = 2\omega_{BL}^2\, c^2/(\omega_{pL}^2\, v_0^2)$, $\omega_p^2/\omega_B = \omega_{pL}^2/\omega_{BL} = \text{const}$. According to (7.60), u_* is related to ξ by

$$u_* = 1 - \alpha\xi. \tag{7.65}$$

The roots $u_{1,2}$ are equal to

$$u_{1,2} = -Q \pm \sqrt{Q^2 - q/\alpha - u_g^2/2}, \qquad 2Q \equiv \frac{3}{2} u_g + q(\eta/u_g - \alpha^{-1}). \tag{7.66}$$

The condition for equal roots is that the expression under the square root sign in (7.66) is equal to zero:

$$q(\alpha^{-1} - \eta/u_g) - \frac{3}{2} u_g = \pm 2\sqrt{q\alpha^{-1} + u_g^2/2}\,. \tag{7.67}$$

Thus we have

$$\xi_{\rm st}(\eta) = \xi_0 \equiv \alpha^{-1}(1 - u_*) \equiv \alpha^{-1}\left(1 \mp \sqrt{q\alpha^{-1} + u_{\rm g}^2/2}\right). \tag{7.68}$$

The dependence of $u_{\rm g}$ on η is determined from (7.65). In the case $\omega \ll \omega_B$, the inequality $u_{\rm g}^2 \ll u_0^2$ holds, and we can use a simpler relation to obtain the function $u_{\rm g}(\eta)$:

$$q(\alpha^{-1} - \eta/u_{\rm g}) - \frac{3}{2}u_{\rm g} = \pm 2(q/\alpha)^{1/2}. \tag{7.69}$$

Analysis shows that the lower sign in Equations (7.66)–(7.69) does not exist for this particular case because it corresponds to the generation region with $z > 0$ ($\omega_B < 0$). The relation (7.69) can be written in the form

$$\alpha\eta = \tilde{u}_{\rm g}\left(w_0 - \frac{3}{2}\tilde{u}_{\rm g} - 2\right) \tag{7.70}$$

where $\tilde{u}_{\rm g} = u_{\rm g}/w_0$, $w_0 = (q/\alpha)^{1/2}$.

Now using (7.68) and (7.70), we can find the time τ_ω and place s_ω where a signal with the frequency ω (or, according to (7.63), the group velocity $u_{\rm g} = (2q\omega/\omega_{BL})^{1/2}$) is generated. For that it is necessary to substitute (7.58) into (7.68) and (7.70) and to find τ_ω and s_ω from these two equations for a fixed value of $u_{\rm g}$. We obtain

$$\alpha\tau_\omega = \frac{w_0(1+\tilde{u}_{\rm g}) - \tilde{u}_{\rm g}\left(2 + \frac{3}{2}\tilde{u}_{\rm g}\right) - 1}{w_0 + u_{\rm g}} \geq 0, \quad \alpha s_\omega = -\frac{u_{\rm g}(1+3u_{\rm g})}{w_0 + u_{\rm g}} < 0. \tag{7.71}$$

Using Equations (7.63) and (7.71) we can find the dynamical spectrum of the generated discrete whistler-mode wave signal. It is shown in Fig. 7.1(a). The beam moves in the $\vec{z}_0$-direction corresponding to a decrease of the magnetic field; the wave packet generated propagates in the opposite direction, its frequency decreasing with increasing time. This is termed a faller.

It is necessary to stress that, in the case considered here, second-order resonance effects (namely, a substantial increase of the amplification) take place only if v_* decreases with ξ (7.60); the inequality $w_0 > 1$ has to be satisfied. If $v_*(\xi)$ is increasing with increasing ξ (i.e. if $\alpha < 0$ in (7.60)), the same effects take place for the electron beam propagating in the direction of increasing magnetic field ($\omega_B = +\,\omega_{BL}\,s$). In the latter case we have a signal whose dynamical frequency spectrum is a rising tone (Fig. 7.1b). The quantitative characteristics of such a signal are described by the same formulae (7.71), but α must be changed to $-\alpha$.

Now we have derived all the parameters to calculate the amplification Γ_m using (7.44). With the new variables (7.58) and ψ defined according to (7.64), we

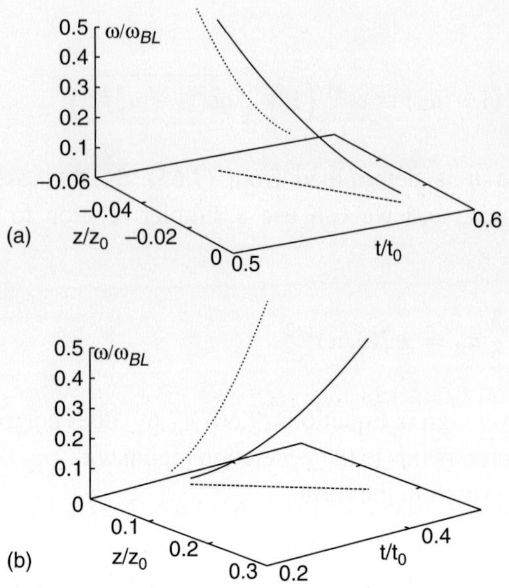

Figure 7.1 The dynamical spectra of whistler-mode discrete signals in a linearly varying magnetic field; fallers (risers) are generated when the electron beam moves toward decreasing (increasing) magnetic field.

write Γ_m in the form

$$\Gamma_m = \frac{\alpha_{\text{eff}}^2\, a^2\, \Gamma_0^2(1/3)}{v_* v_{\text{g}}} \left[3 \left(\frac{p_0\, \alpha^2 u_{\text{g}}}{q u_\Sigma^2} \right)^2 \left(\frac{\partial \xi_0}{\partial \eta} \right) \right]^{-1/3} . \tag{7.72}$$

From (7.68) and (7.70), we obtain

$$\left| \frac{\partial \xi_0}{\partial \eta} \right| = \frac{2u_{\text{g}}}{\alpha w_0 \, |\partial \eta / \partial u_{\text{g}}|} \simeq \frac{u_{\text{g}}}{|2 - w_0 + 3u_{\text{g}} w_0^{-1}|} . \tag{7.73}$$

For whistler-mode waves with $\omega \ll \omega_B$, the interaction coefficients α_1 and α_2 which come into the expression (7.25) can be written in the form

$$\alpha_1 = \frac{2\pi \omega_{BL}\, \omega}{\omega_{\text{pL}}^2} , \qquad \alpha_2 = \frac{e^2 n_{\text{b}}}{m^2 v_{*0}^2} \frac{\omega_B J_0}{\omega B^{1/2}} . \tag{7.74}$$

Substitution of (7.74) into (7.25) gives

$$\alpha_{\text{eff}}^2 \simeq \frac{n_{\text{b}}}{2n_{\text{c}} w_0^2} \frac{p_0^2}{ma^2} \frac{v_{\text{g}} v_*}{v_\Sigma^2} J_0 B_* \tag{7.75}$$

and the final value of Γ_m is equal to

$$\Gamma_m \simeq \frac{n_{\text{b}}}{n_{\text{c}}} p_0^{4/3} \delta \tag{7.76}$$

where the dimensionless parameter $\delta \sim 1$ is equal to:

$$\delta \simeq \Gamma_0^2(1/3)\,\frac{J_0 B_*}{2mv_0^2}\,q\left(\frac{\omega_{BL}}{2\omega}\right)^{1/2}\left(\frac{2}{3\alpha^8}\right)^{1/3} \tag{7.77}$$

where α is determined in (7.60). Thus $\Gamma_m \simeq 10^4\,\frac{n_\mathrm{b}}{n_\mathrm{c}}$ up to about $2\cdot 10^5\,\frac{n_\mathrm{b}}{n_\mathrm{c}}$.

7.6 Summary

The hydrodynamic cyclotron beam–plasma instability in an inhomogeneous magnetic field differs essentially from the case of a homogeneous medium. For small beam densities, the instability is of the convective type and amplification of the wave occurs as it passes through the cyclotron resonance region. The duration of this interaction is determined by the corresponding spatial and temporal dependencies of the gyrofrequency, wave frequency and beam velocity. The width of the cyclotron resonance (CR) region in this case is rather narrow, so that the effects of the mutual compensation of CR mismatch by suitable dependencies of ω_B and ω on z and t are very important. These are the so-called second-order CR effects. Such a compensation occurs at different moments of time and at different places for different frequencies, which results in the generation of quasi-monochromatic wave packets (wavelets) with different frequency spectra. In the general case, the compensation in the case of a step distribution cannot be complete because, as a rule, the gyrofrequency depends only on z, while the wave frequency depends on the combination $\eta = z + v_\mathrm{g}t$. Typical values for the extra amplification due to this compensation can be as much as $p_0^{1/3} \gtrsim 10$ for typical p_0 values of 10^3 to 10^4. Second-order CR effects determine the starting conditions for the instability. As such, they can be very important, in particular in connection with the generation of triggered ELF/VLF emissions in the Earth's magnetosphere.

When the beam density increases, some qualitatively new wave generation regimes appear. In particular, the convective instability can be changed into an absolute instability, when multiple signals with discrete frequency spectra are generated. An example of such a transition to the backward wave oscillator generation regime has been considered in a finite-length system with a homogeneous magnetic field (see Section 5.6). This transition obviously occurs for the inhomogeneous magnetic field situation too. This problem is very interesting because it could be the key to another important problem of magnetospheric physics, namely the origin of chorus emissions. We shall return to that issue in Section 11.2.

Chapter 8

Quasi-linear theory of cyclotron masers

In previous chapters the cyclotron interaction of charged particles with whistler (or Alfvén) mode waves of a given amplitude was considered, as was the amplification of monochromatic waves and wavelets using a linear approach. At the same time, such important questions as the dynamics of the radiation belts, pitch-angle diffusion and the precipitation of energetic charged particles, their acceleration, the origin of different types of electromagnetic emissions in the Earth's and other magnetospheres demand the development of a self-consistent approach to the problem of wave–particle interactions; this should be based on a nonlinear theory of the cyclotron instability. This theory includes the influence of feedback of the waves generated on the energetic charged particle distribution. The formulation of such a nonlinear theory is non-trivial in the case of space CMs.

Experiments show a huge diversity of wave generation regimes in CMs, including both noise-like and quasi-monochromatic regimes, and complicated interactions between these two regimes. A general theory which permits us to describe this diversity of generation regimes from a single point of view does not yet exist.

However, it is helpful to classify the observed natural electromagnetic signals into one of two groups – noise-like emissions or signals with a discrete frequency spectrum. A dominant example of the first group is ELF or VLF hiss; representatives of the second group are ELF or VLF chorus. Similar groupings exist in the ion cyclotron frequency range; these are the short-period geomagnetic pulsations of IPDP (irregular pulsations of diminishing period) and Pc1 types, respectively. An adequate theory for the description of noise-like (broadband) emissions seems to be the well-known quasi-linear (QL) theory of plasma physics of wave–particle interactions.

The basis of this theory in application to the cyclotron wave–particle interactions was developed by Vedenov *et al.* (1962), Andronov and Trakhtengerts (1963, 1964),

Kennel and Engelman (1966), Galeev (1967), Trakhtengerts (1966, 1967), Kennel (1969), Bespalov and Trakhtengerts (1976a,b, 1979) and others. The results of these investigations were summarized in the review by Trakhtengerts (1984). This theory has turned out to be very effective when applied to magnetospheric cyclotron masers, and has been developed to a rather sophisticated degree (see Bespalov and Trakhtengerts, 1980, 1986b and references therein). The basis of this theory is given in Chapters 8 and 9.

8.1 Transition to a noise-like generation regime. Energy transfer equation

The main assumption of the quasi-linear approach is that there is an ensemble of wave packets, closely spaced in frequency and having random phases. The typical frequency width $\Delta\omega$ of this ensemble should be much greater than the trapping width of a particle in the potential well of a single wave packet

$$\Delta\omega \gg \Omega_{\mathrm{tr}1} = (k\omega_{\mathrm{h}1} v_\perp)^{1/2} > \Delta\omega_{\mathrm{i}} \tag{8.1}$$

where the expression (5.14) has been used for trapping frequency $\Omega_{\mathrm{tr}1}$, $\omega_{\mathrm{h}1} = \dfrac{eb_1}{mc}$, b_1 is the magnetic field amplitude of a separate wave packet, and $\Delta\omega_i$ is the frequency interval between wave packets. A qualitative diagram of charged particle motions in phase space ($v_\parallel$, ψ) is shown in Fig. 8.1, to be compared with Fig. 5.1 and Fig. 6.4.

Gyroresonances overlap in this case, and the random phases of the wave packets lead to stochastic (Brownian) motion of a charged particle in this wave field. Actually, the particle motion is much more complicated in such a situation, because many additional ‘cat's eyes’ appear in the phase plane, Fig. 5.1 (Sagdeev *et al.*, 1989, Zaslavsky, 1985).

As we have seen in Chapter 3, the generation of wave emissions over a wide frequency band takes place in the case of energetic particle distributions with a broad velocity dispersion. In the case of an inhomogeneous magnetic field there are additional factors which lead to the breaking of particle trapping by the wave (see Chapter 6). Moreover, the drift (due to magnetic gradient and curvature forces,

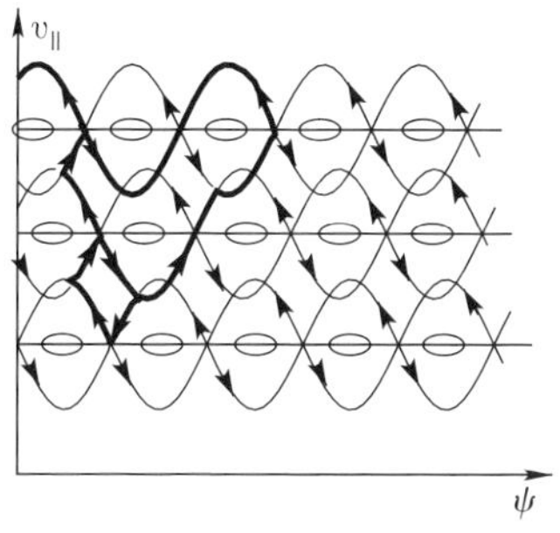

Figure 8.1 A qualitative diagram of resonant electron motions in phase space ($v_\parallel$, ψ) in the case of many waves satisfying (8.1).

see Walt, 1994) of the energetic particles through a wave generation region can mix the phase relations of energetic charged particles undergoing wave–particle interactions.

Thus, we shall suggest further that the wave magnetic field can be represented as

$$\vec{B}_{\sim}(\vec{r}, t) = \sum_k \vec{B}_k \exp\{i\omega_k t - i\vec{k}\vec{r}\} \tag{8.2}$$

where the frequency ω_k and the wave vector of the k-th wave packet are connected by the dispersion relation (see Chapter 2). The complex amplitudes $\vec{B}_k$ have a random phase distribution that gives:

$$\sum_{k'} \langle \vec{B}_{\vec{k}} \, \vec{B}_{\vec{k}} \rangle = b_{\vec{k}}^2 \, \delta(\vec{k} - \vec{k}') \tag{8.3}$$

where $b_{\vec{k}}^2$ is spectral intensity of the wave magnetic field, and $\delta(x)$ is a delta-function.

The same relations are valid for the electric field amplitude, which is connected with B_k by the vector cross-product relation (for $\vec{k} \| \vec{B}$):

$$\vec{B}_k = \frac{c}{\omega} \left[\vec{k} \times \vec{A}_k \right], \quad B_k = N_k A_k \tag{8.4}$$

where N_k is the refractive index.

The self-consistent approach in quasi-linear plasma theory considers the simultaneous usage of two equations: the energy transfer equation for the waves, and the kinetic equation for the averaged (over the wave ensemble) distribution function of the particles. It is suggested that this function changes with time slowly in comparison with the wave period. The energy transfer equation for the wave spectral energy density $\mathcal{E}_{\vec{k}}$ is obtained from the wave equation for the complex amplitude (2.132) in a standard way. For that it is necessary to multiply (2.132) by the complex conjugate amplitude $A_{\vec{k}}^*$, to add it to the complex conjugate equation and, after that, to average both parts of the equation so obtained over the wave ensemble. Then, for the wave spectral energy density we have:

$$\frac{\partial \mathcal{E}_{\vec{k}}}{\partial t} + \vec{v}_{gk} \frac{\partial \mathcal{E}_{\vec{k}}}{\partial \vec{r}} - \frac{\partial w_k}{\partial \vec{r}} \frac{\partial \mathcal{E}_{\vec{k}}}{\partial \vec{k}} = 2\gamma_{\vec{k}} \, \mathcal{E}_{\vec{k}} \tag{8.5}$$

where the third term on the left hand side of the equation has been added; this takes the plasma inhomogeneity into account. The growth rate $\gamma_{\vec{k}}$, in the case of a smooth distribution function, when the quasi-linear approach is valid, is determined by the relation (3.15), where $\mathcal{F}$ is the slowly changing part of the distribution function.

For whistler-mode waves or Alfvén waves the spectral energy density $\mathcal{E}_{\vec{k}}$ associated with the wave is related to $b_{\vec{k}}^2$ by the relation (see Chapter 2 and Ginzburg, 1970):

$$\mathcal{E}_{\vec{k}} = \frac{b_{\vec{k}}^2}{8\pi} + \frac{\partial(\omega \varepsilon_{ij} a_i a_j^*)}{8\pi \, \partial\omega} A_{\vec{k}}^2$$

which leads to

$$\mathcal{E}_k = \frac{\omega}{4\pi k v_g} b_k^2. \tag{8.6}$$

8.2 Derivation of the quasi-linear equation for a distribution function

The inhomogeneity of the magnetic field in a space CM is a most important factor; we therefore need to use the initial kinetic equation in the form (2.36). To apply the QL approach to CMs in the clearest possible way we shall consider the simplest case of low-frequency whistler-mode waves (with $\vec{k} \| \vec{B}$), i.e. when

$$\omega \ll \omega_{BL} \tag{8.7}$$

where ω_{BL} is the minimum value of the electron (or ion) gyrofrequency which for a dipole approximation of an ambient magnetic field is reached in the equatorial plane of every magnetic flux tube.

In accordance with (6.7), we can in this case neglect the change of particle energy and consider only variations of the magnetic moment $J_\perp$ under the action of the waves. The drift term $\vec{v}_D \dfrac{\partial \mathcal{F}}{\partial \vec{R}_{\perp c}}$ in (2.36) gives an additional contribution to the change of the averaged function $\langle \mathcal{F} \rangle$; we shall take this into account as and when necessary. Finally, the initial kinetic equation takes the form:

$$\frac{\partial \mathcal{F}}{\partial t} + v_\| \frac{\partial \mathcal{F}}{\partial z} + \dot{\mu} \frac{\partial \mathcal{F}}{\partial \mu} + \dot{\varphi} \frac{\partial \mathcal{F}}{\partial \varphi} = 0 \tag{8.8}$$

where φ gives the phase of charged particle rotation (2.11); $\mu = \dfrac{\sin^2\theta}{B} B_L$ (2.16). The derivatives $\dot{\mu}$ and $\dot{\varphi}$ can be written with the help of (2.122), (2.123) in the following form, so that

$$\dot{\mu} \equiv \sum_k 2\mu^{1/2}(1 - \mu\tilde{B})^{1/2} \tilde{B}^{-1/2} \omega_{Bk} \cos\psi_k \equiv \sum_k M_k \cos\psi_k \tag{8.9}$$

$$\begin{aligned} \dot{\varphi} = \dot{\varphi}^{(0)} + \dot{\varphi}^{(1)} &\equiv -\omega_B - \sum_k \frac{(1-\mu\tilde{B})^{1/2}}{(\mu\tilde{B})^{1/2}} \omega_{Bk} \sin\psi_k \\ &\equiv -\omega_B - \sum_k d_k \sin\psi_k \end{aligned} \tag{8.10}$$

where $\tilde{B} = \dfrac{B}{B_L}$, $v_\| = \sqrt{\dfrac{2}{m} W(1 - \mu\tilde{B})}$, $\omega_{Bk} = \dfrac{eB_k}{mc}$, and

$$\psi_k = \varphi - \Theta_k = \varphi - \int^z k\,dz' + \omega_k t.$$

In accordance with the physical picture described in Section 7.1, changes of μ_k and $\psi_k = \varphi - \Theta_k$ occur under the interaction with a single wave packet, for which k' is small, and the considerable evolution of $\mathcal{F}$ is determined by an accumulation of these small additions, which are of Brownian character for stochastic phase relations between the different wave packets. In this case we can write the solutions of (8.8)–(8.10) in the form of a series of terms involving the small parameter $\varepsilon = \frac{B_k}{B}$:

$$\mathcal{F} = \langle\mathcal{F}\rangle + \mathcal{F}_{\sim}^{(1)} + \ldots, \quad \psi_k = \psi_k^{(0)} + \psi_k^{(1)} + \ldots, \quad \mu_k = \mu_k^{(1)} + \ldots \tag{8.11}$$

where $\langle\mathcal{F}\rangle$ is the slowly changing part of the distribution function, $\mathcal{F}_{\sim}$ is the rapidly changing random addition, and $\langle\mathcal{F}_{\sim}\rangle = 0$ (the angle brackets mean averaging over the rapid oscillations). Substituting (8.11) into (8.8) and averaging over time, we obtain:

$$\frac{\partial\langle\mathcal{F}\rangle}{\partial t} + v_{\parallel}\frac{\partial\langle\mathcal{F}\rangle}{\partial z} + \left\langle \dot{\mu}^{(1)}\frac{\partial\mathcal{F}_{\sim}}{\partial\mu}\right\rangle + \left\langle \dot{\varphi}^{(1)}\frac{\partial\mathcal{F}_{\sim}}{\partial\varphi}\right\rangle = 0. \tag{8.12}$$

We should remember that, without the wave field, $\dot{\mu} = 0$, $\frac{\partial\mathcal{F}}{\partial z} = 0$ and $\frac{\partial\mathcal{F}}{\partial\varphi} = 0$. Because $\mathcal{F}_{\sim}$ includes only linear terms in the wave amplitude, we can look for the solution of $\mathcal{F}_{\sim}$ in the form of a Fourier series:

$$\mathcal{F}_{\sim} = \sum \mathcal{F}_k = \sum_k (S_k \sin\psi_k + C_k \cos\psi_k) \tag{8.13}$$

with new variables $\psi_k = \varphi - \Theta_k$. In these variables the combination

$$\frac{\partial\mathcal{F}_{\sim}}{\partial t} + v_{\parallel}\frac{\partial\mathcal{F}_{\sim}}{\partial z} + \dot{\varphi}^{(1)}\frac{\partial\mathcal{F}_{\sim}}{\partial\varphi} = \sum_k \dot{\psi}_k\frac{\partial\mathcal{F}_k}{\partial\psi_k} \tag{8.14}$$

where

$$\dot{\psi}_k = \dot{\psi}_k^{(0)} + \dot{\psi}_k^{(1)} = \omega - \omega_B - kv_{\parallel} + \dot{\psi}_k^{(1)}. \tag{8.15}$$

$\dot{\psi}_k^{(1)}$ is determined by the term $d_k \sin\psi_k$ in the sum over k in (8.10).

Subtracting (8.12) from (8.8), and using (8.11), we obtain an equation for $\mathcal{F}_k$ in the form:

$$\dot{\psi}_k^{(0)}\frac{\partial\mathcal{F}_k}{\partial\psi_k} + \dot{\mu}_k^{(1)}\frac{\partial\langle\mathcal{F}\rangle}{\partial\mu} = 0 \tag{8.16}$$

where $\psi_k^{(0)}$ is determined by (8.15), and $\dot{\mu}_k^{(1)}$ is the separate term in the sum (8.9) with

$$\psi_k = \psi_k^{(0)} = \int_{-\ell/2}^{z}\frac{dz'}{v_{\parallel}}(\omega - \omega_B - kv_{\parallel});$$

the coordinate $-\ell/2$ is at the foot of the magnetic flux tube, on the planet's surface. We find from (8.16) that

$$\mathcal{F}_k = M_k \frac{\partial \langle \mathcal{F} \rangle}{\partial \mu} \int_{-\ell/2}^{z} \cos \psi_k^{(0)} \frac{dz}{v_\parallel}. \tag{8.17}$$

Now we can find the third and fourth terms in (8.12). Omitting the brackets for $\mathcal{F}$, they are:

$$\begin{aligned}
&\left\langle \dot{\mu} \frac{\partial \mathcal{F}_\sim}{\partial \mu} \right\rangle + \left\langle \dot{\varphi}^{(1)} \frac{\partial \mathcal{F}_\sim}{\partial \varphi} \right\rangle \\
&\quad = -\sum_{k,k'} \left\langle M_k \cos \varphi_k \frac{\partial}{\partial \mu} M_{k'} \frac{\partial \mathcal{F}}{\partial \mu} \int_{-\ell/2}^{z} \cos \psi_k^{(0)} \frac{dz'}{v_\parallel} \right\rangle \\
&\qquad - \sum_{k,k'} \left\langle d_k \sin \psi_k^{(0)} M_{k'} \left(\frac{\partial \mathcal{F}}{\partial \mu} \right) \frac{\partial}{\partial z} \int_{-\ell/2}^{z} \cos \psi_k^{(0)} \frac{dz'}{v_\parallel} \right\rangle \\
&\quad = -\sum_{k} \left\{ M \frac{\partial}{\partial \mu} \left(M Q_k \omega_{bk}^2 \frac{\partial \mathcal{F}}{\partial \mu} \right) + d \left(M Q_k \omega_{bk}^2 \frac{\partial \mathcal{F}}{\partial \mu} \right) \right\}.
\end{aligned} \tag{8.18}$$

Here the second term has been transformed by integrating over z by parts. In (8.18) we have introduced the following definitions:

$$Q_k = \left\langle \cos \left(\int_{-\ell/2}^{z} \frac{\Delta dz'}{v_\parallel} \right) \int_{-\ell/2}^{z} \frac{dz'}{v_\parallel(z')} \cos \left(\int_{-\ell/2}^{z'} \frac{\Delta dz''}{v_\parallel} \right) \right\rangle \tag{8.19}$$

and

$$\omega_{bk}^2 = \frac{e^2}{m^2 c^2} b_k^2, \qquad M = 2\mu^{1/2} (1 - \mu \tilde{B})^{1/2} (\tilde{B})^{-1/2},$$
$$d = \frac{(1 - \mu \tilde{B})^{1/2}}{(\mu \tilde{B})^{1/2}}, \qquad \Delta = \omega - \omega_B - k v_\parallel. \tag{8.20}$$

We can see that the function $Q_k(\mu, W)$ coincides with the function $\mathcal{I}$ which was introduced in Chapter 3 during calculations of the cyclotron amplification in an inhomogeneous plasma. For smooth distributions of electrons (or ions) this function can be considered as a delta-function $\mathcal{I} \to \pi \delta(\omega - \omega_B - k v_\parallel)$. The same applies for $Q(\mu, W, \omega)$ if the frequency spectrum of the waves is a smoothly varying function of ω:

$$Q_k \to \pi \delta(\omega - \omega_B - k v_\parallel). \tag{8.21}$$

Finally the QL equation (8.12), with (8.18) and (8.19) being taken into account, is written as:

$$\frac{\partial \mathcal{F}}{\partial t} + v_\parallel \frac{\partial \mathcal{F}}{\partial z} = \frac{v_\parallel}{v} \frac{\partial}{\partial \mu} \left(\mu \sqrt{1 - \mu \tilde{B}}\, D_{\mu\mu} \frac{\partial \mathcal{F}}{\partial \mu} \right) \tag{8.22}$$

where

$$D_{\mu\mu} = \int d^3\vec{k}\, \delta(\omega - \omega_B - kv_{\parallel}) \frac{8\pi^2 e^2 k v_g}{\omega m^2 c^2 \tilde{B}} \mathcal{E}_{\vec{k}}. \tag{8.23}$$

In (8.22) we have come to the continuous wave spectrum $\left(\sum_k \to \int d^3\vec{k}\right)$ and have introduced the spectral energy density with the help of relation (8.6). It is easy to see that the diffusive operator on the right hand side of (8.22) takes its well-known form (Schulz and Lanzerotti, 1974), if we consider the case of a homogeneous plasma:

$$\frac{1}{\sin\theta} \frac{\partial}{\partial\theta} \left(\sin\theta \cdot D_{\mu\mu} \frac{\partial \mathcal{F}}{\partial\theta} \right) \tag{8.24}$$

where θ is the charged particle pitch-angle. When $\omega \ll \omega_B$, and $\vec{k} \| \vec{B}$, the refractive index N_{w} (2.72) for whistler-mode waves is given by

$$N_{\mathrm{w}}^2 = \frac{\omega_{pe}^2}{\omega\omega_B} = \frac{k^2 c^2}{\omega^2}. \tag{8.25}$$

In the Earth's magnetosphere this may often be considered to be a constant along a magnetic flux tube. We should also take into account that, for propagation in the $(+z)$-direction along a duct in the form of a magnetic flux tube, the energy transfer equation (8.5) has to be written in curvilinear coordinates. The equation for the energy flux $\mathcal{E}_\omega$ through the cross-section of a duct in unit frequency band has the simple form:

$$\frac{1}{v_g} \frac{\partial \mathcal{E}_\omega}{\partial t} + \frac{\partial \mathcal{E}_\omega}{\partial z} = æ_\omega \mathcal{E}_\omega \tag{8.26}$$

where z is the coordinate along the magnetic field line (in a curvilinear coordinate system), the amplification coefficient $æ_\omega = \gamma / v_g$, and the group velocity $v_g = 2\omega/k$ for the case of $\omega \ll \omega_B$. The energy flux at any position along the flux tube (where the geomagnetic field is B) $\mathcal{E}_\omega$ is connected with $\varepsilon_{\vec{k}}$ via the relation

$$\mathcal{E}_\omega = \sigma e_\omega = \sigma \mathcal{E}_{k_z} = \sigma \int \mathcal{E}_{\vec{k}}\, dk_x\, dk_y \tag{8.27}$$

where $\sigma = B_M / B$ is the cross-sectional area of the magnetic flux tube, of unit cross-section at its foot where B_M is the value of the magnetic field, so $\mathcal{E}_\omega$ is the wave spectral energy density flux into the atmosphere (Fig. 8.2).

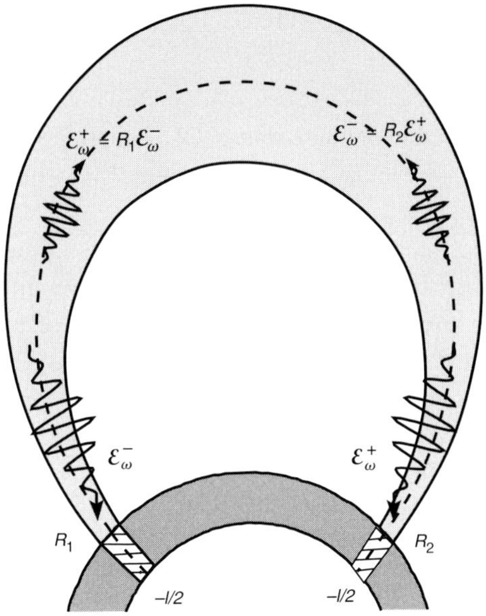

Figure 8.2 A schematic picture of the wave energy flux propagation inside a magnetic flux tube; waves are reflected from the conjugate ionospheres with reflection coefficients R_1 and R_2, respectively.

In accordance with (3.62), the amplification coefficient $æ_\omega$ is:

$$æ_\omega = \frac{4\pi^3 e^2 \tilde{B}^2 \omega}{m^2 k c^2} \int_0^\infty \int_0^{\frac{B_L}{B}} \frac{W\,dW\,\mu\,d\mu}{1-\mu B}\left[\left(\frac{\omega_{BL}}{\omega} - \mu\right)\frac{1}{W}\frac{\partial \mathcal{F}}{\partial \mu} + \frac{\partial \mathcal{F}}{\partial W}\right] \times \delta\left\{k\left(\frac{v_{\mathrm{R}}}{v_\parallel} - 1\right)\right\} \tag{8.28}$$

where v_{R} is the gyroresonance velocity. Making the transition from (3.62) to (8.28) we have taken into account that $\mu = \frac{J_\perp}{W} B_L$, and $B_L\,dJ_\perp\,dW = W\,d\mu\,dW$.

The diffusion coefficient $D_{\mu\mu}$ in (8.23) can be rewritten, with (8.27) taken into account, as:

$$\sqrt{1-\mu B}\,D_{\mu\mu} \equiv D = \frac{8e^2\pi^2\mu_c}{m^2c^2v}\int \frac{d\omega\,k}{\tilde{B}\omega}\,\mathcal{E}_\omega\,\delta\left[k\left(\frac{v_{\mathrm{R}}}{v_\parallel} - 1\right)\right]. \tag{8.29}$$

The system of equations (8.22) and (8.26), together with relations (8.28) and (8.29), is just the self-consistent set of equations of QL plasma theory. When initial and boundary conditions are supplied it can be used to solve particular problems of wave–particle interactions in the Earth's and other magnetospheres. Such topics will be discussed in Chapters 10–13.

8.3 Simplification of QL equations for cyclotron masers

It is very difficult to solve the system of quasi-linear (QL) equations (8.22) and (8.26) in the general case. And there are factors applicable in real CMs which can simplify this system considerably.

Experimental data on electromagnetic ELF/VLF emissions in the Earth's CMs show that there are two distinct types of signals, which have very different dynamical frequency spectra. One group consists of different types of broadband hiss emissions – time stationary hiss, quasi-periodic and burst-like noise emissions. The other group includes different types of discrete emissions having fine (narrowband) structure in the frequency–time domain; these consist of a succession of quasi-monochromatic signals whose frequency changes with time. The interval of time between successive signals is comparable with, or less than, a few seconds, the typical period for bounce oscillations of energetic electrons between northern and southern magnetic mirrors and/or the time for the propagation (at the group velocity) of a wave packet between conjugate ionospheres (the one-hop whistler-mode travel time). The full periods (double bounce time, and two-hop travel time) are, respectively:

$$T_B = 2\int_{-z_*}^{z_*} \frac{dz}{\sqrt{\frac{2W}{m}\left(1-\mu\tilde{B}\right)}} \tag{8.30}$$

and

$$T_{\rm g} = 2\int_{-\ell/2}^{\ell/2} \frac{dz}{v_{\rm g}} \tag{8.31}$$

where $\pm z_*$ locate the electron mirror points where the pitch-angle $=\pi/2$ (where $\mu\tilde{B}(\pm z_*)=1$), $\pm(\ell/2)$ are the field-aligned coordinates of the conjugate ionospheres where wave reflection takes place, and 2ℓ is the total length of the magnetic trap, $\tilde{B}=B/B_L$.

For the first group of signals with noise-like spectra the following two inequalities are valid:

$$\frac{\Delta\omega}{\omega} \sim 1 \text{ and } \tau \gg T_{\rm g},\ T_B \tag{8.32}$$

where $\Delta\omega$ is the frequency bandwidth of the hiss, and τ is the characteristic scale of its temporal variations. The first inequality in (8.32) permits QL theory to be applied. The second inequality in (8.32) permits Equations (8.22) and (8.26) to be averaged over the periods (2.30) and (2.31) of fast oscillations during which changes of the distribution function $\mathcal{F}$ and the spectral flux density $\mathcal{E}_\omega$ are small. Averaging over time actually means integration over z using the zero-order approximation solutions $dz=v_{\rm g}\,dt$ and $dz=v\left(1-\mu\tilde{B}\right)^{1/2}dt$.

Next, we have to take into account the boundary conditions over z. For the particle distribution function, the loss cone is a natural boundary. The particles are either trapped, outside the loss cone given by

$$\mu > \mu_c = \left|\frac{B_L}{B_0}\right|,$$

in which case the boundary condition over z is a periodic one:

$$\mathcal{F}^{\downarrow}(\pm z_*) = \mathcal{F}^{\uparrow}(\pm z_*)$$

where the superscript $\downarrow$ ($\uparrow$) means particle motion towards (away from) the mirror point; or the particles within the loss cone (with $\mu \le \mu_c$) are precipitated into the upper atmosphere and lost from the radiation belts. Thus, the following boundary conditions apply:

$$\begin{aligned} &\text{for } z = \pm\ell/2: \quad \mathcal{F}^{\downarrow}(\pm\ell/2) \simeq \frac{1}{2}\mathcal{F} \\ &\text{for } \mu \le \mu_c: \quad \mathcal{F}^{\uparrow}(\pm\ell/2) = 0. \end{aligned} \tag{8.33}$$

Here $\mathcal{F}$ is the non-averaged distribution function.

The averaging of (8.22) over T_B together with boundary condition (8.33) gives the equation for the changing distribution function $\mathcal{F}$ in the following form:

$$\frac{\partial\langle\mathcal{F}\rangle}{\partial t} = \frac{1}{T_B}\frac{\partial}{\partial\mu}\mu\overline{D}\frac{\partial\langle\mathcal{F}\rangle}{\partial\mu} - \delta\langle\mathcal{F}\rangle + J. \tag{8.34}$$

For simplicity in what follows, the averaging sign is omitted.

For losses

$$\delta = \begin{cases} 0 & \text{for } \mu \ge \mu_c = \dfrac{B_L}{B_0} \\ \delta_0 \equiv \dfrac{2}{T_B(\mu_c)} & \text{for } \mu \le \mu_c. \end{cases} \tag{8.35}$$

Here J represents the source of energetic particles injected into a particular flux tube, and

$$\overline{D} = \frac{16\pi^2 e^2 \mu_c}{m^2 c^2 v^2}\left(\int \frac{d\omega}{\tilde{B}\omega}\,\mathcal{E}_{\omega\Sigma}\ell_{\text{eff}}\right)_{z_R}, \tag{8.36}$$

the averaging being performed over z. Here $\mathcal{E}_{\omega\Sigma} = \mathcal{E}_\omega^+ + \mathcal{E}_\omega^-$, the superscript $\pm$ means waves propagating in ($\pm z$)-directions, z_R is the root of the exact CR condition:

$$\omega - \omega_B - kv_\parallel = 0 \tag{8.36a}$$

and $\ell_{\text{eff}} = k\alpha_B^{-1}$; α_B is determined by the relation (6.34). For the specific case $\omega \ll \omega_B$, the relation for ℓ_{eff} can be written as

$$\ell_{\text{eff}} = \left[\left(1 + \mu\,\frac{k^2 v^2}{\omega_B \omega_{BL}}\right)\frac{d \ln \omega_B}{dz}\right]^{-1}_{z_{\text{R}}}. \tag{8.37}$$

We shall investigate the particular dependence of z_{R} and ℓ_{eff} on (μ, v, ω) when the theory is applied to particular situations in the Universe in later chapters.

Equation (8.34) has to be solved for particular boundary conditions on μ, which can be written in the form:

$$(\mu \overline{D})\left(\frac{\partial \mathcal{F}}{\partial \mu}\right) = 0 \text{ at } \mu = 0 \text{ and } \mu_m. \tag{8.38}$$

The relation (8.38) means that the diffusive fluxes of charged particles through the boundaries $\mu = 0$ and $\mu = \mu_m$ of the physically determined region are zero.

Together with (8.38), the continuity conditions for the distribution function $\mathcal{F}$ and its derivative $\partial\mathcal{F}/\partial\mu$ have to be met on the boundary of the loss cone, i.e. for $\mu = \mu_c$:

$$\mu = \mu_c: \quad \mathcal{F}_{\text{i}}(\mu = \mu_c - \varepsilon) = \mathcal{F}_{\text{e}}(\mu = \mu_c + \varepsilon); \quad \frac{\partial \mathcal{F}_{\text{i}}}{\partial \mu} = \frac{\partial \mathcal{F}_{\text{e}}}{\partial \mu} \tag{8.38a}$$

where $\mathcal{F}_{\text{i (e)}}$ is the distribution function inside (outside, external to) the loss cone. In real magnetic traps in space, $\mu_c = (B_L/B_0) \ll 1$, and the coefficient δ in (8.34) can be put as

$$\delta = \frac{2}{T_B(\mu_c)} \simeq \frac{v}{\ell}.$$

We now consider a simplification of Equation (8.26) for small values of amplification. We introduce $R_{1,2}$ as the reflection coefficient of waves reflected from the two conjugate ionospheres (Fig. 8.2). The boundary conditions for the flux of wave energy with spectral density $\mathcal{E}_\omega$ can be written in the form:

$$\begin{aligned} z = -\ell/2 \qquad & \mathcal{E}_\omega^+\left(-\frac{\ell}{2}\right) = R_1\,\mathcal{E}_\omega^-\left(-\frac{\ell}{2}\right) \\ z = \ell/2 \qquad & \mathcal{E}_\omega^-\left(\frac{\ell}{2}\right) = R_2\,\mathcal{E}_\omega^+\left(\frac{\ell}{2}\right) \end{aligned} \tag{8.39}$$

where $z = \pm\ell/2$ are the coordinates of the conjugate ionospheres, and superscripts $\pm$ determine the propagation direction of the wave energy flux in the $(\pm z)$-direction, respectively (Fig. 8.2). We average (8.26) over t, suggesting that there are only small variations of $\mathcal{E}_\omega$ on a time scale of the two-hop propagation time T_g.

Carefully considering the direction of wave propagation, it is convenient to write (8.26) for the amplification coefficient in the form:

$$\frac{1}{v_{\mathrm{g}}} \frac{\partial \ln \mathcal{E}_{\omega}^{\pm}}{\partial t} \pm \frac{\partial \ln \mathcal{E}_{\omega}^{\pm}}{\partial z} = æ_{\omega}^{\pm} \tag{8.40}$$

where $v_{\mathrm{g}} = \mathrm{const}(z) > 0$. Integrating both Equations (8.40) over $-\ell/2 \leq z \leq \ell/2$ to find the one-hop amplification and summing them, we obtain for the time rate of change of the wave energy flux density $\mathcal{E}_{\omega}$,

$$\frac{d\mathcal{E}_{\omega}}{dt} = \frac{1}{T_{\mathrm{g}}} (\Gamma_{\omega} + \ln R_1 R_2)\mathcal{E}_{\omega} \tag{8.41}$$

where $\mathcal{E}_{\omega} = \frac{1}{2} (\mathcal{E}_{\omega}^{+} + \mathcal{E}_{\omega}^{-})$, $\Gamma_{\omega} = \Gamma_{\omega}^{+} + \Gamma_{\omega}^{-}$ is the total cyclotron amplification over the two-hop propagation path:

$$\Gamma_{\omega} = \oint æ_{\omega}\, dz = \int_{-\ell/2}^{\ell/2} (æ^{+} + æ^{-})\, dz. \tag{8.42}$$

Substituting (8.28) into (8.42), we find that in the limit $\omega/\omega_B \ll 1$:

$$\Gamma_{\omega} = \frac{8\pi^3 e^2}{m^2 c^2 \omega_{BL}} \int_{W_{\min}}^{\infty} v^2\, dW \int_0^{1 - \frac{W_{\min}}{W}} \left(\mu \frac{\partial \mathcal{F}}{\partial \mu} - \frac{\omega}{\omega_{BL}} \mathcal{F} \right) \ell_{\mathrm{eff}}\, d\mu \tag{8.43}$$

where $W_{\min} = \frac{m}{2} \frac{\omega_{BL}^2}{k^2}$ is the minimum energy for gyroresonance, which occurs in the equatorial plane of a certain magnetic flux tube. The limits of integration in (8.43) are determined by the inequalities:

$$\mu \leq 1 - \frac{\omega_{BL}^2}{k^2 v^2}, \qquad W > W_{\min} \tag{8.44}$$

which follow directly from the cyclotron resonance condition when $\omega \ll \omega_{BL}$. From this condition we can find the value of the magnetic field at the resonance point, which is $\tilde{B}_{z_{\mathrm{R}}} = B(z_{\mathrm{R}})/B_L$:

$$\begin{aligned} \tilde{B}_{z_{\mathrm{R}}} &= \left(\sqrt{\frac{k^2 v^2}{\omega_{BL}^2} + \mu^2 \frac{k^4 v^4}{4\omega_{BL}^4}} - \frac{\mu k^2 v^2}{2\omega_B^2} \right) \\ &= \frac{kv}{\omega_{BL}} \left(\sqrt{1 + \mu^2 \frac{k^2 v^2}{4\omega_B^2}} - \frac{\mu k v}{2\omega_B} \right) \geq 1. \end{aligned} \tag{8.45}$$

Under the condition $\mu^2 \dfrac{k^2 v^2}{4\omega_B^2} \ll 1$, i.e. for electrons with low μ, it follows from (8.45) that

$$\tilde{B}_{zR} \simeq \frac{kv}{\omega_{BL}} \geq 1\,. \tag{8.46}$$

We can find from (8.44) the integration limits over ω in (8.36), taking into account (8.23):

$$\omega_{BL} \gg \omega_m \geq \omega \geq \frac{\omega_{BL}^3}{\omega_{pL}^2 \beta^2} (1-\mu)^{-1} \tag{8.47}$$

where $\beta = v/c$, and ω_m is the maximum frequency in the wave spectrum.

8.4 Regimes of pitch-angle diffusion

We now investigate some common properties of CM operation within the framework of the QL approach. The simplified system of QL equations (8.34) and (8.41) is valid when

$$T_D \gg T_B \sim \frac{2\ell}{v} \tag{8.48}$$

where $T_D \sim (T_B/\overline{D})$ is the characteristic time for pitch-angle diffusion under the action of whistler-mode waves. There are two particularly important cases, which describe the regimes of so-called weak and strong pitch-angle diffusion (Kennel, 1969).

The regime of weak diffusion is determined by the inequality

$$T_D \gg T_B/\mu_c. \tag{8.49}$$

Physically this means that the loss cone remains empty all the time during CM operation. From the mathematical point of view it is a very important simplification of the problem, because it makes it possible to look for the solution of Equation (8.34) only outside the loss cone $\mu > \mu_c$, applying the boundary condition that $\mathcal{F}(\mu_c) = 0$.

In the other limit, corresponding to the limit of strong diffusion, the inequality

$$T_B \ll T_D \ll T_B/\mu_c. \tag{8.49a}$$

is fulfilled. It is evident that the distribution function is almost isotropic over μ in this case.

To elucidate more carefully the special features of these two regimes we consider the general properties of the solution to the systems of equations (8.34) and (8.41).

First we suggest that the diffusion coefficient $\overline{D}$ in (8.36) does not depend on μ. We consider the stationary solution of (8.34), when the precipitation of particles into the loss cone is compensated by a generous supply of particles from the source J. For the situation when $\mu_c \ll 1$, the source inside the loss cone is negligible. Putting $\partial\mathcal{F}/\partial t = 0$ in (8.34), we obtain an expression for $\mathcal{F}$ in the region $0 \le \mu \le \mu_c$, inside the loss cone:

$$\frac{\partial}{\partial\mu}\,\mu\,\frac{\partial\mathcal{F}_{\mathrm{i}}}{\partial\mu} - \frac{2}{\overline{D}}\,\mathcal{F}_{\mathrm{i}} = 0. \tag{8.50}$$

The bounded solution of (8.50) is

$$\mathcal{F}_{\mathrm{i}} = C_1 I_0\left(2\sqrt{2\mu/\overline{D}}\right) \tag{8.51}$$

where I_0 is the Bessel function of imaginary argument, and C_1 is a constant which can be found from the continuity conditions (8.38). The stationary solution of (8.34) outside the loss cone is written as

$$\mathcal{F}_{\mathrm{e}} = \mathcal{F}_{\mathrm{e}1} + C_2 \tag{8.52}$$

$$\mathcal{F}_{\mathrm{e}1} = \int_{\mu_c}^{\mu}\left[(\mu'\,\overline{D})^{-1}d\mu' \int_{\mu'}^{\mu_m} T_B J\,d\mu''\right]. \tag{8.53}$$

The constants $C_{1,2}$, can be easily found from (8.38a). We deduce that

$$\mathcal{F}_{\mathrm{i}} = \frac{I_0\left(2\sqrt{\frac{2\mu}{\overline{D}}}\right)\cdot 2}{\sqrt{2\overline{D}\mu_c}\,I_1\left(2\sqrt{\frac{2\mu_c}{\overline{D}}}\right)}\int_{\mu_c}^{\mu_m} T_B J\,d\mu, \quad \mu \le \mu_c \tag{8.54}$$

and

$$\mathcal{F}_{\mathrm{e}} = \mathcal{F}_{\mathrm{e}1}(\mu) + \mathcal{F}_{\mathrm{i}}(\mu_c), \quad \mu \ge \mu_c. \tag{8.55}$$

In the limit of weak diffusion (8.49), when $2\sqrt{\frac{2\mu_c}{\overline{D}}} \sim 2\left(\frac{2T_D\,\mu_c}{T_B}\right)^{1/2} \gg 1$, we have from (8.54):

$$\mathcal{F}_{\mathrm{i}}(\mu) \approx \left[\left(\frac{2T_D}{\mu_c T_B}\right)^{1/2}\int_{\mu_c}^{\mu_m} T_B J\,d\mu\right]\exp\left\{2\sqrt{\frac{2T_D}{T_B}}\left(\mu^{1/2} - \mu_c^{1/2}\right)\right\}. \tag{8.56}$$

According to (8.55) and (8.56), the distribution function decreases exponentially inside the loss cone and increases monotonically in the region $\mu > \mu_c$. This result (Fig. 8.3) confirms our qualitative considerations of weak diffusion with an empty loss cone.

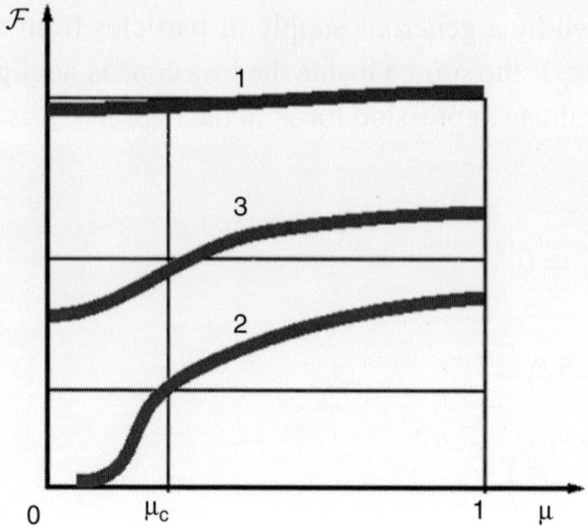

Figure 8.3 Regimes of pitch-angle diffusion:
1. Strong diffusion (SD);
2. weak diffusion (WD);
3. intermediate case.

In the limit of strong diffusion $2\left(\frac{2T_D\,\mu_c}{T_B}\right) \ll 1$ one obtains

$$I_0 \sim 1, \quad I_1 \sim \left(\frac{2T_D\mu_c}{T_B}\right)^{1/2} \ll 1$$

and we have from (8.54):

$$\begin{aligned}\mathcal{F}_{\rm i}(\mu) &\simeq \frac{1}{\mu_c}\int_{\mu_c}^{\mu_m} T_B J\,d\mu' \\ &= \text{const.} \gg \mathcal{F}_{\rm el} \sim \frac{T_D}{T_B}\int_{\mu_c}^{\mu}\frac{d\mu'}{\mu'}\int_{\mu'}^{\mu_m} T_B J\,d\mu''. \qquad (8.57)\end{aligned}$$

In this case the loss cone is continuously filled, and the distribution function is almost isotropic (Fig. 8.3).

Using (8.53)–(8.55), it is possible to estimate the lifetimes T_ℓ of trapped particles. With the definition

$$T_\ell = \left|\frac{1}{N_{\rm h}}\frac{dN_{\rm h}}{dt}\right|^{-1}_{J=0} \qquad (8.58)$$

where $N_{\rm h}$ is the number of energetic (hot) particles inside the magnetic flux tube having unit cross-sectional area in the ionosphere (see Fig. 8.2):

$$N_{\rm h} = \pi\mu_c^{-1}\int_0^{\infty}\int_{\mu_c}^{1}\frac{T_B W\,dW}{m^2}\,d\mu\cdot\mathcal{F}(\mu, W); \qquad (8.59)$$

$dN_{\rm h}/dt$ is a rate which determines changes of $N_{\rm h}$ with time when the source J is switched off. Integrating (8.34) over velocity space and the volume of the magnetic flux tube with $J=0$, we obtain:

$$\left.\frac{dN_{\text{h}}}{dt}\right|_{J=0} = -\pi\mu_c^{-1}\int_0^\infty\left(\int_0^{\mu_c}\mathcal{F}T_B\cdot\delta_0\,d\mu\right)\frac{W\,dW}{m^2}. \tag{8.60}$$

Substitution of (8.54) into (8.60) and usage of (8.59) give us the following expression for the lifetime of the electrons:

$$T_\ell = \frac{\int_0^\infty\left(\int_{\mu_c}^1 T_B\mathcal{F}\,d\mu\right)W\,dW}{\int_0^\infty\left(\int_{\mu_c}^1 T_B J\,d\mu\right)W\,dW} = \frac{N_{\text{h}}}{J_\Sigma} \tag{8.61}$$

where

$$J_\Sigma = \pi\mu_c^{-1}\int_0^\infty\left(\int_{\mu_c}^1 T_B J\,d\mu\right)\frac{W\,dW}{m^2} \tag{8.62}$$

is the total source power. The physical sense of (8.61) is very clear. Indeed the number of particles lost per unit time is, in the stationary state, equal to the number of new particles supplied by the source per unit time.

According to (8.57) and (8.61) the electron lifetime T_ℓ in the regime of strong diffusion does not depend on the source power. Rather it is determined by the universal formula:

$$T_\ell^{\text{s}} = \overline{T}_B/\mu_c. \tag{8.63}$$

Here $\overline{T}_B$ is the value of T_B averaged over μ, and the superscript 's' indicates strong diffusion.

In the case of weak diffusion we need to know the wave energy flux $\mathcal{E}$. A rough estimation from (8.41) in the stationary case gives us the following value:

$$\mathcal{E} \approx \int \mathcal{E}_\omega\,d\omega \simeq \frac{J_\Sigma}{|\ln R_1 R_2|}\,\frac{m\omega_0\omega_{BL}}{k_0^2}\left(\tilde{B}^2\right)_{z_{\text{R}}}. \tag{8.64}$$

If $\mathcal{E}_\omega$ is narrowband with a central frequency ω_0,

$$\tilde{B}^2_{z_{\text{R}}} = \frac{B^2(z_{\text{R}})}{B_L^2}.$$

In the case (8.46) we have:

$$\mathcal{E} \simeq \frac{2J_\Sigma}{|\ln R_1 R_2|}\cdot\left(\frac{\omega_0}{\omega_{BL}}\right)\cdot W_0 \tag{8.65}$$

where W_0 is the characteristic energy of the particles.

This result is valid for both diffusion regimes. For strong diffusion the anisotropic part of $\mathcal{F}_e$, which determines $\mathcal{E}_\omega$, is a small addition to the full distribution function (8.55). However, for weak diffusion $\mathcal{F}_e$ is the main part of the distribution function.

The particle lifetime in the weak diffusion regime can be found from (8.61), substituting (8.53). This estimate coincides with the direct estimation from Equation (8.34) which gives:

$$T_\ell^{\mathrm{w}} \sim \frac{\overline{T}_B}{\overline{\overline{D}}} \sim \frac{n_{cL}\, v_{\mathrm{g}} \ln|R_1 R_2|}{J_\Sigma\, \omega_{BL}\, \mu_c} \gg \overline{T}_B/\mu_c. \tag{8.66}$$

The lifetime depends directly upon n_{cL}, the cold plasma density; the superscript 'w' means weak diffusion.

The different regimes of pitch-angle diffusion are illustrated in Fig. 8.4. The lifetime and the number of trapped particles are shown as functions of the source intensity. Besides the regimes of weak (drawn on the left) and strong (centre) diffusion, a new regime of super strong (right) diffusion exists when the diffusion time T_D is much less than the bounce oscillation period T_B: $T_D \ll T_B$.

A peculiar effect of particle locking in the magnetic trap exists, when the lifetime grows due to strong pitch-angle scattering (Fig. 8.4). The particle can move backwards, as well as forwards, in a random walk fashion. Some details of such a regime can be found in Bespalov and Trakhtengerts (1986b). In space conditions this regime can be realized for huge magnetic traps and very intense sources, in particular in the solar corona.

For a long magnetic trap, other effects which include energy exchange between cold and hot plasmas can prevail. Some of these are considered in Chapter 13.

8.5 The case of an arbitrary ratio ω/ω_B

In Section 8.3 the simplest application of QL theory to a CM was considered in the limit that $\omega/\omega_{BL} \ll 1$. Indeed, this is a most important case, realized both in planetary magnetospheres and in the solar corona. Having an arbitrary ω/ω_{BL} ratio is more applicable to real life conditions, especially for regions of low cold plasma density. This case introduces some new features which can be critical for CM dynamics. In particular, with any arbitrary ratio of ω/ω_{BL}, diffusion over energy can be switched on. Such a situation is clearly extremely important for charged particle acceleration.

The derivation of the QL system of equations for an arbitrary ratio of ω/ω_{BL} is similar to the previous case $(\omega/\omega_B) \ll 1$ but is more cumbersome. To avoid this difficulty, we use the well-known system of QL equations for a homogeneous plasma, and the fact that, according to Section 8.2, these work for an inhomogeneous plasma if we deal with smooth distributions of particles in velocity phase space.

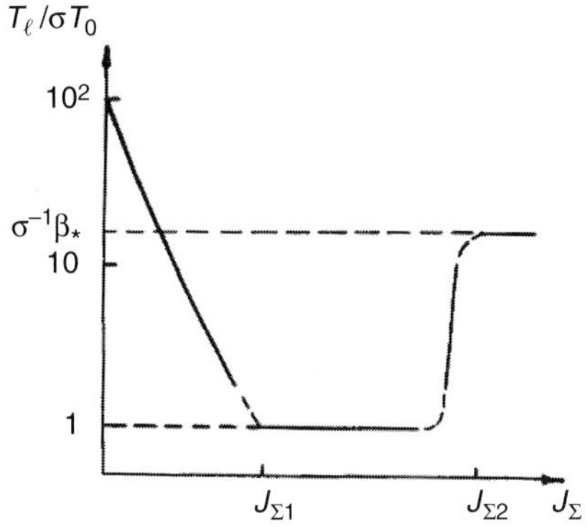

(a)

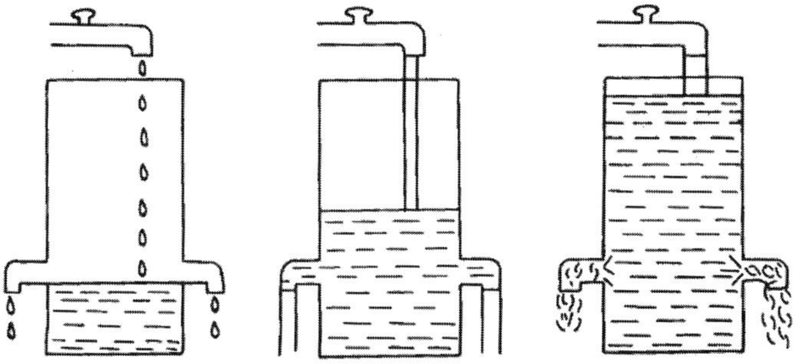

(b)

Figure 8.4 (a) Diagram illustrating the dependence of the lifetime (T_l) and the number of trapped particles as functions of the source intensity J_Σ; (b) The different regimes of pitch-angle diffusion as they can be compared with the outflow of liquid from a vessel (Kennel, 1969). (Copyright American Geophysical Union, reproduced with permission.)

For this system, the equivalent equation to (8.22), (8.23) has the form:

$$\frac{d\mathcal{F}}{dt} = \frac{8\pi^3}{m^2} \sum_{s=-\infty}^{\infty} \int d^3\vec{k}\, (\hat{\hat{\Lambda}}_s G_{\vec{k},s} \mathcal{E}_{\vec{k}} \hat{\Lambda}_s \mathcal{F}) \tag{8.67}$$

and the wave energy equation is the same as (8.5):

$$\frac{\partial \mathcal{E}_{\vec{k}}}{\partial t} + \frac{\partial \omega}{\partial \vec{k}} \frac{\partial \mathcal{E}_{\vec{k}}}{\partial \vec{\tau}} - \frac{\partial \omega}{\partial \vec{\tau}} \frac{\partial \mathcal{E}_{\vec{k}}}{\partial \vec{k}} = 2\gamma \mathcal{E}_{\vec{k}}, \tag{8.68}$$

with the growth rate

$$\gamma = \frac{4\pi^3}{m} \sum_{s=-\infty}^{\infty} \int (G_{\vec{k},s} \hat{\Lambda}_s \mathcal{F})\, d^3\vec{v}. \tag{8.69}$$

Here

$$\hat{\hat{\Lambda}}_s = \left(\frac{1}{v_\perp}\right)\frac{\partial}{\partial v_\perp}\left(\frac{s\omega_B}{\omega}\right) + \frac{\partial}{\partial v_\parallel}\left(\frac{k_\parallel}{\omega}\right), \tag{8.69a}$$

$$\hat{\Lambda}_s = \frac{s\omega_B}{\omega v_\perp}\frac{\partial}{\partial v_\perp} + \frac{k_\parallel}{\omega}\frac{\partial}{\partial v_\parallel}, \tag{8.69b}$$

$G_{\vec{k},s}$ is the spectral power of radiation from a single charged particle at the s-th harmonic of its gyrofrequency. We are especially interested in the case of quasi-longitudinal propagation $\vec{k}\parallel\vec{B}$, when only the term $G_{\vec{k},1}$ with $s=1$ is non-zero. This term is equal to:

$$G_{\vec{k},1} = \frac{e^2\omega v_g}{4\pi c^2 k}\, v_\perp^2\, \delta(\omega - kv_\parallel - \omega_B). \tag{8.70}$$

To generalize this system for an inhomogeneous plasma, we rearrange (8.67) and (8.69) with new variables:

energy $\quad W = \dfrac{mv^2}{2}$ and

sine of equatorial pitch-angle, squared $\quad \mu = \dfrac{B_L}{B}\sin^2\theta.$

After that we obtain:

$$\begin{aligned}\frac{\partial\mathcal{F}}{\partial t} &+ v_\parallel\frac{\partial\mathcal{F}}{\partial z}\\ &= 8\pi^3\int dk\left\{\left[\frac{1}{W^{1/2}}\frac{\partial}{\partial W}W^{1/2} + \frac{1}{W}\sqrt{1-\mu\tilde{B}}\right.\right.\\ &\quad\left.\left.\times\frac{\partial}{\partial\mu}\left(\frac{\omega_{BL}/\omega-\mu}{\sqrt{1-\mu\tilde{B}}}\right)\right]G_{k,1}\mathcal{E}_k\left[\frac{\partial\mathcal{F}}{\partial W} + \frac{1}{W}\left(\frac{\omega_{BL}}{\omega}-\mu\right)\frac{\partial\mathcal{F}}{\partial\mu}\right]\right\},\end{aligned} \tag{8.71}$$

with the growth rate

$$\begin{aligned}\gamma = \frac{4\sqrt{2}\pi^4}{m^{3/2}}\tilde{B}\int_0^\infty\int_0^{\tilde{B}-1} G_{k,1}\,d\mu\,dW\,\frac{W^{1/2}}{\sqrt{1-\mu\tilde{B}}}\\ \times\left[\frac{\partial\mathcal{F}}{\partial W} + \frac{1}{W}\left(\frac{\omega_{BL}}{\omega}-\mu\right)\frac{\partial\mathcal{F}}{\partial\mu}\right]\end{aligned} \tag{8.72}$$

where, as earlier, $\tilde{B} = B/B_L$, $v_\parallel = \pm\, v\sqrt{1-\mu\tilde{B}}$.

In this case, when the diffusion time (over μ and W) $T_D \gg T_B, T_g$, Equations (8.71) and (8.72) plus the energy transfer Equation (8.68) are simplified similarly to the case for $\omega/\omega_{BL} \ll 1$ by averaging over time. The averaged system of equations has the form:

$$\frac{\partial \mathcal{F}}{\partial t} = \frac{1}{T_B}\left\{\frac{1}{W}\frac{\partial}{\partial W}(W^2\mu)\,\overline{D}\,\frac{\omega}{\omega_{BL}} + \frac{\partial}{\partial \mu}\mu \overline{D}\right\}\hat{\Lambda}_\perp \mathcal{F} - \delta\mathcal{F} + J \qquad (8.73)$$

where $W = \dfrac{mv^2}{2}$, the operator $\hat{\Lambda}_\perp$ is

$$\hat{\Lambda}_\perp = \frac{\omega}{\omega_{BL}} W \frac{\partial}{\partial W} + \left(1 - \frac{\mu\omega}{\omega_{BL}}\right)\frac{\partial}{\partial \mu} \qquad (8.74)$$

and $\overline{D}$, which is determined by the relation (8.36), should be considered as an integral operator acting on functions to the right of it. In the limit $(\omega/\omega_{BL}) \ll 1$, Equation (8.73) reduces to Equation (8.71).

The important modification to the QL equation (8.73), in comparison with (8.34), is the appearance of diffusion over energy. This changes the initial (source) energy spectrum of particles to a new spectrum. It is clear from the general structure of Equation (8.73) that such a modification to the energy spectrum will be severe when the ratio ω/ω_{BL} become large, and when the loss cone region μ_c goes to zero.

In a dipole magnetic field, $\mu_c = L^{-3}\left(4 - \dfrac{3}{L}\right)^{-1/2}$ (see Chapter 2), and μ_c decreases with increasing L-shell. According to the analysis of Chapter 3, the ratio ω/ω_{BL} for maximum growth rate of unstable waves increases with a decrease of the cold plasma density n_c. So, in the case of the Earth's magnetosphere, we can expect that such a stochastic acceleration would be effective outside the plasmasphere.

Here we illustrate some general features of the stochastic acceleration for the example of the strong diffusion regime, when the distribution function is almost isotropic. In this case, we can average Equation (8.73) over μ, taking $\mathcal{F}$ out of the integral. In the stationary case ($\partial\mathcal{F}/\partial t = 0$) we obtain:

$$\frac{1}{W}\frac{\partial}{\partial W}\left(W^2 D_1 \frac{\partial \mathcal{F}}{\partial W}\right) - 2\mu_c\mathcal{F} + \langle T_B J\rangle = 0 \qquad (8.75)$$

where the angle brackets mean averaging over μ, and

$$D_1 = \int_0^1 d\mu\,\mu\,\frac{8\pi^2 e^2 \mu_c}{c^2 m}\int \frac{d\omega\,\omega}{\omega_{BL}^2}\,\mathcal{E}_{\omega\Sigma}\ell_{\text{eff}}. \qquad (8.76)$$

Here D_1 in general depends on W, because ℓ_{eff} and the limits of integration over ω depend on W. We suggest that this dependence is weak. In Equation (8.75) we use the dimensionless variable

$$x = \frac{v}{v_0} = \left(\frac{W}{W_0}\right)^{1/2}, \tag{8.77}$$

where v is the modulus of the velocity and $W_0 = mv_0^2/2$ is the characteristic energy of the energetic particles (of mass m) in the source region. Equation (8.75) is then written as

$$\frac{d^2\mathcal{F}}{dx^2} + \frac{3}{x}\frac{d\mathcal{F}}{dx} - \frac{8\mu_c W_0}{D_1}\mathcal{F} + \left\langle\frac{4\tau_B W_0 J}{D_1}\right\rangle = 0. \tag{8.78}$$

The boundary conditions for Equation (8.78) are as follows:

$$\mathcal{F} \to 0, \quad x \to \infty; \qquad \frac{\partial \mathcal{F}}{\partial x} = 0, \quad x = 0. \tag{8.79}$$

The solution of (8.78), satisfying conditions (8.79), has the form:

$$\mathcal{F} = \frac{4}{D_1}\left\{\frac{I_1(\alpha x)}{x}\int_x^{\infty}\left\langle T_B W_0 x'^2 J\right\rangle \mathcal{K}_1(\alpha x')\,dx' + \frac{\mathcal{K}_1(\alpha x)}{x}\int_0^x \left\langle T_B W_0 x'^2 J(x')\right\rangle I_1(\alpha x')\,dx'\right\}, \tag{8.80}$$

where $I_1(\alpha x)$ and $\mathcal{K}_1(\alpha x)$ are modified Bessel functions, and

$$\alpha = (8\mu_c W_0/D_1)^{1/2}. \tag{8.81}$$

Being interested in the conditions for the effective acceleration of charged particles, when the final energy spread is much larger than the characteristic energy W_0 in the source, we take $J(x)$ in the following form:

$$J = J_0\, v_0\, \delta(v - v_0) \equiv J_0\, \delta(x - 1). \tag{8.82}$$

After substitution of (8.82) into (8.80), we obtain:

$$\mathcal{F} = \frac{4T_{B0} W_0 J_0}{D_1} \cdot \begin{cases} \mathcal{K}_1(\alpha)\, I_1(\alpha x)/x, & x < 1 \\ I_1(\alpha)\, \mathcal{K}_1(\alpha x)/x, & x \geq 1. \end{cases} \tag{8.83}$$

The distribution function (8.83) is shown in Fig. 8.5 for different values of α. It is clear from (8.83) and Fig. 8.5 that acceleration is possible, if $\alpha \ll 1$.

To estimate α we take the value of the integral over ω in D_1 (8.76) at some middle point $\omega = \omega_0$:

$$D_1 \sim \frac{4\pi^2 e^2 \mu_c}{mc^2}\frac{\ell_{\rm eff}\,\omega_0}{\omega_{BL}^2}\mathcal{E}_\Sigma(\omega_0). \tag{8.84}$$

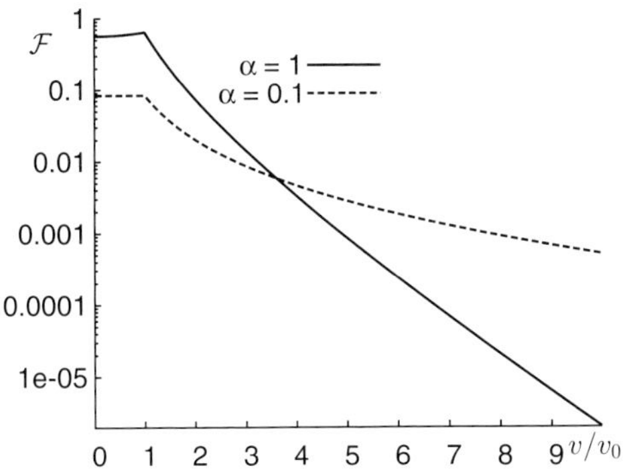

Figure 8.5 Diagram illustrating the efficiency of charged particle stochastic acceleration as a function of the parameter α.

In the self-consistent approach and for $\omega_0 \ll \omega_{BL}$ the wave energy flux density $\mathcal{E}_\Sigma$ is determined by (8.65). After substitution of (8.65) and (8.83) into (8.81) we find that

$$\alpha = \frac{2B_L^2 |\ln R| \omega_{BL}}{\pi^2 J_\Sigma \omega_0^2 m \ell_{\text{eff}}} . \tag{8.85}$$

The minimum gyroresonant energy of electrons, which can be involved in the process of stochastic acceleration, is:

$$W_0 \geq m \left(\frac{\omega_{BL}^2}{2k_{\omega L}^2} \right) = \frac{\omega_{BL}^3}{\omega_{pL}^2 \omega_0} \cdot \frac{mc^2}{2} . \tag{8.86}$$

It follows from (8.85) that effective stochastic acceleration ($\alpha < 1$) is possible on high L-value geomagnetic flux tubes and for rather high frequencies.

8.6 Summary

Feedback of the waves generated in a cyclotron resonance instability influences parameters of the energetic charged particle distribution appreciably. The simplest theoretical representation of this process involves quasi-linear theory; here the different wave packets have random phases which produce stochastic charged particle motions.

An approach is developed which leads, for the simplest case of extremely low-frequency ducted whistler-mode waves, to a self-consistent set of equations for the changing electron distribution function (8.22), the changing wave energy flux along the duct (8.26), the wave amplification coefficient (8.28) and the charged particle diffusion coefficient (8.29). For broadband hiss emissions, with the changes being both small and slow, the changing distribution function is given by (8.34) and the changing wave energy flux density by (8.41). When losses from the radiation belts via pitch-angle diffusion into the loss cone match the magnetospheric sources of energetic charged particles exactly, a stationary state of the radiation belts is achieved. For the intense sources when the inequality (8.49a) is fulfilled, strong pitch-angle diffusion takes place with an almost isotropic distribution function and filling of the loss cone. In this case the radiation belt's particle lifetime is given by the relation (8.63). If the loss cone is almost empty, the situation corresponds to weak diffusion, and the lifetime is given by (8.66).

For the higher frequencies of whistler-mode emissions, besides diffusion over pitch-angle there is also diffusion over energy (8.73), i.e. stochastic acceleration. According to (8.85) and (8.86) this acceleration is effective on high L-value geomagnetic flux tubes and for rather high frequencies.

Chapter 9

Non-stationary CM generation regimes, and modulation effects

Experimental data show that the generation of ELF/VLF waves and energetic charged particle dynamics in the magnetosphere are non-stationary processes. They operate over different time scales ranging from a fraction of a second up to some tens or even hundreds of seconds. This time interval includes the characteristic period of bounce oscillations, T_B, of energetic electrons between magnetic mirror points in the northern and southern hemispheres, as well as the period of a whistler-mode wave packet, T_g, between reflection points (the conjugate ionospheres); both of these are of the order of 1 s and or a few seconds. It is clear that the quantitative description of these non-stationary processes demands different approaches in the CM theory.

In particular, for slow processes with a characteristic time $\tau \gg (T_B, T_g)$, we can use the system of quasi-linear equations (8.34) and (8.41). Fast processes in CMs, for which $\tau \leq (T_B, T_g)$, as a rule are associated with discrete ELF/VLF emissions having a narrow dynamic frequency spectrum. In this case the nonlinear theory of wave–particle interactions, developed in Chapters 5 and 6, is more suitable. An example of such an approach is the nonlinear theory of the BWO generator, developed in Chapter 5. Here we shall concentrate our attention on the slow dynamics of CMs in the frame of quasi-linear theory, following in the main the review by Bespalov and Trakhtengerts (1986b) and later papers by Trakhtengerts *et al.* (1986), Demekhov and Trakhtengerts (1994), Trakhtengerts *et al.* (1996) and Pasmanik *et al.* (2002).

Sometimes it is difficult to select either slow or fast processes in CMs. Indeed, the multiple reflection of whistler-mode or Alfvén mode waves from the ionospheric mirrors can lead to slow changes of a wave packet envelope and also the formation of short electromagnetic pulses. This is similar to the regime of mode locking in lasers. This regime as applied to whistler CMs was first considered by

Bespalov (1984), and was applied further to Alfvén sweep masers by Trakhtengerts *et al.* (2000). Below we consider this regime in more detail.

9.1 Balance approach to the QL equations, the two-level approximation

In this chapter we use the self-consistent system of QL equations (8.34) and (8.41), adapted to the space CM, for investigation of the relatively slow dynamic situation with characteristic temporal scales $\tau \gg T_g, T_B$. In the general case this system of equations is rather complicated and demands numerical simulations. Such solutions will be analysed in Sections 9.3 and 9.4. At the same time, many qualitative features of the major characteristics of CM dynamics can be elucidated from the analytical solution of (8.34) and (8.41) in some important particular cases.

Such a case in point is the approach of weak pitch-angle diffusion (8.49), when it is possible to omit the loss term $(-\delta\mathcal{F})$ in (8.34) and to consider its solution in the region $\mu_c < \mu < 1$, taking into account the energetic electron precipitation with the help of the boundary condition $\mathcal{F}(\mu_c) = 0$. An analytical solution can be obtained in this case, if we suggest a narrow spectrum $\Delta\omega$ of whistler waves $\Delta\omega \ll \omega_0$, where the characteristic frequency ω_0 can be associated with the frequency of the maximum growth rate in (8.41). Then we can write (8.34) and the boundary conditions in the following form:

$$\frac{\partial \mathcal{F}}{\partial t} = \frac{\mathcal{E}_\Sigma(t)}{T_B} \frac{\partial}{\partial \mu} \mu\, D_0(\mu, w) \frac{\partial \mathcal{F}}{\partial \mu} + J \tag{9.1}$$

$$\mu = \mu_m, \quad \left(\mu D_0 \frac{\partial \mathcal{F}}{\partial \mu}\right)_{\mu_m} = 0, \quad \mu = \mu_c, \quad \mathcal{F}(\mu_c) = 0 \tag{9.2}$$

where the total wave energy flux density $\mathcal{E}_\Sigma = \mathcal{E}^+_{\omega 0} + \mathcal{E}^-_{\omega 0}$ is determined from the equation:

$$\frac{1}{\mathcal{E}_\Sigma} \frac{d\mathcal{E}_\Sigma}{dt} = \left\{ (\Gamma_\omega + \ln R_1 R_2)\, T_g^{-1} \right\}_{\max}. \tag{9.3}$$

Actually, for (9.1)–(9.2) we are dealing with the Sturm–Liouville problem, when the function $\mathcal{F}(\mu, t)$ can be presented as the expansion

$$\mathcal{F}(\mu, t) = \sum_{n=1}^{\infty} Z_n(\mu)\, N_n(t), \tag{9.4}$$

where the eigenfunctions $Z_n(\mu)$ are found from the equations:

$$\frac{1}{T_B(\mu)} \frac{d}{d\mu} \left(D_0 \mu \frac{dZ_n}{d\mu} \right) = -p_n^2 Z_n \tag{9.5}$$

and the eigenvalues p_n^2 follow from the boundary conditions:

$$Z_n\Big|_{\mu=\mu_c} = 0, \quad D_0\left(\frac{dZ_n}{d\mu}\right)\Big|_{\mu=\mu_m} = 0. \tag{9.6}$$

The eigenfunctions Z_n corresponding to different eigenvalues are orthogonal, that is

$$\int_{\mu_c}^{\mu_m} Z_n Z_m \, d\mu = 0 \text{ for } n \neq m \tag{9.7}$$

and any function can be expanded in the eigenfunctions of (9.5). So we can present the source $J(\mu, t)$ as the sum

$$J(\mu) = \sum_{n=1}^{\infty} J_n(t)\, Z_n(\mu) \tag{9.8}$$

and obtain, using the orthogonality condition (9.7), the system of ordinary differential equations for the amplitudes $N_n(t)$:

$$\frac{dN_n}{dt} = -p_n^2 \mathcal{E}_\Sigma N_n + J_n, \quad n = 1, 2, \ldots \tag{9.9}$$

which should be solved jointly with (9.3).

There is a close analogy between the system of equations (9.9) and the multi-level balance approach in the theory of quantum generators (Khanin, 1995). This analogy, developed in more detail by Bespalov and Trakhtengerts (1986b), turned out to be very useful for the prediction of such important generation regimes in space CMs as self-sustained oscillations of the whistler wave intensity (Bespalov, 1981, 1982, and Bespalov and Koval, 1982) and the regime of passive mode synchronization (Bespalov, 1984). Unfortunately, the simplifications used for the transition to the multi-level approach (9.9), such as weak pitch-angle diffusion and a narrow frequency spectrum, are not appropriate for the quantitative analysis of experimental data, and it is necessary to use more sophisticated calculations, based on a computer simulation (Sections 9.3 and 9.4).

In this section we shall consider in more detail the so-called two-level approximation, when only one equation from the chain (9.9) – that with the lowest eigenvalue p_1^2 – is considered together with (9.3). The justification for this approach is the fact that the eigenvalues p_n^2 grow quickly with increasing n, and the high 'levels' N_n (with $n \geq 2$) can be neglected for slow processes and for a smooth distribution of $J(\mu)$.

The two-level approximation is the key to understanding the basic physical features of slow dynamical regimes in space CMs. It serves as the starting point for a consideration of more complicated CM dynamics, when external factors such as a temporal modulation of either the ionospheric mirrors or the geomagnetic field play an important role. We have to generalize (9.3) to include the dependence of the reflection coefficient $R_{1,2}$ in the flux of precipitated energetic particles.

The reflection coefficient $R_{1,2}$ in (9.3) for the whistler or Alfvén wave depends on the plasma density in the lower ionosphere which, in turn, depends on the flux of the precipitated radiation belt (RB) electrons or protons. Therefore, we add to (9.1) and (9.3) the balance equation for ionization in the lower ionosphere, particularly in the E region, which includes an extra production term due to energetic charged particle precipitation:

$$\left.\begin{aligned} \frac{\partial n_E}{\partial t} &= q_E + \eta_E S_E - \alpha_E n_E^2, \\ S_E &= \frac{1}{2}\left(J_\Sigma - \frac{\partial N}{\partial t}\right). \end{aligned}\right\} \tag{9.10}$$

Here q_E is a constant source (e.g. solar ultraviolet) determining the mean ionization rate in the ionosphere, η_E is the mean number of the electron–ion pairs per unit volume produced by an energetic electron, α_E is the recombination coefficient, N is the total number of energetic charged particles in a certain flux tube (see Fig. 9.1), with a cross-sectional area of 1 cm^2 at ionospheric altitude, and n_E is the electron density in the ionosphere. Here S_E is the energetic electron (proton) flux density into the ionosphere, and J_Σ is the total source of energetic particles on the same magnetic flux tube; the factor of a half indicates the precipitated particle flux to one of the two conjugate ionospheres.

The flux of precipitated RB particles can easily be expressed in terms of $\mathcal{F}$ and D (see also (8.58)–(8.60)). To do this we integrate both sides of (9.1) over the phase-space volume in velocity space and over the volume of the magnetic flux tube with 1 cm^2 cross-sectional area in the ionosphere:

$$\begin{aligned} S_E &= \frac{\pi\sigma}{2}\int_0^\infty \int_{\mu_0}^1 T_B\left(J - \frac{\partial \mathcal{F}}{\partial t}\right) v^3\, dv\, d\mu \\ &= \frac{\pi\sigma}{2}\int_0^\infty \left(\mu D \frac{\partial \mathcal{F}}{\partial \mu}\right)\bigg|_{\mu=\mu_0} v^3\, dv. \end{aligned} \tag{9.11}$$

We remember that $\sigma = \mu_c^{-1}$, and that the case $\omega \ll \omega_{BL}$ is considered.

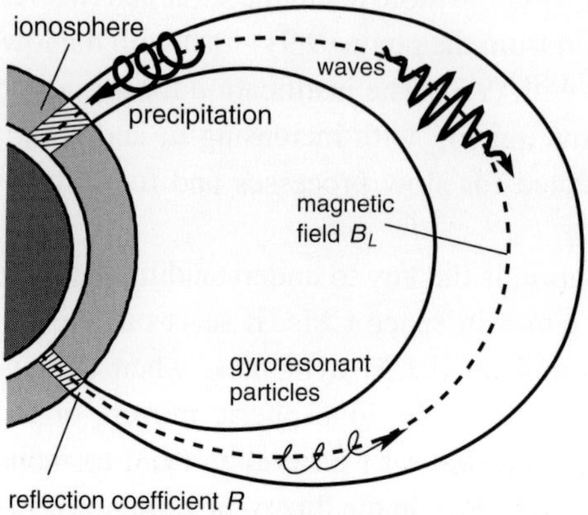

Figure 9.1 Diagram illustrating the dynamical cyclotron interaction situation considered, which increases the ionospheric electron density in both hemispheres due to the precipitation of energetic electrons.

For the sake of simplicity we assume that the spread of energies is small ($v \sim v_0$, where v_0 is a characteristic velocity), and we put $\mu^{1/2} = æ$. Then we can use (8.34), (8.41)–(8.46) and (9.10)–(9.11) to derive a complete system of equations describing the CI dynamics for weak pitch-angle diffusion and for a narrowband wave spectrum ($\omega \sim \omega_0$, $\Delta\omega \ll \omega_0$):

$$\frac{\partial \mathcal{F}_*}{\partial t} = \Delta_1 \mathcal{E}_\Sigma \frac{\partial}{\partial æ}\left(æ \frac{\partial \mathcal{F}_*}{\partial æ}\right) + J_*(t, æ),$$

$$\frac{\partial \mathcal{E}_\Sigma}{\partial t} = \Delta_2 \left(\int_{æ_c}^{æ_m} æ^2 \frac{\partial \mathcal{F}_*}{\partial æ}\, dæ\right) \mathcal{E}_\Sigma - \nu(n_E)\mathcal{E}_\Sigma,$$

$$\frac{\partial n_E}{\partial t} = q_E + \Delta_3 \mathcal{E}_\Sigma \left(æ \frac{\partial \mathcal{F}_*}{\partial æ}\right)\Bigg|_{æ_c} - \alpha_E n_E^2, \tag{9.12}$$

where

$$\begin{Bmatrix} \mathcal{F}_* \\ J_* \end{Bmatrix} = \pi\sigma \int_0^\infty v^3 \begin{Bmatrix} \mathcal{F} \\ J \end{Bmatrix} dv,$$

$$\Delta_1 \approx \frac{2\pi e^2}{m^2 c^2 \omega_0 v_0}, \qquad \Delta_2 = 2\pi\Delta_1 \frac{m v_g\, v_0\, \omega_0}{\sigma \omega_{BL}},$$

$$\Delta_3 = \frac{2\pi a \eta_E}{\omega_0} \Delta_1, \qquad \nu = \frac{2}{T_g} |\ln R(n_E)| \tag{9.12a}$$

and we have used the parabolic approximation for $B(z)$ away from the magnetic equator ($z = 0$), and a is the characteristic scale of $B(z)$ (see (2.8)).

Considering the two-level approximation of (9.12) we take the source $J_*(t, æ)$ in the form:

$$J_*(t, æ) = J_\Sigma(t)\, Z_1(æ), \tag{9.13}$$

where $Z_1(æ)$ is the solution of the Sturm–Liouville problem for the first equation of (9.12), corresponding to the minimum eigenvalue p_1. Z_1 and p_1 are found from the equations:

$$\frac{d}{dæ}\left(æ \frac{dZ_1}{æ}\right) = -p_1^2 Z_1, \tag{9.14}$$

$$\left(Z_1\right)_{æ=æ_c} = 0, \qquad \left(\frac{dZ}{dæ}\right)_{æ=æ_m} = 0. \tag{9.14a}$$

The general solution of (9.14) is:

$$Z(æ) = C_1 J_0(2pæ^{1/2}) + C_2 N_0(2pæ^{1/2}), \tag{9.15}$$

where J_0 and N_0 are the zeroth-order Bessel and Neumann functions, respectively. Using the boundary conditions (9.14a) and the fact that $Z > 0$, we can find the lowest eigenvalue

$$p_1 = 0.15\pi\left(æ_m^{1/2} - æ_c^{1/2}\right)^{-1}. \tag{9.16}$$

Then we have:

$$Z_1(æ) = \frac{v_0 p_1^2}{4a}\left[N_0(2p_1 æ_c^{1/2})\, J_0(2p_1 æ^{1/2}) - J_0(2p_1 æ_c^{1/2})\, N_0(2p_1 æ^{1/2})\right] \tag{9.17}$$

where we have used the normalization condition

$$\frac{4\pi a}{v_0}\int_{æ_c}^{æ_m} Z_1(æ)\, dæ = 1.$$

Now we can write the solution of the first equation of (9.12) in the form

$$\mathcal{F}_*(t, æ) = N_{\rm h}(t)\, Z_1(æ), \tag{9.18}$$

where $N(t)$ is the total number of energetic charged particles in the flux tube with unit cross-sectional area at its base. We thus have the following system of ordinary differential equations for $N_{\rm h}(t)$, $\mathcal{E}(t)$ and $n_E(t)$:

$$\left.\begin{aligned} \frac{dN_{\rm h}}{dt} &= -\delta\mathcal{E} N_h + J_\Sigma, \\ \frac{d\mathcal{E}}{dt} &= h\mathcal{E} N_{\rm h} - \nu(n_E)\,\mathcal{E}, \\ \frac{dn_E}{dt} &= q_E + \tau\mathcal{E} N_{\rm h} - \alpha_E n_E^2. \end{aligned}\right\} \tag{9.19}$$

Here

$$\delta = \Delta_1 p_1^2, \quad h = \Delta_2 \int_{æ_c}^{æ_m} æ^2 \frac{dZ_1}{dæ}\, dæ, \ \text{and}\ \tau = \frac{\Delta_3 v_0 p_1^2}{4\pi a}, \tag{9.19a}$$

and J_Σ is the total number of energetic charged particles coming per unit time into the flux tube with unit cross-sectional area at its ionospheric base. We remind ourselves that $\nu(n_E)$ is the damping of whistler waves (9.12a).

Using (9.17) and the inequality $æ_c \ll æ_m < 1$, we obtain

$$h = \frac{\Delta_2 v_0}{4\pi a}\,(æ_m - æ_c).$$

9.1.1 CI evolution, with initial conditions given

First we treat the CI evolution without a charged particle source, and we ignore the time variation of the ionospheric properties. Then Equations (9.19) reduce to

$$\frac{dN_{\text{h}}}{dt} = -\delta\mathcal{E}N_{\text{h}}, \quad \frac{d\mathcal{E}}{dt} = h\mathcal{E}N - \nu\mathcal{E} \tag{9.20}$$

with the initial conditions

$$t = 0, \quad N_{\text{h}} = \tilde{N} > \nu/h, \quad \mathcal{E} = \tilde{\mathcal{E}}. \tag{9.21}$$

Physically, this means that the CI threshold is exceeded at $t = 0$.

Equations (9.20) then have the integral of motion

$$\delta(\mathcal{E} - \tilde{\mathcal{E}}) + h(N_{\text{h}} - \tilde{N}) = \nu \ln(N_{\text{h}}/\tilde{N}). \tag{9.22}$$

The function $N(t)$ is determined by the relation:

$$t = \int_{\ln(N_{\text{h}}/\tilde{N})}^{0} \left[\delta\tilde{\mathcal{E}} + \nu\xi + \tilde{\gamma}(1 - \exp\xi)\right]^{-1} d\xi, \tag{9.23}$$

where $\tilde{\gamma} = h\tilde{N}$ is the initial growth rate of the instability. The energy flux density $\mathcal{E}$ of the whistler-mode waves as a function of time has its maximum value

$$\mathcal{E}_{\text{max}} = \tilde{\mathcal{E}} + \delta^{-1}\left[\tilde{\gamma} - \nu - \nu\ln(\tilde{\gamma}/\nu)\right], \tag{9.24}$$

and the characteristic time of the process is

$$T \simeq (\tilde{\gamma} - \nu)^{-1} \ln(\mathcal{E}_{\text{max}}/\tilde{\mathcal{E}}). \tag{9.25}$$

The density of energetic electrons on the flux tube decreases monotonically with increasing energy flux $\mathcal{E}$, and the flux of precipitated electrons has the same time dependence as does the total energy of the whistler-mode waves. Significantly, in the final state when $\mathcal{E} \to 0$, the density of trapped particles can be considerably lower than the threshold density N corresponding to the instability threshold $hN = \nu$, especially if the energetic charged particle density was initially considerably higher than ν/h. The consequences of this situation are explored further in Chapter 11.

9.1.2 Relaxation oscillations of radiation belt parameters

Now we assume that the geomagnetic trap has a charged particle source with J_{Σ} constant in time. Then the system of equations (9.19) has an equilibrium state described by the following parameters:

$$n_{E0} = \left(\frac{\delta q_E + \tau J_{\Sigma}}{\delta\alpha_E}\right)^{1/2}, \quad N_0 = \frac{\nu(n_{E0})}{h}, \quad \mathcal{E}_0 = \frac{hJ_{\Sigma}}{\delta\nu(n_{E0})}. \tag{9.26}$$

Here, all the symbols are as defined earlier in this section. We now analyse the CI dynamics near this equilibrium state. We shall linearize Equations (9.19), writing all quantities such as $a = a_0 + a_\sim \exp(\lambda t)$, where λ is complex. Using the linearized equations (9.19), we can easily write the characteristic equation:

$$(\lambda + 2\alpha_E n_{E0})\left(\lambda^2 + 2\nu_J\lambda + \Omega_J^2\right) - \chi\lambda = 0, \tag{9.27}$$

where the following notation has been introduced:

$$\Omega_J^2 = hJ_\Sigma, \quad 2\nu_J = \frac{hJ_\Sigma}{\nu(n_{E0})}, \quad \chi = \frac{2\tau J_\Sigma}{T_g\, R_0\, \delta}\left(\frac{\partial R}{\partial n_E}\right)_0.$$

The subscript J indicates that the modulation of the electron fluxes and the wave energy with frequency Ω_J is due to the action of the source J_Σ, with χ showing the change in the whistler-mode wave reflection coefficient of the ionosphere due to the production of ionospheric plasma by energetic charged particle precipitation. ν_J is the damping rate of these oscillations.

The equilibrium state is unstable if:

$$4\alpha_E^2 n_{E0}^2 + 4\alpha_E n_{E0}\,\nu_J + \Omega_J^2 < \left(\frac{\alpha_E n_{E0}}{\nu_J} + 1\right)\chi. \tag{9.28}$$

We see that, in the presence of a constant source of energetic electrons, periodic CI modes can be generated only if $\chi > 0$, that is, if $\partial R/\partial n_E > 0$. An increase in the ionization (n_E) in the lower ionosphere typically leads to increased absorption of the whistler-mode waves ($\chi < 0$). But, in some cases, the sign of χ may be reversed if the precipitation of the energetic charged particles leads primarily to the temporal growth of the spatial gradients of the ionospheric electron density and to enhanced reflection of the whistler-mode waves from the ionosphere. Clearly, the situation is more unstable if ν_J is small, i.e. if $\nu_J \ll \alpha_E n_{E0}$.

We now discuss in more detail the stable periodic CI modes ($\chi < 0$) which are more typical for the particle sources described by (9.13). If the variation of the Q factor of the CI resonator due to charged particle precipitation is negligibly small ($\chi \to 0$), (9.27) has the following solution:

$$\lambda_{1,2} = -\nu_J \pm i\left(\Omega_J^2 - \nu_J^2\right)^{1/2}, \quad \lambda_3 = -2\alpha n_{E0}, \tag{9.29}$$

where

$$\Omega_J = (\nu J_\Sigma/N_0)^{1/2}, \quad \nu_J = J_\Sigma/N_0, \tag{9.30}$$

and N_0 is the number of energetic electrons in the flux tube with unit cross-sectional area at its base, determined by (9.26). If the energetic electron source is described by $J_\Sigma = N_0/T_\ell$, where T_ℓ is their lifetime, then we obtain the frequency of the relaxation oscillation $\Omega_J = (\nu/T_\ell)^{1/2} = (2\gamma/T_\ell)^{1/2}$, where $\gamma = \Gamma/T_g$ is the total whistler-mode growth Γ per one-hop whistler-mode propagation time, i.e. the averaged CI growth rate.

Then weakly damped oscillations of the electromagnetic radiation intensity and (weak) fluxes of precipitated particles occur if the following inequality is satisfied:

$$\Omega_J T_g < 4 |\ln R|. \tag{9.31}$$

If we take into account the temporal modulation of R, the eigenfrequencies described by (9.29) are changed. For instance, when $\chi \gg \Omega_J$ (the weak flux situation), the eigenvalues corresponding to the quasi-periodic regime are:

$$\lambda_{1,2} \simeq \pm i |\chi|^{1/2} - (\nu_J + \alpha_E n_{E0}), \quad |\chi| \gg (\nu_J - \alpha_E n_{E0})^2. \tag{9.32}$$

Equations (9.29)–(9.32) indicate that the eigenfrequency increases with the source amplitude as $J_\Sigma^{1/2}$, while the Q factor varies as $J_\Sigma^{-1/2}$; it decreases with decreasing absorption of the whistler-mode waves at the two ends of the flux tube on which the cyclotron interactions are taking place.

Now we shall analyse the system of equations (9.19) far from the equilibrium state. In particular, if we ignore the variation of R due to charged particle precipitation ($\chi \to 0$), we obtain quasi-periodic CI modes in the form of individual spikes (see Fig. 9.2) described by the nonlinear equation:

$$\frac{d^2\zeta}{dt^2} + 2\nu_J \frac{d\zeta}{dt} \exp\{\zeta\} + \Omega_J^2(\exp\{\zeta\} - 1) = 0, \tag{9.33}$$

where $\mathcal{E} = \mathcal{E}_0 \exp\{\zeta\}$, $\mathcal{E}_0$ is the energy flux density of the whistler-mode waves in the equilibrium state, and $d\zeta/dt = hN - \nu$; $\zeta = \ln(\mathcal{E}/\mathcal{E}_0)$.

For weakly damped oscillations the 'spike' shape is determined by

$$\sqrt{2}\,\Omega_J(t - t_0) = \pm \int_{\zeta(t_0)}^{\zeta} (\exp\{\zeta_{\max}\} - \zeta_{\max} - \exp\{\zeta'\} + \zeta')^{-1/2} d\zeta'. \tag{9.34}$$

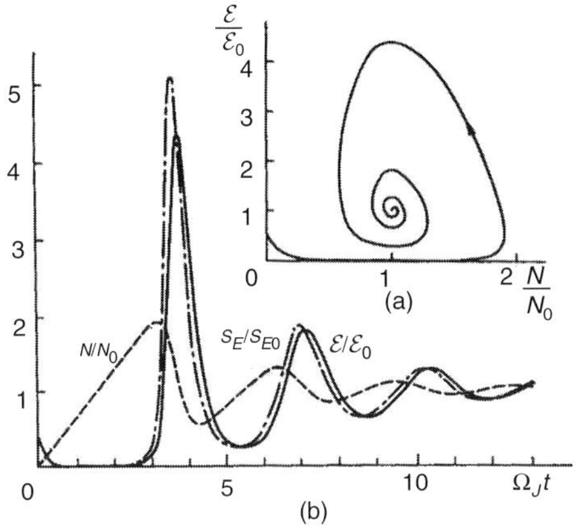

Figure 9.2 Relaxation CI oscillations. (a) Inset, the phase plane (energy space) corresponding to (9.19), and (b) the relaxation of three parameters towards their equilibrium values in terms of the characteristic time $\Omega_J t$.

Equations (9.34) can be readily analysed in the highly nonlinear case when $\mathcal{E}/\mathcal{E}_0 \gg 1$. The repetition frequency of the spikes is:

$$\Omega_{JN} \simeq \pi \left(\frac{\mathcal{E}_0}{2\mathcal{E}_{\max}} \right)^{1/2} \Omega_J . \tag{9.35}$$

In Fig. 9.2, the repetition period is given by $\Omega_J t \simeq 3.5$, which means that $\mathcal{E}_{\max} \simeq \mathcal{E}_0$. The duration of each spike is:

$$\Delta t \simeq \Omega_J^{-1} \left(\frac{2\mathcal{E}_0}{\mathcal{E}_{\max}} \right)^{1/2} \ln \frac{4\mathcal{E}_{\max}}{\mathcal{E}_0} . \tag{9.36}$$

Interestingly, under these conditions the maximum amplitude $\mathcal{E}_{\max}$ decays exponentially with increasing time, as is the case near the equilibrium state, and

$$\nu_{JN} = (2/3)\, \nu_J . \tag{9.37}$$

Alternatively, if we ignore the decay (i.e. for $t \ll \nu_{JN}^{-1}$), we can express $\mathcal{E}_{\max}$ in terms of the initial RB parameters $\tilde{\mathcal{E}}$ and $\tilde{N}$:

$$\frac{\mathcal{E}_{\max}}{\mathcal{E}_0} \simeq \ln \frac{\mathcal{E}_0}{\tilde{\mathcal{E}}} + \frac{1}{2} \left[\frac{\nu - h\tilde{N}}{\Omega_J} \right]^2 . \tag{9.38}$$

Substituting (9.38) into (9.35), we obtain

$$\Omega_{JN} = \begin{cases} \pi \left(2 \ln \dfrac{\mathcal{E}_0}{\tilde{\mathcal{E}}} \right)^{-1/2} \Omega_J, & \dfrac{(\nu - h\tilde{N})^2}{\Omega_J^2} \ll 2 \ln \dfrac{\mathcal{E}_0}{\tilde{\mathcal{E}}}, \\ \pi (\nu - h\tilde{N})^{-1} \Omega_J^2, & \dfrac{(\nu - h\tilde{N})^2}{\Omega_J^2} \gg 2 \ln \dfrac{\mathcal{E}_0}{\tilde{\mathcal{E}}}, \quad \nu > h\tilde{N}. \end{cases} \tag{9.38a}$$

We see that, in the nonlinear case far from equilibrium, the CI behaviour retains the main features exhibited near equilibrium, such as the condition for quasi-periodic modes, and the growth of eigenfrequencies with the source power. These equations are discussed further in their magnetospheric application in Chapter 11.

9.1.3 CI modulation by hydromagnetic waves

The relaxation oscillations treated above are especially significant when periodic disturbances due to magnetohydrodynamic waves or drift waves act on the radiation belts. Hydromagnetic oscillations modulating both the density and the degree of pitch-angle anisotropy of trapped particles give rise to an oscillating component in the source J_Σ. For instance, for a fast small-amplitude magnetoacoustic wave with

$k \perp B$ and amplitude b, we can describe its effect by introducing the effective source

$$J_{\Sigma_\sim} = \left(\frac{\partial b}{\partial t}\Big/2B\right)\left[2(1-\mu)\,\mu\,\frac{\partial \mathcal{F}}{\partial \mu} - \mu v\,\frac{\partial \mathcal{F}}{\partial v}\right], \tag{9.39}$$

in which we can substitute $\mathcal{F}$ corresponding to the equilibrium state given by (9.26) if the ratio b/B is sufficiently small.

In the presence of such disturbances, an external force $\Delta\Omega_J^2 \cos\Omega t$, where $\Delta = \max(J_{\Sigma_\sim}/J_\Sigma)$, appears on the right-hand side of (9.33). If the frequency of the external force coincides with the frequency of the relaxation oscillations, $\Omega \simeq \Omega_J$, a resonance occurs for which a small disturbance can result in the deep modulation of the intensity of the wave radiation and also in the fluxes of energetic charged particles. Using the linear approximation with respect to the equilibrium state, we can easily find the depth of modulation for the following parameters:

$$\max\left(\frac{\mathcal{E}_\sim}{\mathcal{E}_0}\right) \simeq \max\left(\frac{S_{E_\sim}}{S_{E_0}}\right) = Q_J^2 \max\left(\frac{N_\sim}{N_0}\right) = Q_J^2\Delta. \tag{9.40}$$

Here $Q_J = \Omega_J/2\nu_J$ is the Q factor for the relaxation oscillations, and S_E is the flux density of the precipitated particles. Equation (9.40) indicates that, for a high Q factor $Q_J \gg 1$, the depth of modulation of the whistler wave intensity and the precipitated electron flux is much greater than the depth of modulation of the trapped particles.

Of course, modulation of the reflection coefficient R should lead to similar effects. Such a modulation can be artificially produced, for instance, using a high-power ground-based shortwave radio transmitter to change the ionospheric conditions in a periodic manner.

The above results have been obtained for a specific dependence of the diffusion coefficient on the pitch-angle and the energy (see (9.12)) and a specific source of energetic charged particles. The more detailed analysis of Bespalov and Trakhtengerts (1986b) shows that relaxation oscillations in the RB occur for a wide range of particle sources and for an arbitrary dependence of the diffusion coefficient on the pitch-angle and the energy. Thus, such relaxation oscillations are expected to be a common feature of the RB in the presence of particle sources. This conclusion is confirmed by computer simulations of the system of equations (9.12) (see Section 9.3).

9.2 Self-modulation of waves in a CM

The description of cyclotron maser (CM) dynamics on the basis of the two-level approximation is valid in the case of a weak source of energetic particles with a smooth distribution over μ, i.e. a smooth pitch-angle distribution. These conditions often do not hold in 'the real world'. The charged particle source can be non-monotonic in μ (e.g. for the 'butterfly'-like distribution) and quite strong. Then the weak pitch-angle diffusion regime is changed during the process of CI dynamics into strong diffusion.

Moreover, the plasma pressure changes in a magnetic flux tube due to the energetic particle precipitation in the process of pitch-angle diffusion, as well as a nonlinear modulation of the wave reflection coefficient by the ionosphere, can lead to some additional effects in CM dynamics. These result in a self-modulation of the cyclotron wave intensity and of the energetic charged particle flux.

Analytical investigations of CM dynamics for an arbitrary source and for the general case of pitch-angle diffusion are extremely complicated, and we leave their discussion to Section 9.3, where the results of computer simulations of the full system of QL Equations (8.34) and (8.41) are presented. In this section we analyse self-modulation effects due to the connection of CI with MHD waves through the modulation of plasma pressure and to the nonlinearity of the ionospheric reflection.

First we consider the interaction between the pulsations of magnetic flux tubes and CI relaxation oscillations. In Section 9.1 we assumed that the relaxation oscillations were caused by geomagnetic pulsations due to external factors. However, magnetohydrodynamic waves can be generated in the same process as that which generates the relaxation oscillations. Indeed, in the geomagnetic trap, the pressure disturbances in the RB are linked to the magnetic field disturbances via the effects of diamagnetism and 'frozen in' effects. If these disturbances are appropriately linked, an instability can occur and the relaxation oscillations can go over to a self-oscillatory regime accompanied by relatively large-amplitude micropulsations of the geomagnetic field. Further, the case of small plasma pressure $P_\perp$, $\beta = 8\pi P_\perp / B^2 \ll 1$, is considered, which is typical for the inner magnetosphere.

We shall first consider two MHD wave modes, i.e. both fast magnetoacoustic and Alfvén modes. For weak MHD waves, which are of great interest in the magnetosphere when the wave amplitude $|b| \ll |B|$, the external magnetic field, the plasma pressure change for a fast magnetoacoustic wave and the connection with CI relaxation oscillations is much stronger than for Alfvén waves. A significant change of the plasma pressure for an Alfvén wave only takes place when the Alfvén wavelength along the magnetic field line is comparable with its radius of curvature. At the same time the phase and group velocities of the Alfvén mode across the magnetic field are much less than for the magnetoacoustic wave.

This circumstance is important for 'real world' conditions, when the interaction region is localized in space. It is necessary to have a resonator for magnetoacoustic waves to prevent the leakage of MHD energy away from a generation region due to group propagation. Such a problem does not occur for Alfvén waves.

The role of whistler-mode turbulence in the generation of MHD waves in the magnetosphere was first considered by Pilipenko and Pokhotelov (1975) and Mikhailovskii and Pokhotelov (1975). They showed that the drift instability connected with the energetic electron pressure gradient is strongly modified. Pitch-angle scattering of these electrons due to the cyclotron interaction with whistler turbulence causes the additional dissipation of MHD waves and leads to the development of the dissipative MHD drift instability. The same generation mechanism operates when the development of the CI and the excitation of MHD waves are considered in a self-consistent approach. But the drift instability can be strongly amplified when the eigenfrequency of magnetic flux tube oscillations is close to the frequency of the relaxation oscillations of the CI. Below we separately analyse the cases of magnetoacoustic waves and Alfvén waves.

9.2.1 Modulational instability associated with magnetoacoustic waves

We assume that the magnetosphere contains a resonator for magnetohydrodynamic waves of the magnetoacoustic type. Such a resonator can be an individual magnetic flux tube with an enhanced cold plasma density. These often appear in the magnetosphere during periods of higher than usual geomagnetic activity. As noted in Chapter 3 and Section 9.3, such a flux tube with higher cold plasma density n_c also provides favourable conditions for the development of the CI. A source of new particles in the magnetic flux tube can be the flux of electrons drifting longitudinally in the geomagnetic field which do not interact with the whistler-mode waves outside the tube owing to the low n_c there (an analogy with gas-transport lasers can be seen here). If the transverse size of the resonating flux tube is $\ell_\perp \ll R_c$, the radius of curvature of the geomagnetic field line, we can ignore the curvature of the field lines and approximate the resonator as an almost rectangular rod (Fig. 9.3). In this case, the non-uniformity of the magnetic field along z determines the longitudinal structure of the hydromagnetic wave field. We shall analyse the simplest case when the resonator is uniform along z (i.e. the Alfvén velocity $v_A(z) = \text{const.}$) and $\beta = 8\pi P_\perp / B^2 \ll 1$ so that the fast magnetoacoustic waves can be treated independently of other wave types. Under such conditions the wave equation for magnetoacoustic eigenmodes is:

$$\frac{\partial^2 b}{\partial x^2} + \frac{\partial^2 b}{\partial y^2} + \frac{\partial^2 b}{\partial z^2} - \frac{1}{v_A^2} \frac{\partial^2 b}{\partial t^2} = -\frac{4\pi}{B_L} \left(\frac{\partial^2 P_\perp}{\partial x^2} + \frac{\partial^2 P_\perp}{\partial y^2} \right) \tag{9.41}$$

where b is the magnetoacoustic wave magnetic field component along the geomagnetic field $\vec{B}$, v_A is the Alfvén velocity, and the RB electron pressure across the magnetic field is equal to:

$$P_\perp = \frac{1}{2} m \int_0^\infty v_\perp^2 \, \mathcal{F} \, d^3\vec{v}. \tag{9.42}$$

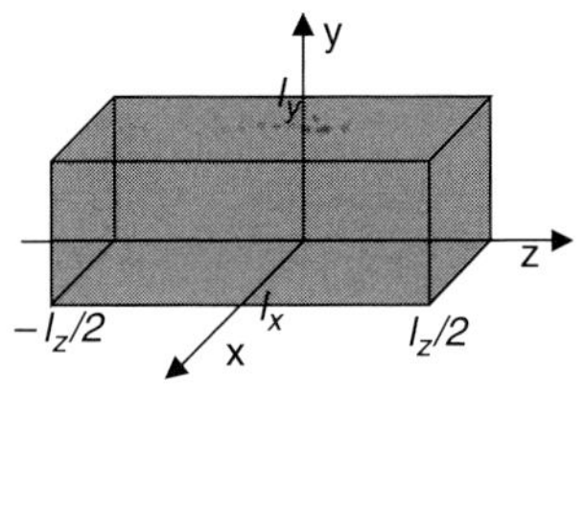

Figure 9.3 Simple model for a magnetoacoustic resonator along the z-axis, from $-\ell_z/2$ to $+\ell_z/2$.

The condition that the perturbations at the ends of the geomagnetic trap vanish yields the longitudinal eigenvalues $æ_s = s\pi/\ell_z$ and the respective eigenfunctions $\sin æ_s(z + \ell_z/2)$; $s = 1, 2, \ldots$ In the case of a rectangular resonator with cross-field scales ℓ_x and ℓ_y (Fig. 9.3) and the cold plasma density $n_{\rm ci}$ inside much larger than $n_{\rm ce}$ outside (external to) the resonator, it is possible to neglect the damping of the magnetoacoustic waves (due to leakage of wave energy from the resonator) and to present the spatial structure of the eigenmodes in the form:

$$b = \sum_{n,m,s}^{\infty} b_{nms}(t) \sin(k_{nx}x) \sin(k_{ny}y) \sin æ_s \left(z + \frac{\ell_z}{2}\right), \tag{9.43}$$

where the eigenvalues

$$k_{nx} = n\,\pi\,\ell_x \text{ and } k_{my} = m\,\pi\,\ell_y, \tag{9.44}$$

with $n, m = 1, 2, \ldots$, and the eigenvalues satisfy the dispersion equation, determining the eigenfrequency Ω_M:

$$k_{nx}^2 + k_{my}^2 + æ_s^2 = \Omega_M^2/v_{\rm A}^2. \tag{9.45}$$

Without loss of generality we can suggest that $æ_s^2 \ll k_{nx}^2, k_{my}^2$ and neglect the dependence of P on z. Using the orthogonality conditions

$$\begin{aligned} &\frac{2\pi}{\ell_x}\int_0^{\ell_x} dx \sin\left(\frac{n_1\pi x}{\ell_x}\right)\sin\left(\frac{n_2\pi x}{\ell_x}\right) = \delta n_1 n_2 \\ &\frac{2\pi}{\ell_y}\int_0^{\ell_y} dy \sin\left(\frac{m_1\pi y}{\ell_y}\right)\sin\left(\frac{m_2\pi y}{\ell_y}\right) = \delta m_1 m_2, \end{aligned} \tag{9.46}$$

where, for example, $\delta n_1 n_2$ is the Kronecker delta, we can obtain from (9.41) the following equation for $b_{nm}(t)$:

$$\begin{aligned} \frac{d^2 b_{nm}}{dt^2} + \Omega_M^2 b_{nm} = \frac{(4\pi)^2 v_A^2 \pi}{B_L \ell_x \ell_y}\int_0^{\ell_x}\int_0^{\ell_y} dx\,dy \sin\left(\frac{n\pi x}{\ell_x}\right) \\ \times \sin\left(\frac{m\pi y}{\ell_y}\right)\Delta_\perp P_{\perp\sim} \end{aligned} \tag{9.47}$$

where Ω_M is determined by (9.45) and $\Delta_\perp = \dfrac{\partial^2}{\partial x^2} + \dfrac{\partial^2}{\partial y^2}$; B_L is the magnetic field along the length of the resonator. The pressure disturbance $P_{\perp\sim}$ has two terms; one is due to the CI, and the other is connected with MHD waves. To take both effects into account we should add two terms to the left-hand side of (8.34)

$$\mu \frac{\partial b}{B_L\,\partial t}\left[W\frac{\partial \mathcal{F}}{\partial W} - (1-\mu)\frac{\partial \mathcal{F}}{\partial \mu}\right] \tag{9.48}$$

and

$$\frac{\mu W}{m\omega_{BL} B_L}\left[\frac{\partial b}{\partial x}\frac{\partial \mathcal{F}}{\partial y}-\frac{\partial b}{\partial y}\frac{\partial \mathcal{F}}{\partial x}\right] \tag{9.49}$$

where $b(x, y, t)$ is described by (9.43). The first term in (9.48) gives the effects of wave magnetic compression, and the second term describes the effects of the drift instability (Mikhailovskii, 1992).

We shall consider the case of a weak source J of the energetic electrons, when the two-level approach can be used. In the linear approach over b ($|b|/B \ll 1$) it is possible to present $\mathcal{F}$ as the sum:

$$\mathcal{F}=\mathcal{F}_0(\mu;\ W, x, y)+\mathcal{F}_*+\mathcal{F}_{M\sim} \tag{9.50}$$

where $\mathcal{F}_*$ is due to the dynamics of the CM itself, and $\mathcal{F}_{M\sim}$ determines the contribution of the MHD oscillations. In the case of a small range of electron energies and a narrow band whistler wave spectrum, the first equation in (9.3) is supplemented on its right-hand side by the two terms:

$$\frac{\mu}{B_L}\frac{\partial b}{\partial t}\left[\mathcal{F}_*+(1-\mu)\frac{\partial \mathcal{F}_*}{\partial \mu}\right] \tag{9.48a}$$

and

$$-\frac{\mu W_0}{m\omega_{BL} B_L}\left[\frac{\partial b}{\partial x}\frac{\partial \mathcal{F}_*}{\partial y}-\frac{\partial b}{\partial y}\frac{\partial \mathcal{F}_*}{\partial x}\right]. \tag{9.49a}$$

We shall consider a possible role of $\nu(n_E)$ changes associated with the precipitation of energetic electrons due to their cyclotron interaction with whistler-mode waves in subsection 9.2.3, suggesting below in this subsection that $\nu(n_c)=\text{const}$. Now we linearize the system of equations (9.3) relative to the state of CM stationary generation with $\nu=\text{const.}$, and putting $\mathcal{E}=\mathcal{E}_0+\mathcal{E}_\sim$ and $\mathcal{F}_*=\mathcal{F}_0+\mathcal{F}_\sim$ we obtain:

$$\begin{aligned}\frac{\partial \mathcal{F}_\sim}{\partial t}=&-\frac{\mathcal{E}_\sim}{\mathcal{E}_0}J_0+\Delta_1\mathcal{E}_0\frac{\partial}{\partial æ}\left(æ\frac{\partial \mathcal{F}_\sim}{\partial æ}\right)+\frac{\mu}{B_L}\frac{\partial b}{\partial t}\left[\mathcal{F}_0+(1-\mu)\frac{\partial \mathcal{F}_0}{\partial \mu}\right]\\&-\frac{\mu W_0}{m\omega_{BL} B_L}\left(\frac{\partial b}{\partial x}\frac{\partial \mathcal{F}_0}{\partial y}-\frac{\partial b}{\partial y}\frac{\partial \mathcal{F}_0}{\partial x}\right),\end{aligned} \tag{9.51}$$

$$\frac{\partial \mathcal{E}_\sim}{\partial t}=\Delta_2\mathcal{E}_0\int_{æ_c}^{æ_m}æ^2\frac{\partial \mathcal{F}_\sim}{\partial æ}\,dæ, \tag{9.52}$$

where J_0 is the source of energetic electrons, and $\mathcal{E}_0$ and $\mathcal{F}_0$ are found from the joint solution of the relations:

$$\Delta_1 \mathcal{E}_0 \frac{\partial}{\partial æ}\left(æ \frac{\partial \mathcal{F}_0}{\partial æ}\right) = -J_0(æ), \tag{9.53}$$

$$\Delta_2 \int_{æ_c}^{æ_m} æ^2 \left(\frac{\partial \mathcal{F}_0}{\partial æ}\right) dæ = \nu. \tag{9.54}$$

ν determines the threshold of the cyclotron instability, the wave magnetic field b in (9.51) is given by (9.43), and Δ_1 and Δ_2 are defined by (9.12a). In the case of a weak, and constant in time, source $J_0(\mu)$, when the two-level approach (9.13) is valid, relations (9.51) and (9.52) are transformed (after integration over æ) to the system of equations

$$\frac{\partial N_\sim}{\partial t} = -\frac{\mathcal{E}_\sim}{\mathcal{E}_0} J_\Sigma - \delta \mathcal{E}_0 N_\sim + \delta_1 N_0 \frac{\partial (b/B_L)}{\partial t} - \frac{\delta_2 W_0}{m\omega_{BL} B_L}\left(\frac{\partial b}{\partial x}\frac{\partial N_0}{\partial y} - \frac{\partial b}{\partial y}\frac{\partial N_0}{\partial x}\right), \tag{9.55}$$

$$\frac{\partial \mathcal{E}_\sim}{\partial t} = h \mathcal{E}_0 N_\sim, \tag{9.56}$$

where N_0 corresponds to the stationary state (see (9.26)); δ and h are determined by the relations (9.19a), and $\delta_{1,2} \sim 1$. The last term in (9.55) characterizes the drift instability effects.

Now we consider the magnetic compression effects, neglecting the last term in (9.55). Differentiating (9.56) with respect to time and using (9.55), we obtain:

$$\frac{\partial^2 \mathcal{E}_\sim}{\partial t^2} + 2\nu_J \frac{\partial \mathcal{E}_\sim}{\partial t} + \Omega_J^2 \mathcal{E}_\sim = h\delta_1 N_0 \mathcal{E}_0 \frac{\partial (b/B_L)}{\partial t} = (\delta_1 \nu/\mathcal{E}_0) \frac{\partial (b/B_L)}{\partial t} \tag{9.57}$$

where Ω_J and ν_J are the frequency and damping rate of the relaxation oscillations, respectively.

For a process varying as $\exp(i\Omega t)$, this gives:

$$\left(-\Omega^2 + 2i\nu_J \Omega + \Omega_J^2\right) \frac{\mathcal{E}_\sim}{\mathcal{E}_0} = i\delta_1 \nu \Omega\, (b/B_L). \tag{9.58}$$

Taking into account relations (9.42) and (9.56) we find from (9.58) that the time varying energetic electron pressure perpendicular to the geomagnetic field is:

$$P_{\perp\sim} = W_0 N_\sim = -\delta_1 N_0 W_0 \frac{\Omega}{\Omega_J^2 - \Omega^2} (b/B_L). \tag{9.59}$$

If the plasma pressure $P_{\perp 0} = N_0 W_0$ is changing in space more slowly than b, we find from (9.59) that

$$\Delta_\perp P_{\perp\sim} \simeq -\delta_1 P_{\perp 0} \frac{\Omega^2}{\Omega_J^2 - \Omega^2} \Delta_\perp (b/B_L)$$

$$= \frac{\delta_1 P_{\perp 0}}{B_L} \frac{\Omega^2}{\Omega_J^2 - \Omega^2} \sum_{m,n} b_{nm} \frac{\Omega_M^2}{v_A^2} \sin\frac{n\pi x}{\ell_x} \sin\frac{m\pi y}{\ell_y}. \tag{9.60}$$

Substitution of (9.60) into (9.47) gives, for a process varying as $\exp(i\Omega t)$, Ω being the modulation frequency:

$$\left(\Omega_M^2 - \Omega^2\right)\left(\Omega_J^2 + 2i\nu_J\Omega - \Omega^2\right) = \Omega^2\Omega_M^2\beta \tag{9.61}$$

where the ratio of energetic electron pressure (perpendicular to the magnetic field) to the magnetic pressure $\beta = 4\pi\delta_1 P_{\perp 0}/B_L^2 \ll 1$. It is evident that there is no positive feedback between MHD oscillations and pressure modulation due to the CI in (9.61); there is only a correction to the eigenfrequencies Ω_J and Ω_M.

However, the situation is different if we take into account the second term in (9.55), which is usually responsible for drift instabilities. In the case of magneto-acoustic modes with spatial structure (see (9.43)), the effects of the drift instability are very much weakened by interference effects, when the background pressure $P_{\perp 0}$ changes slowly (in space) on the resonator scale. The situation is more favourable for Alfvén waves as considered in the next subsection.

9.2.2 Modulational instability due to Alfvén waves

We have considered the self-modulational instability in the case of fast magneto-acoustic waves. We have seen that, in the approximation of straight magnetic field lines, an instability appears under very special conditions. Now we shall treat the self-modulation associated with Alfvén waves which are greatly affected by magnetic field curvature, and by the non-uniformity of both the magnetic field and the hot plasma density. We shall proceed from the results obtained for the excitation of Alfvén waves in a non-uniform magnetic field of variable curvature (Mikhailovskii, 1978).

In problems closely related to that under consideration here, the analysis of low-frequency drift instabilities takes into account the effective collision frequency due to the presence of whistler-mode turbulence (Mikhailovskii and Pokhotelov, 1975b; Pilipenko and Pokhotelov, 1975). In contrast to these analyses, we shall use a self-consistent approach which takes the effect of magnetohydrodynamic oscillations on the whistler turbulence into account (Bespalov and Trakhtengerts, 1978b). This feedback effect is important as it significantly affects both the spectrum and the growth rate of the drift instabilities under consideration. Feedback leads to the deep modulation of the cyclotron wave intensity and also of the fluxes of energetic electrons precipitated from the magnetosphere.

The mathematical consideration of this problem is similar to the previous case of magnetoacoustic waves, but for Alfvén waves it is not necessary to have the resonator. This is because the component of group velocity across the magnetic field is equal to zero. However, as we have already stated, the interaction of Alfvén waves with the CI is essential, if compression effects due to the curvature of the magnetic field lines are taken into account. Moreover, we cannot use the two-level approximation. And we have to take into account the magnetic drift of the RB electrons in longitude ψ, with a mean drift angular velocity $\bar{\Omega}_{\rm d}$. Using the variables v (energetic electron velocity) and $J_\perp = \dfrac{v_\perp^2}{2B}$ (the first adiabatic invariant), Equation (9.51) for $\mathcal{F}_\sim$ in the case of Alfvén waves is written as follows:

$$\frac{\partial \mathcal{F}}{\partial t} + \bar{\Omega}_{\rm d} \frac{\partial \mathcal{F}}{\partial \psi} = \frac{1}{T_B} \frac{\partial}{\partial J_\perp} \left[G(J_\perp, v)\, \mathcal{E} \frac{\partial \mathcal{F}}{\partial J_\perp} \right] + J + \frac{\partial}{\partial t} (\Delta \mathcal{F}_A), \tag{9.62}$$

$$\frac{\partial \mathcal{E}}{\partial t} = \left\{ \int_0^\infty \left[\int_{J_{\perp c}}^{J_{\perp m}} K(J_\perp, v) \frac{\partial \mathcal{F}}{\partial J_\perp}\, dJ_\perp \right] dv \right\} \mathcal{E} - \nu \mathcal{E}. \tag{9.63}$$

Here, the quantities $G(J_\perp, v)$ and $K(J_\perp, v)$ are determined by the relations for $\mathcal{E}_k = \mathcal{E}\delta(k - k_0)$, where k is a general wavenumber and k_0 its value for the gyroresonant condition (equation (8.36a)). We should note that, since G and K depend only on the magnitude of B, we can ignore their variation under the effect of the Alfvén waves which is of second order in the wave amplitude, b. $\Delta \mathcal{F}_{\rm A}$ is the term added to the distribution function determined by the Alfvén waves and averaged over the bounce oscillation period T_B.

The boundary conditions for $J_{\perp \rm c} = \dfrac{v^2}{2B_L \sigma}$ and $J_{\perp \rm m} = \dfrac{\mu_m v^2}{2B_L}$, at the edge of the loss cone and at the point corresponding to the maximum value $\mu = \mu_{\rm m}$, respectively, are:

$$\mathcal{F}\Big|_{J_\perp = J_{\perp \rm c}} = 0, \qquad \frac{\partial \mathcal{F}}{\partial J_\perp}\Big|_{J_\perp = J_{\perp \rm m}} = 0,$$

and the normalization condition is such that:

$$N = 2\pi B_E \int_0^\infty \left(\int_{J_{\perp c}}^{J_{\perp m}} T_B\, \mathcal{F}\, dJ_\perp \right) v\, dv.$$

Here N is the total number of energetic electrons in the magnetic flux tube with unit cross-sectional area at its base, and $\mathcal{F}(t, J_\perp, v)$ is the distribution function averaged over the bounce period of the oscillations of the charged particles between mirror points in the coordinate system fixed to the instantaneous position of the magnetic field lines.

We now consider small perturbations of $\mathcal{F}$ and wave energy $\mathcal{E}$ near the equilibrium state corresponding to a stationary CI regime. The stationary state is described by $\mathcal{F}_0$ (9.53), $\mathcal{E}_0$ (9.54) and the Alfvén wave amplitude $b_A = 0$. Linearization of (9.62) and (9.63) for harmonic processes ($a = a_0 + a_\sim \exp(-i\Omega t)$) yields

$$-i(\Omega - s\bar{\Omega}_\mathrm{d})\,\mathcal{F} = -\frac{\mathcal{E}_\sim}{\mathcal{E}_0}\,J + \frac{1}{T_B}\frac{\partial}{\partial J_\perp}\left(G\mathcal{E}_0\,\frac{\partial \mathcal{F}}{\partial J_\perp}\right) - i\Omega\,\Delta\mathcal{F}_\mathrm{A}, \qquad (9.64)$$

$$-i\Omega\mathcal{E}_\sim = \mathcal{E}_0 \int_0^\infty \left(\int_{J_{\perp\mathrm{c}}}^{J_{\perp\mathrm{m}}} K\,\frac{\partial \mathcal{F}_\sim}{\partial J_\perp}\,dJ_\perp\right) dv. \qquad (9.65)$$

As previously, the boundary conditions are:

$$\mathcal{F}_\sim\Big|_{J_\perp = J_{\perp c}} = 0, \qquad \frac{\partial \mathcal{F}_\sim}{\partial J_\perp}\Big|_{J_\perp = J_{\perp m}} = 0.$$

Now we introduce the system (x^1, x^2, x^3) of curvilinear coordinates associated with the magnetic field lines in which $x^2 = \psi$ is the longitude (azimuthal) angle, $x^3 = z$ is the distance from the magnetic equator along the field line, and $x^1 = \Phi/2\pi g^{1/2}$ is the third coordinate to complete the right-handed system, where Φ is the magnetic flux within a given surface. This system of coordinates (see Fig. 9.4) is described by a metric tensor g_{ik} using which we can write down the elements of length and volume in the following form:

$$d^2r = g_{11}(dx^1)^2 + g_{22}(dx^2)^2 + g_{33}(dx^3)^2, \quad d^3r = \sqrt{g}\,dx^1\,dx^2\,dx^3. \qquad (9.66)$$

We have $B = (0, 0, B)$, $B = (\partial\Phi/\partial x^1)/2\pi g^{1/2}$, and $g = g_{11}\,g_{22}\,g_{33}$. If the geomagnetic trap is taken to be axisymmetric, the term $\Delta\mathcal{F}_A$ added to the distribution function due to the Alfvén waves can be written, in this system of coordinates, in

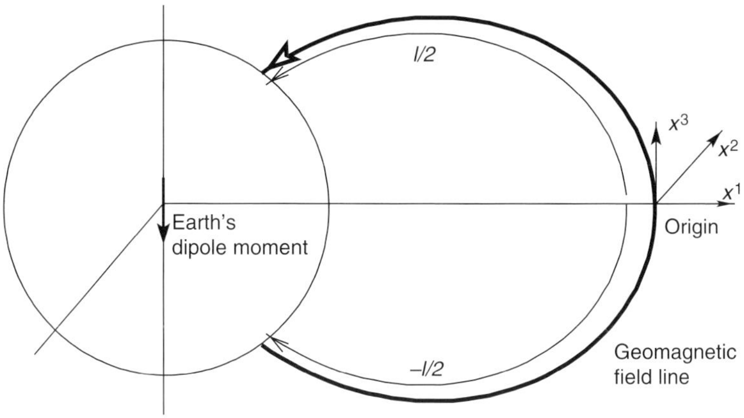

Figure 9.4 Diagram illustrating the magnetic coordinate system used.

the form:

$$\Delta\mathcal{F}_{\mathrm{A}} = \Delta\mathcal{F}_{\mathrm{A}}^{(1)} + \Delta\mathcal{F}_{\mathrm{A}}^{(2)}, \tag{9.67}$$

where

$$\Delta\mathcal{F}_{\mathrm{A}}^{(1)} = -\xi\,\frac{\partial\mathcal{F}_0}{\partial x^1} \tag{9.68}$$

and

$$\Delta\mathcal{F}_{\mathrm{A}}^{(2)} = \frac{\Omega - s\Omega_*}{\Omega}\,\frac{1}{v}\,\frac{\partial\mathcal{F}_0}{\partial v}\,\frac{1}{T_B}\oint\frac{v_\parallel^2 + J_\perp B}{R}\,\xi\,\frac{dz}{v_\parallel}. \tag{9.69}$$

When we derived (9.69) we assumed that the dependence on the coordinate ψ had the form $\exp(is\psi)$, $s = 0, \pm 1, \pm 2, \ldots$ The mean angular velocity of the azimuthal drift in a non-uniform magnetic field is

$$\Omega_{\mathrm{d}} = -\frac{v_\parallel^2 + J_\perp B}{R}\,2\pi mc\sqrt{g_{11}}/e\,\frac{\partial\Phi}{\partial x^1}, \qquad \bar{\Omega}_{\mathrm{d}} = \frac{1}{T_B}\oint\Omega_{\mathrm{d}}\,\frac{dz}{v_\parallel}, \tag{9.70}$$

the Larmor drift frequency (Mikhailovskii, 1992) is

$$\Omega_* = -2\pi mcv\,\frac{\partial\mathcal{F}_0}{\partial x^1}\Big/\frac{\partial\mathcal{F}_0}{e\,\partial v}\,\frac{\partial\Phi}{\partial x^2},$$

ξ is the displacement of the medium along x^1 due to the Alfvén waves, and R is the radius of curvature of the magnetic field lines.

The term (9.68) is due to the displacement of the magnetic flux tube as a whole; therefore, it does not make a contribution to (9.61). This is explained by the fact that group propagation of the whistler-mode waves is determined by the geomagnetic field lines, and is displaced together with them when they are oscillating due to the magnetohydrodynamic waves. The growth rate of the whistler-mode waves is determined by the distribution function in the coordinate system fixed to the geomagnetic field lines.

We seek a solution to (9.64) using an expansion in the eigenfunctions of the diffusion operator (see Section 9.1). We obtain

$$\mathcal{F}_\sim = \sum_p \left(i\Delta\Omega - \nu p^2\right)^{-1}\left(\frac{\mathcal{E}_\sim}{\mathcal{E}_0}\int_{J_{\perp\mathrm{c}}}^{J_{\perp\mathrm{m}}} T_B Z_p J\,dJ_\perp \right.$$
$$\left. + \,i\Omega\int_{J_{\perp\mathrm{c}}}^{J_{\perp\mathrm{m}}} T_B Z_p\,\Delta\mathcal{F}_{\mathrm{A}}^{(2)}\,dJ_\perp\right) Z_p, \tag{9.71}$$

where $\Delta\mathcal{F}_{\mathrm{A}}^{(2)}$ is given by (9.69), $\Delta\Omega = \Omega - s\Omega_{\mathrm{d}}$, and $Z_p(J_\perp, v)$ are the eigenfunctions of the self-adjoint operator

$$\frac{1}{T_B}\frac{\partial}{\partial J_\perp}\left[\frac{\mathcal{E}_0}{\nu}\, G(J_\perp, v)\,\frac{\partial Z_p}{\partial J_\perp}\right] = -p^2(v)\, Z_p, \tag{9.72}$$

$$Z_p\Big|_{J_\perp = J_{\perp\mathrm{c}}} = 0, \qquad \frac{\partial Z_p}{\partial J_\perp}\Big|_{J_\perp = J_{\perp\mathrm{m}}} = 0,$$

which correspond to the eigenvalue p^2 and satisfy the normalization condition

$$\int_{J_{\perp\mathrm{c}}}^{J_{\perp\mathrm{m}}} T_B Z_p^2\, dJ_\perp = 1.$$

Using (9.65) and (9.71), we can express $\mathcal{E}_\sim$ in terms of $\Delta\mathcal{F}_{\mathrm{A}}^{(2)}$:

$$\begin{aligned}\frac{\mathcal{E}_\sim}{\mathcal{E}_0} = -i\Omega &\left\{\int_0^\infty dv \int_{J_{\perp\mathrm{c}}}^{J_{\perp\mathrm{m}}} dJ_\perp\, K\, \frac{\partial}{\partial J_\perp}\right.\\ &\times\left.\left[\sum_p Z_p(i\Omega - \nu p^2)^{-1} \int_{J_{\perp\mathrm{c}}}^{J_{\perp\mathrm{m}}} T_B Z_p\, \Delta\mathcal{F}_{\mathrm{A}}^{(2)}\, dJ_\perp\right]\right\}\\ &\times\left\{i\Omega + \int_0^\infty dv \int_{J_{\perp\mathrm{c}}}^{J_{\perp\mathrm{m}}} dJ_\perp\, K\, \frac{\partial}{\partial J_\perp}\right.\\ &\times\left.\left[\sum_p Z_p(i\Omega - \nu p^2)^{-1} \int_{J_{\perp\mathrm{c}}}^{J_{\perp\mathrm{m}}} T_B Z_p J\, dJ_\perp\right]\right\}^{-1}.\end{aligned} \tag{9.73}$$

We shall analyse only the most interesting case, that of weakly damped RB relaxation oscillations, when $p_{\mathrm{max}}^2 \nu \ll |\Delta\Omega|$. Under these conditions we can replace (9.73) by

$$\frac{\mathcal{E}_\sim}{\mathcal{E}_0} \simeq i\Omega\left(\Omega_J^2 - \Omega\,\Delta\Omega\right)^{-1} \int_0^\infty \left(\int_{J_{\perp\mathrm{c}}}^{J_{\perp\mathrm{m}}} K\, \frac{\partial \Delta\mathcal{F}_{\mathrm{A}}^{(2)}}{\partial J_\perp}\, dJ_\perp\right) dv, \tag{9.74}$$

where

$$\Omega_J = \left\{\int_0^\infty \left[\int_{J_{\perp\mathrm{c}}}^{J_{\perp\mathrm{m}}} K(J_\perp, v)\, \frac{\partial J(J_\perp, v)}{\partial J_\perp}\, dJ_\perp\right] dv\right\}^{1/2} \tag{9.75}$$

is the eigenfrequency of the relaxation oscillations.

Equations (9.71) and (9.74) yield the following expression for the perturbation of the distribution function in a stationary system of coordinates:

$$\mathcal{F}_{\sim} = -\xi \frac{\partial \mathcal{F}_0}{\partial x^1} + \frac{\Omega}{\Delta\Omega} \Delta\mathcal{F}_{\mathrm{A}}^{(2)} + J \left(\frac{\Omega}{\Delta\Omega}\right) \left(\Omega_J^2 - \Omega\,\Delta\Omega\right)^{-1}$$
$$\times \int_0^{\infty} \left(\int_{J_{\perp c}}^{J_{\perp m}} K \frac{\partial \Delta\mathcal{F}_{\mathrm{A}}^{(2)}}{\partial J_{\perp}} \, dJ_{\perp} \right) dv. \tag{9.76}$$

To complete the system of equations, we have to find the efficiency of generation of the Alfvén waves. We analyse the equations describing the generation of the Alfvén waves under magnetospheric conditions. For an axisymmetric plasma distribution and a poloidal (e.g. centred dipolar) geomagnetic field, the equations for a small-amplitude Alfvén wave with frequency Ω can be written in integral form as given by Mikhailovskii and Pokhotelov (1975a):

$$A_1\Omega^2 - A_2 + A_3 = 0. \tag{9.77}$$

The scale of the hydromagnetic perturbation in the coordinate x^1 must be small in comparison with the scale of the transverse non-uniformity of the system. The parameter coupling Alfvén waves and magnetoacoustic waves is of the order of $(æ^1 R)^{-1}$, where $æ^1$ is the transverse wavenumber of the Alfvén wave.

In (9.77)

$$A_1 = \int \left(g^{11} \left| \frac{\partial \xi}{\partial x^1} \right|^2 + s^2 g^{22} |\xi|^2 \right) v_{\mathrm{A}}^{-2} \, d^3 r,$$
$$A_2 = \int \left(g^{11} \left| \frac{\partial^2 \xi}{\partial x^1 \partial x^3} \right|^2 + s^2 g^{22} \left| \frac{\partial \xi}{\partial x^3} \right|^2 \right) (g_{33})^{-1} \, d^3 r, \tag{9.78}$$
$$A_3 = 32\pi^3 s^2 \int \left(\frac{\partial \Phi}{\partial x^1} \right)^{-2} \xi^* P_{\sim} R^{-1} (g_{11})^{1/2} \, d^3 r.$$

Here v_{A} is the Alfvén speed, and

$$P_{\sim} = \frac{m}{2} B \int_0^{\infty} \left[\int_{J_{\perp c}}^{J_{\perp m}} \frac{v_{\parallel}^2 + J_{\perp} B}{|v_{\parallel}|} \mathcal{F}_{\sim} \, dJ_{\perp} \right] v \, dv \tag{9.79}$$

is the variable (at the frequency Ω) pressure component.

It is assumed that the ratio of the gas kinetic pressure to the magnetic pressure is very small, $\beta \equiv 8\pi P / B_L^2 \ll 1$, B_L being the geomagnetic field strength in the equatorial plane in the flux tube of interest.

Equations (9.76) and (9.79) yield

$$P_{\sim} = -\xi\,\frac{\partial P_0}{\partial x^1} + \frac{m}{2}\,B\int_0^{\infty}\left[\int_{J_{\perp c}}^{J_{\perp m}}\frac{v_{\parallel}^2 + J_{\perp}B}{|v_{\parallel}|}\,\frac{\Omega}{\Delta\Omega}\,\Delta\mathcal{F}_{\mathrm{A}}^{(2)}\,dJ_{\perp}\right]v\,dv$$
$$+\frac{m}{2}\,B\int_0^{\infty}\left\{\int_{J_{\perp c}}^{J_{\perp m}}\frac{v_{\parallel}^2 + J_{\perp}B}{|v_{\parallel}|}\,\frac{\Omega}{\Delta\Omega\left(\Omega_J^2 - \Omega\,\Delta\Omega\right)}\right.$$
$$\left.\times\,J\left[\int_0^{\infty}\left(\int_{J_{\perp c}}^{J_{\perp m}}K\,\frac{\partial\Delta\mathcal{F}_{\mathrm{A}}^{(2)}}{\partial J_{\perp}}\,dJ_{\perp}\right)dv\right]dJ_{\perp}\right\}dv. \tag{9.80}$$

The first term in (9.80) is related to $\mathcal{F}^{(1)}$ and makes a contribution to the frequency correction. The second term describes the drift instability of the Alfvén waves due to the non-uniform curvature of the magnetic field, and the third term describes the CI effects.

We analyse the conditions for the modulational instability at the double resonance when the frequency Ω_J of the relaxation oscillations given by (9.75) is close to the Alfvén wave resonant frequency $\Omega_{\mathrm{A}}^2 = A_2/A_1$ (see (9.77)):

$$\Omega \sim \Omega_J \simeq \Omega_{\mathrm{A}}. \tag{9.81}$$

This frequency is much higher than the azimuthal drift frequency Ω_{d}. Under such conditions, the main contribution to (9.80) is made by the third term, which is proportional to $(\Omega_J^2 - \Omega^2)^{-1}$. Substituting (9.80) into (9.77) and (9.78), we obtain the dispersion equation (as $\Omega_d \to 0$)

$$\left(\Omega^2 - \Omega_{\mathrm{A}}^2\right)\left(\Omega^2 - \Omega_J^2\right) = -Y, \tag{9.82}$$

where

$$Y = \frac{32\pi^3\,s^2\,m}{A_1}\int_{\Phi_{\min}}^{\Phi_{\max}} d\Phi\left(\frac{\partial\Phi}{\partial x^1}\right)^{-2}\int_0^{\infty} v^3\,dv\int_{J_{\perp c}}^{J_{\perp m}} dJ_{\perp}\,T_B J$$
$$\times\int_0^{\infty} dv'\int_{J_{\perp c}}^{J_{\perp m}} dJ'_{\perp}\,K(J'_{\perp}, v')\,\frac{\partial}{\partial J'_{\perp}}\left\{(s\Omega_* - \Omega)\,v'\,\frac{\partial\mathcal{F}_0}{\partial v'}\right.$$
$$\times\left[\frac{1}{T_B(J_{\perp}, v)}\oint (v_{\parallel}^2 + J_{\perp}B)\,\xi^* R v^2\,\frac{dz}{v_{\parallel}}\right]$$
$$\left.\times\left[\frac{1}{T_B(J'_{\perp}, v')}\oint\frac{\left(v_{\parallel}'^2 + J'_{\perp}B\right)\xi}{Rv'^2}\,\frac{dz'}{v'_{\parallel}}\right]\right\}. \tag{9.83}$$

If (9.82) has an unstable solution, then the stationary state described by (9.53) and (9.54) is unstable as the depth of modulation of the whistler-mode intensity and the energetic electron fluxes are increasing as time increases. The Alfvén wave amplitude is growing at the same time. Equations (9.82) and (9.83) enable us to find a criterion for such a modulational instability to occur, and to estimate its growth rate.

First, we shall show that this instability is universal if the double resonance condition given by (9.81) is satisfied. If Y is a complex quantity, (9.82) always has an unstable solution. For a real $Y > 0$, there is also an unstable solution; if $Y < 0$, an instability is also found but for the mode with the opposite sign of s.

To estimate the growth rate of the modulational instability, we shall simplify (9.83) assuming that the quantity $T_B^{-1} \oint (v_\parallel^2 + J_\perp B)\, \xi (R v^2 v_\parallel)^{-1}\, dz$ depends on $J_\perp$ and v more smoothly than other functions in the integrand in (9.83). Then we find that, for $\Omega_J = \Omega_A$, the instability growth rate

$$\gamma_{\mathrm{A}} = \frac{2s}{\Omega_J} \left\{ \frac{\pi^3 m}{A_1} \int_{\Phi_{\min}}^{\Phi_{\max}} d\Phi \left(\frac{\partial \Phi}{\partial x^1} \right)^{-2} \int_0^\infty dv \int_{J_{\perp \mathrm{c}}}^{J_{\perp \mathrm{m}}} dJ_\perp\, T_B J v^{-1} \right.$$
$$\times \left| \frac{1}{T_B} \oint \frac{(v_\parallel^2 + J_\perp B)\xi}{R} \frac{dz}{v_\parallel} \right|^2$$
$$\left. \times \int_0^\infty dv' \int_{J_{\perp \mathrm{c}}}^{J_{\perp \mathrm{m}}} dJ'_\perp\, K v' \frac{\partial}{\partial J'_\perp} \left[\left(\frac{s\Omega_*}{\Omega} - 1 \right) \frac{\partial \mathcal{F}_0}{\partial v'} \right] \right\}^{1/2}. \tag{9.84}$$

After a number of simplifications, we find that

$$\frac{\gamma_{\mathrm{A}}}{\Omega} \simeq \frac{1}{R} \left\{ \frac{\beta s^2}{(1 + s^2 \ell_\perp^2 / L^2 R_E^2)} \left(\frac{s\Omega_*}{\Omega} - 1 \right) \right\}^{1/2}, \tag{9.85}$$

where $\Omega \simeq (v_{\mathrm{A}}/\ell) q$, with $q = 1, 2, \ldots$, $\beta = 8\pi P_0 / B^2$, P_0 is the pressure of the energetic electrons, $\Omega_* \simeq v^2 / \omega_B \ell_\perp^2$, and $\ell_\perp$ is the transverse size of the plasma region. Equation (9.82) is valid for a sufficiently small s. The azimuthal wavenumber is limited primarily by the fact that the inequality $|s\Omega_{\mathrm{d}}| \ll \Omega_J$ must be satisfied to allow us to ignore Ω_{d} in (9.82). The additional restriction on the value of s is due to the frequency dispersion of the refractive index of the Alfvén waves, $|s| \ll \Omega_B / \Omega$, where Ω_B is the gyrofrequency for protons.

Summarizing the results of subsections 9.2.1 and 9.2.2, we conclude that, for a smooth spatial dependence of the stationary pressure $P_{\perp 0}$ of the RB electrons, a remarkable self-modulation of the CI by MHD waves is possible for the Alfvén mode, whose wavelength along the magnetic field line is comparable with its curvature. The self-modulation is connected with the drift instability, the latter being strongly amplified when the magnetic flux tube eigenfrequency Ω_{A} is close to the frequency

Ω_J of the CI relaxation oscillations. The growth rate is a factor $\beta^{-1/2}$ ($\gg 1$) greater than that for the ordinary drift instability.

9.2.3 Self-modulation regime in an Alfvén sweep maser

We have so far discussed the CI dynamics of the electron component of the radiation belts. It should be noted, however, that many of the above results are also valid for the proton radiation belt, that is, the stationary states and various dynamic CI regimes, though, of course, on a different time scale. For instance, the characteristic Alfvén wave frequencies Ω are lower than the proton gyrofrequency Ω_{BL} in the equatorial cross-section of the magnetic flux tube. The highest frequency proton CI in the magnetosphere occurs in the geomagnetic pulsation range $0.1\text{ Hz} < \Omega_A/2\pi < 10\text{ Hz}$. This frequency range corresponds to a significantly greater role of the ionosphere as a nonlinear mirror in largely determining the dynamics of the Alfvén maser (AM); see Fig. 9.5. Indeed, the characteristic periods of the relaxation CI oscillations for the proton component are of the same order as the relaxation time for the ionosphere. In addition, for the 0.1–10 Hz frequency range the Alfvén wavelengths are comparable with the thickness of the ionosphere from which the wave is reflected. The coefficient of reflection of the ionosphere, as a function of frequency, has pronounced peaks due to the resonant properties of the ionosphere. Figure 2.8 diagrammatically shows the frequency dependence of the magnitude of the reflection coefficient $R(\Omega)$ and also the characteristic frequency variation of the cyclotron amplification $\Gamma(\Omega)$ in the AM. The excitation threshold of the AM is determined by the condition that

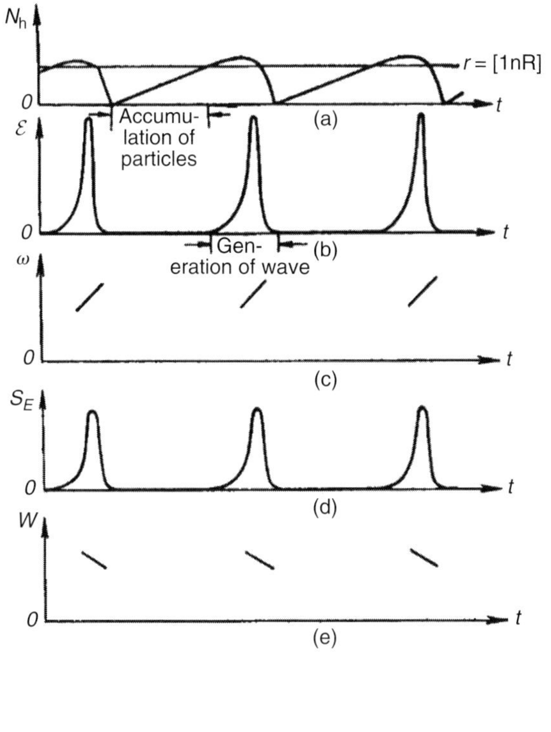

Figure 9.5 Dynamics of the Alfvén maser (AM): (a) the number of energetic protons N_h in a flux tube with a cross-sectional area of 1 cm^2 in the ionosphere; (b) the wave energy $\mathcal{E}$ per unit volume; (c) the dynamic frequency spectrum (up to $\sim$ 10 Hz) of the wave generated; (d) the flux of energetic protons precipitated into the E region of the ionosphere, and (e) the mean energy of precipitated protons, W, all as a function of time t.

the total growth Γ along the flux tube exceeds a certain value:

$$\Gamma(\Omega) = \int_{-\ell/2}^{+\ell/2} æ_{\mathrm{A}}(\Omega)\, dz = |\ln R(\Omega)|, \tag{9.86}$$

where $æ_{\mathrm{A}}(\Omega)$ is the amplification coefficient of the Alfvén waves.

According to (9.86) and Fig. 2.8 wave generation occurs in a narrow frequency range near one of the peaks of the reflection coefficient $R(\Omega)$ where and when $\Gamma(\Omega)$ is sufficiently high.

The additional ionization of the ionosphere caused by the precipitation of the energetic RB protons alters the reflection coefficient and results in the drift of $R(\Omega)$ with respect to the amplified wave frequency $\Gamma(\Omega)$. If the wave generation occurs on the positive slope of the curve $\Gamma(\Omega)$ (where $\partial\Gamma/\partial\Omega > 0$) and is accompanied by a frequency sweep towards increasing frequency, the effective amplification which is equal to $\left(\Gamma(\Omega) + \ln(\Omega)\right)_{\max}$ can increase, rather than decrease, with increasing wave intensity. Under such conditions an 'explosive' growth of the wave amplitude can occur. This corresponds to the transition of the Alfvén sweep maser to the autooscillation generation regime.

Adiabatic approximation. A quantitative theory of the Alfvén sweep maser (Bespalov and Trakhtengerts, 1986a) can be comparatively easily developed in the adiabatic approximation when

$$T_{\mathrm{g}}/T \ll 1, \tag{9.87}$$

where T is the characteristic time of variation of the wave intensity, and T_{g} is the group propagation time of the Alfvén wave packet through the magnetospheric resonator. In this approximation the mean radiation frequency Ω_0 at each moment during the phase of spike development ($\partial\mathcal{E}/\partial t > 0$) can be found from

$$\frac{\partial}{\partial\Omega}\left[\Gamma(t,\Omega) + \ln R(t,\Omega)\right]\Big|_{\Omega=\Omega_0} = 0. \tag{9.88}$$

Then the system of equations (9.10) can be used, replacing the electron mass m with the proton mass M and the whistler wavenumber with the Alfvén wavenumber. For low frequencies ($\Omega \ll \Omega_B$), the Alfvén wave growth rate is half that given by the above procedure for the electron growth rate due to a different dependence of N_{A} on ω (see (3.11)). Taking this into account, we obtain

$$\begin{aligned}
\frac{dN_{\mathrm{h}}}{dt} &= -\delta N_{\mathrm{h}}\mathcal{E} + J_{\Sigma},\\
\frac{d\mathcal{E}}{dt} &= hN_{\mathrm{h}}\mathcal{E} - \nu\mathcal{E},\\
\frac{dn_E}{dt} &= q_E + \tau_E N_{\mathrm{h}}\mathcal{E} - \alpha_E n_E^2,
\end{aligned} \tag{9.89}$$

where

$$\delta = \frac{2.8\,\psi(\Omega_0)}{\Omega_{BL}}\left(\frac{e}{Mc}\right)^2, \qquad h = \delta\,\frac{\Omega_0}{\Omega_{BL}}\,\frac{Mv_i^2}{2\sigma\ell},$$

$\tau_E = \dfrac{\eta_E \delta}{2}$, v_i is the mean velocity of the energetic protons, $æ_0 = \left(\dfrac{\Omega_0 \Omega_{pL}}{c\Omega_{BL}}\right)$ is the wavenumber of the Alfvén disturbances, $\nu = 2\,|\ln R(\Omega_0)|/T_g$, and $\psi(\Omega)$ is the normalized amplification, $\psi(\Omega) = \Gamma(\Omega)/\Gamma_{max}$; Ω_{pL} and Ω_{BL} are, respectively, the cold plasma frequency and cyclotron frequency in the equatorial plane of a magnetospheric flux tube whose McIlwain parameter is L.

As noted above, the parameters of the lower ionosphere where the electron density is n_E vary during the process of CI development under the effect of the flux of precipitated protons. The $R(\Omega)$ curve is shifted along the frequency axis and the frequencies of its maxima and minima are changed. We can take these changes into account for $\Delta n_E/n_E < 1$ by writing

$$\nu = \nu_0 + \left(\frac{\partial \nu}{\partial n_E}\right)_0 \Delta n_E,$$

$$h = h_0 + \left(\frac{\partial h}{\partial n_E}\right)_0 \Delta n_E, \tag{9.90}$$

where ν_0 and h_0 correspond to the equilibrium state of the system described by (9.89), $\partial\nu/\partial n_E$ describes the variation of R (for a fixed frequency), and $\partial h/\partial n_E$ takes into account the shift of the $R(\Omega)$ curve along the frequency axis.

According to condition (9.19), we can write the condition that the system described by (9.89) operates in the spike regime in the form ($\nu_J T_r \ll 1$)

$$\frac{2T_r\,\tau_E\,\nu_0}{\left(1 + \Omega_J^2 T_r^2\right)\delta T_g h_0}\left[\frac{\partial}{\partial n_E}\ln(h\nu)\right]_0 > 1, \tag{9.91}$$

where $T_r = (2\alpha_e n_{E0})^{-1}$ is the characteristic ionospheric relaxation time, Ω_J and ν_J are the frequency and the damping rate of the relaxation oscillations given by (9.30).

The physical meaning of (9.91) is that the spike regime is realized when the system, during the generation process, goes over into a state with a higher reflection factor compensating for the damping of the relaxation oscillations. Condition (9.91) includes the derivative $(\partial/\partial n_E)\ln(h\nu)$, which is determined by the specific form of $R(\Omega, n_E)$ and condition (9.88).

The reflection coefficient $R(\Omega, n_E)$ was analysed by Polyakov *et al.* (1983). The coefficient of reflection of the Alfvén waves from the ionosphere was found to vary over a wide range, depending on the time of day and the geomagnetic activity. The dependence of R on n_E is associated with the dissipation of the Alfvén waves in the E region of the ionosphere, which is determined by the Pedersen and Hall conductivities there. These conductivities vary with the flux and energy of the precipitating energetic protons. Numerical calculations

(Ostapenko and Polyakov, 1990) show that the condition

$$\frac{\partial}{\partial n_E}\ln(h\nu) \simeq \frac{1}{h}\frac{\partial h}{\partial n_E} > 0$$

needed for satisfying condition (9.91) for the spike mode can be most readily found at the morning side of the magnetosphere under quiet geomagnetic conditions. As noted above, see (9.90), the derivative

$$\frac{\partial h}{\partial n_E} = \left(\frac{1}{\psi}\frac{\partial \psi}{\partial \Omega}\right)\bigg|_{\Omega=\Omega_0} L \cdot \frac{\partial \Omega_0}{\partial n_E} \tag{9.92}$$

describes the sweeping of the radiation frequency in a spike. Under quiet morning conditions, $\partial\Omega_0/\partial n_E > 0$ and the spike regime is realized at the positive slope of the amplification $\Gamma(\Omega)$. Under quiet daytime conditions, the sign can be reversed.

AM radiation in the spike regime. We now analyse some of the amplitude-time characteristics of the AM spike regime. Introducing the dimensionless variables

$$\tau = \Omega_J t, \quad x = \ln\left(\frac{\mathcal{E} N_\mathrm{h}}{\mathcal{E}_0 N_0}\right), \quad y = \frac{N_\mathrm{h}}{N_0} \tag{9.93}$$

in the system of equations (9.89), we obtain, for $\Omega_J T_r \ll 1$,

$$\begin{aligned} &\frac{d^2x}{d\tau^2} - Q_J^{-1}\left(gy - \frac{1}{y}\right)\exp(x)\frac{dx}{dt} + \exp(x) - 1 \\ &\qquad - Q_J^{-2}\left(g + y^{-2}\right)[\exp(x) - 1]^2 = 0, \\ &\frac{dy}{d\tau} = Q_J^{-1}[1 - \exp(x)], \end{aligned} \tag{9.94}$$

where

$$Q_J = \frac{\Omega_J}{2\nu_J} \gg 1,$$

and

$$g = \left(\frac{2T_\mathrm{r}\,\tau_E\,\nu_0}{\delta T_\mathrm{g}\,h_0}\right)\left[\frac{\partial}{\partial n_E}\ln(h\nu)\right]_0 .$$

The system of equations (9.94) for $g \simeq 1$ describes weakly damped (yet growing) oscillations of a nonlinear oscillator. In the time stationary regime, the mean work done on the oscillator in the period τ_J of the nonlinear oscillations must be zero:

$$\int_0^{\tau_J} (gy - y^{-1})\left(\frac{dx}{d\tau}\right)^2 \exp(x)\, d\tau = 0 \tag{9.95}$$

where $x(\tau)$ is found from the equation

$$\frac{d^2x}{d\tau^2} + \exp(x) - 1 = 0 \tag{9.96}$$

and y is found from the second equation (9.94). Equations (9.95) and (9.96) determine the parameters of the spiking regime of the AM. Equation (9.96) was analysed in Section 9.1, see (9.34). Thus, the temporal characteristics of the spiking regime can be described by (9.35)–(9.37) where the electron parameters must be replaced by the proton parameters according to (9.89). The spike amplitude $\mathcal{E}_{\max}$ in (9.35)–(9.37) can be found from (9.95). In particular, slightly above the instability threshold $(g - 1 \ll 1)$ when $x_m \simeq \ln(\mathcal{E}_{\max}/\mathcal{E}_0) < 1$, we obtain

$$x = x_m \cos(\tau + \varphi), \quad y = 1 - \frac{x_m}{Q_J} \sin(\tau + \varphi), \quad x_m = 2Q_J \sqrt{\frac{g-1}{3}}. \tag{9.97}$$

9.3 Results of computer simulations

The analytical solution of the QL self-consistent system of equations (8.34) and (8.41) is very complicated in the case of an arbitrary source J, when the two-level approach discussed in Section 9.1 is not valid. Then it is necessary to consider the change of pitch-angle diffusion regimes during CI development. Some simplified analytical models, taking into account the full diffusion (the multi-level approach), were considered by Bespalov (1981) and Bespalov and Trakhtengerts (1986b). It was shown that the relaxation oscillations can be changed by self-sustained oscillations of the whistler wave intensity and of the energetic electron flux. The physical cause for these is the temporal evolution of the frequency spectrum of the whistler-mode waves generated and involving newly resonant electrons in the cyclotron instability. However, these models describe neither the changes of the diffusion regimes nor the details of the evolution of the frequency spectrum of the whistler waves generated. Both of these are important for the interpretation of experimental ELF/VLF data on the near Earth space plasma environment.

In this section we discuss the results of computer simulations of the system of equations (8.34) and (8.41) provided by Pasmanik *et al.* (2004a). These take into account all the features mentioned here and give us the basis for a quantitative interpretation of observed quasi-periodic (QP) ELF/VLF emissions (see Chapter 10).

Two types of energetic particle sources have been considered. The first is the magnetic drift of energetic electrons into the wave generation region and the second is the local acceleration (due, for example, to magnetic compression). Electrons can be removed from the cyclotron interaction region by two mechanisms. One is precipitation into the upper atmosphere of energetic electrons, via the loss cone, and the other is the drift of electrons away from the interaction region across geomagnetic field lines. Drift removal of gyroresonant electrons is most effective for interactions in a duct of enhanced density that is outside the plasmasphere, or if the cross-section of the interaction region is rather small. This mechanism was taken into account in the so-called flow cyclotron maser (FCM) model developed by Trakhtengerts *et al.* (1986) and Demekhov and Trakhtengerts (1994). In the case of interactions inside the plasmasphere, or in rather large regions of enhanced cold plasma density beyond it, the dominant loss mechanism for energetic electrons is due to precipitation.

So, we assume in our model that whistler-mode waves are generated in a region with enhanced cold plasma density, where the cyclotron resonance condition (Equation 3.66) is satisfied for most of the energetic electrons (see below). Outside this region, development of the cyclotron instability is impossible, and the distribution of energetic electrons may remain anisotropic. The magnetic drift of such particles into the region of dense plasma serves as the source of free energy for wave generation. To include this source in our considerations we add to the left side of Equation (8.34) the drift term $v_{\mathrm{D}}\,\partial\mathcal{F}/\partial x$, where v_{D} is the magnetic drift velocity, and x is the coordinate along the magnetic drift trajectory (x^2 in Fig. 9.4).

We shall restrict our consideration to the case of ducted whistler wave propagation, when the waves propagate parallel to the geomagnetic field (i.e. $\vec{k}\|\vec{B}$). In this case, we can assume the quasi-homogeneous distribution of wave energy flux $\mathcal{E}$ across the interaction region, and average the kinetic equation over the cross-section of this region:

$$\frac{\partial\Phi}{\partial t} = \frac{1}{T_B}\frac{\partial}{\partial\mu}\mu D\frac{\partial\Phi}{\partial\mu} + J - \delta\cdot\Phi, \tag{9.98}$$

$$J = \frac{v_D}{S_0}\int_{y_1}^{y_2}(\mathcal{F}_{\mathrm{in}} - \mathcal{F}_{\mathrm{out}})\,dy, \tag{9.99}$$

where the flux of energetic electrons $\Phi = S_0^{-1}\int\mathcal{F}\,ds$, S_0 is the area of the duct cross-section, J is the effective source of energetic electrons, $\mathcal{F}_{\mathrm{in}}$ and $\mathcal{F}_{\mathrm{out}}$ are the distribution functions of the electrons entering and leaving the duct, respectively, and y_1 and y_2 are the boundaries of the interaction region in the direction transverse to the drift velocity $\vec{v}_D$. The diffusion operator is given by Equations (8.34)–(8.36), $\mathcal{E}_\omega$ is defined by (8.27) (see equation (8.41)), and δ is defined by (8.35).

Noting that energy diffusion is negligible in the frequency range considered, $\omega \ll \omega_B$, we can simplify the analysis of this model by using a distribution with a narrow energy spectrum, parameterized by the characteristic gyroresonant energy $W_0 = mv_0^2/2$. The use of this approximation is made possible by the fact that the inhomogeneous geomagnetic field provides a wide spread of resonant energies: a wave with a given frequency ω interacts with electrons whose velocity is $v > v_{\min}(\omega, \mu)$, where $v_{\min}$ is determined from the cyclotron resonance condition in the equatorial plane (see Equations 3.66). Due to that, the growth rate of ducted whistler-mode waves is determined by integral parameters of the energy distribution, such as the characteristic energy. This approximation allows us to obtain the correct integral characteristics, such as the electron flux and the wave amplification without analysing the evolution of the electron energy spectrum. Hereafter, Φ denotes the distribution function over μ, i.e. the distribution integrated over its narrow energy spread near one resonant energy W_0.

Equation (9.98) should be solved together with Equation (8.41), where Γ_ω is determined by the relation (8.43) with $\mathcal{F}$ replaced by Φ. Using the parabolic approximation for the geomagnetic field close to the magnetic equator (2.43), and putting $(z_R/a) \ll 1$ in (8.37), we obtain for ℓ_{eff} in (8.43):

$$\ell_{\text{eff}} \approx a \frac{1-\xi}{\sqrt{2-\xi}} \frac{1}{\sqrt{\xi-\mu}} \tag{9.100}$$

where

$$\xi = 1 - \frac{\omega_{BL}^2}{k^2 v_0^2} \equiv 1 - \frac{\omega_0}{\omega}, \tag{9.101}$$

$$\omega_0 = \frac{\omega_{BL}}{\beta_*}, \qquad \beta_* = \left(\frac{\omega_{pL} v_0}{\omega_{BL}\, c}\right)^2.$$

Substituting (9.100) and (9.101) into (8.36) and (8.43), we finally obtain:

$$D(\mu, t) = D_* \cdot (2-\mu)^{-1/2} \int_\mu^1 \varepsilon(\xi-\mu)^{-1/2}\, d\xi, \tag{9.102}$$

the one-hop amplification

$$\Gamma(\omega, t) = \Gamma_* \left(\frac{1-\xi}{\sqrt{2-\xi}}\right) \int_0^\xi \left(\mu \frac{\partial \Phi}{\partial \mu} - \frac{\omega}{\omega_{BL}} \Phi\right) (\xi-\mu)^{-1/2}\, \delta\mu \tag{9.103}$$

where

$$D_* = \frac{32\sqrt{2}\,\pi^2 e^2\, L R_0}{3m^2 c^2 v_0 \beta_*}, \qquad \Gamma_* = \frac{\pi a \omega_{pL}^2\, v_0}{\omega_{BL}\, c^2\, n_{\text{cL}}}, \qquad \varepsilon = \frac{\mathcal{E}_\omega}{v_{\text{g}}}, \tag{9.104}$$

and n_{cL} is the cold plasma density in the equatorial plane of the geomagnetic flux tube of a certain-value.

The main attention in the numerical analysis of (8.41), (9.98), (9.102) and (9.103) was to study the dependence of the generation regime on the properties of the energetic electron source, i.e. the characteristics of the distribution $\mathcal{F}_{\text{in}}$ of the longitudinally drifting electrons, such as the shape of their pitch-angle distribution and their flux density, and on the frequency dependence of the wave damping due to the ionosphere $R(\omega)$. The following realistic values for parameters were used in our magnetospheric simulations: $L = 4.4$ ($f_{BL} \approx 10.3$ kHz), $n_{\text{cL}} = 55$ cm^{-3}, and $W_0 = 45$ keV.

As mentioned above, two different mechanisms of energetic electron removal from the generation region are considered in the kinetic equation (9.98). We analyse these two mechanisms separately. Two cases are studied; the first is when only losses due to precipitation are significant (in this case $\mathcal{F}_{\text{out}} \equiv 0$), and the second is when only drift losses are taken into account (in this case $\delta \equiv 0$).

9.3.1 Case of losses due to precipitation

The numerical analysis of this case shows that the crucial parameter determining the generation regime in the system is the intensity of the source, i.e. the number of particles supplied in unit time. Variations of other characteristics of the particle source or of wave damping do not affect the generation regime so much. In the particular examples presented below, a source with moderate pitch-angle anisotropy was used:

$$\mathcal{F}_{\text{in}}(\mu) = C\sqrt{\mu}, \tag{9.105}$$

where the value C is defined from the normalization condition

$$\int_0^1 \mathcal{F}_{\text{in}} \frac{d\mu}{\sqrt{1-\mu}} = n_0, \tag{9.106}$$

and n_0 is the energetic electron density. It is convenient to characterize the intensity of the source by the value

$$J_0 = v_0 \int_0^1 \mathcal{F}_{\text{in}}\, d\mu. \tag{9.107}$$

J_0 is equal to the flux of precipitating electrons provided by the source in the stationary state, i.e. when the number of precipitating electrons equals that supplied by the source. As a dimensionless characteristic of the source intensity, we can use the value $j_0 = (\nu\tau_j)^{-1}$, where $\tau_j = N/J_0$ is the time scale of energetic electron supply, N is the number of energetic electrons in the magnetic flux tube considered, and $\nu = 2\,|\ln R|/T_g$ is the wave damping rate.

In the case of a weak source, the wave generation regime exhibits relaxation oscillations. After the system reaches the threshold of the cyclotron instability, several spikes of wave intensity with diminishing amplitude are generated; eventually the system goes into the stationary generation regime. An example of such a generation regime is presented in Fig. 9.6. The whistler-mode wave spectrogram is shown in the left panel, and 'snapshots' of the electron distribution function corresponding to the minimum and maximum wave intensity are shown in the right panel. It is evident that this generation regime is characterized by almost constant shapes of the wave spectrum and electron pitch-angle distribution, but their amplitudes change with time. With an increase of the source intensity, both the oscillation period and the characteristic relaxation time decrease.

A simplified analytical model for such a generation regime has been developed in Section 9.1. To obtain this model, the so-called two-level approximation

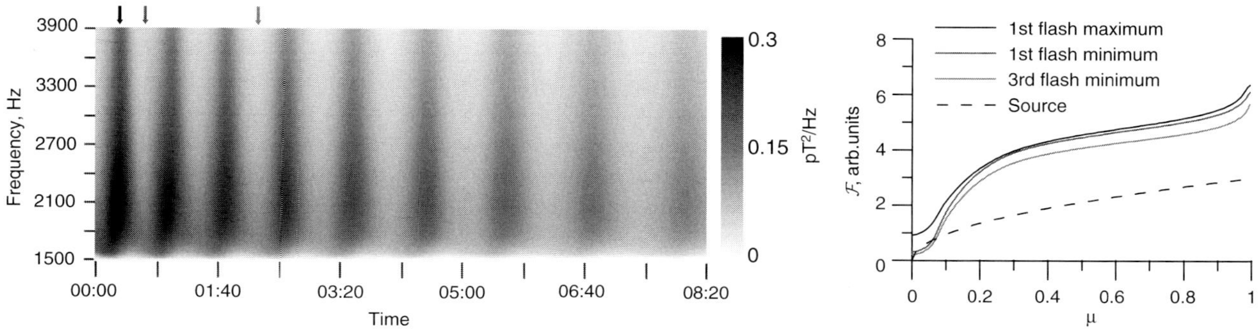

Figure 9.6 Relaxation oscillations in the model with losses due to precipitation; the spectrogram of the wave intensity is given in the left panel, and 'snapshots' of the pitch-angle distribution of energetic electrons in the duct (solid lines) and the source distribution (dashed line) are shown in the right panel. The system parameters are $L = 4.4$, $n_{cL} = 55\,\text{cm}^{-3}$, $W_0 = 45$ keV, $|\ln R| = 2\omega/\omega_0$, $J_0 = 10^6\,\text{cm}^{-2}\text{s}^{-1}$. (Taken from Pasmanik *et al.*, 2004a.)

is used, where the shapes of the wave spectrum and the electron pitch-angle distribution remain constant during the process of cyclotron instability development. This assumption allows us to obtain from (9.98) the system of ordinary differential equations (9.10) in new variables, namely the number of energetic electrons N in the magnetic flux tube, and the wave energy flux density $\mathcal{E}$. Analysis of this simplified model yields the following expression for the period of relaxation oscillations (see (9.21)):

$$T_J = \frac{2\pi}{\Omega_J} = 2\pi\sqrt{\frac{N_0/J_0}{|\ln R|/T_g}}, \tag{9.108}$$

where N_0 is the average number of trapped energetic electrons in the magnetic flux tube with a unit cross-section at the ionospheric level. A comparison of this result from the simplified model with the results obtained from the numerical solution of the full system (9.98) and (8.41) is presented in Fig. 9.7; these two approaches are in a good agreement.

With a further increase in the source of energetic electrons, another generation regime takes over – the regime of self-sustained oscillations. Here, periodic undamped oscillations of wave intensity and precipitating electron flux occur (Fig. 9.8). During a single spike of the wave intensity (see Fig. 9.8, left panel) the wave frequency rises. This happens because new particles with higher pitch-angles become involved in the interaction as the cyclotron instability develops. The time evolution of the energetic electron distribution function differs from that for the relaxation oscillations regime: there is a stronger variation of the pitch-angle distribution shape from the maximum to the minimum of wave intensity (Fig. 9.8, right panel). An increase in the source intensity leads to a decrease of the oscillation period.

In general, the generation regimes in the case of precipitation losses are characterized by a rather small modulation of the wave intensity and a small variation in the energetic electron pitch-angle distribution.

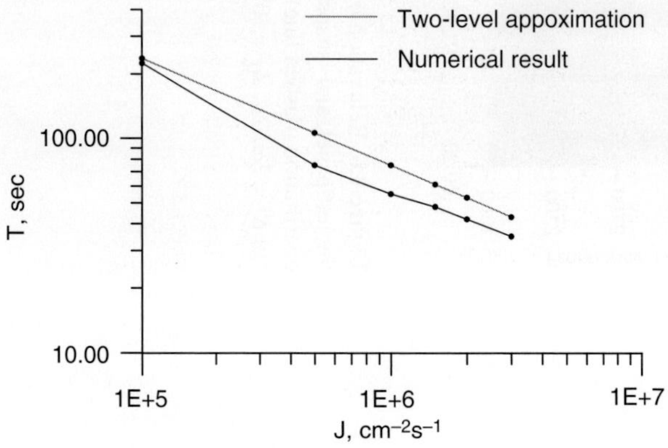

Figure 9.7 Dependence of the period of relaxation oscillations on the intensity of the energetic electron source, for the simplified 'two-level' approximation model and for the result of numerical analysis. (Taken from Pasmanik *et al.*, 2004a.)

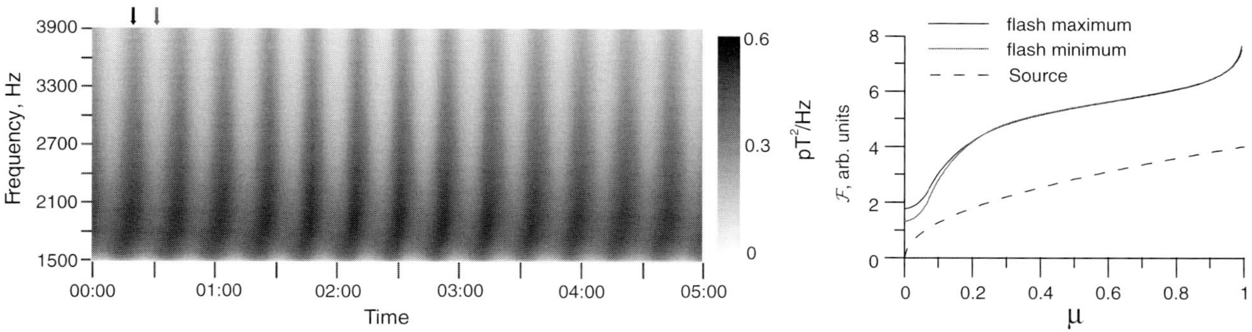

Figure 9.8 Regime of self-sustained oscillations in the case of losses due to precipitation. The source of energetic particles is 10 times greater than in Fig. 9.6 ($J_0 = 10^7\ \mathrm{cm}^{-2}\mathrm{s}^{-1}$); all the other parameters and plots are the same as in Fig. 9.6. (Taken from Pasmanik *et al.*, 2004a.)

9.3.2 Case of drift losses

As mentioned above, this case was studied by Demekhov and Trakhtengerts (1994) and applied to pulsating aurorae. In Chapter 10 quantitative analysis of this phenomenon will be performed on the basis of the results of this section. Here we focus on an analysis of how the dynamic wave spectra and electron pitch-angle distributions depend on the source properties of the energetic electrons.

To solve (9.98) for the distribution function Φ, it is necessary to know the distribution $\mathcal{F}_{\text{out}}$ of energetic electrons as they exit the duct. The shape of this distribution is determined by the evolution of the initial distribution $\mathcal{F}_{\text{in}}$ during the drift of the electrons across the duct, and depends on the dynamics of the development of the cyclotron instability in the duct. We assume that the distribution $\mathcal{F}_{\text{out}}$ is isotropic at all μ and that its amplitude is defined from the particle conservation law:

$$\int \mathcal{F}_{\text{out}}\, T_B\, d\mu = \int \mathcal{F}_{\text{in}}\, T_B\, d\mu. \tag{9.109}$$

Such an assumption can be justified if the loss cone is small enough, the wave intensity is rather high, and the main source of free energy is the pitch-angle anisotropy of trapped electrons.

Wave generation in this case is possible if the intensity of the source is higher than some threshold value, which is determined by the condition that the cyclotron instability threshold is exceeded if the whole duct is filled by energetic electrons:

$$\Gamma_0 \geq |\ln R|, \tag{9.110}$$

where the value of Γ_0 is calculated from (9.103) using the distribution $\mathcal{F}_{\text{in}}$. This is different from the case of a large interaction region (the previous case Section 9.3.1), in which the instability threshold may be reached due to the accumulation of energetic electrons.

In contrast to the previous case, there is a much wider variety of generation regimes in the case of drift losses, and the dependence of their characteristics on the system parameters is much stronger. In particular, the shape of the source (the initial pitch-angle distribution of the drifting electrons) is one of the crucial parameters determining the generation regime. Three different types of pitch-angle distributions in the source have been studied.

The first type corresponds to a monotonic pitch-angle distribution with moderate anisotropy:

$$\mathcal{F}_{\text{in}} = C\mu^{\alpha}, \tag{9.111}$$

where the parameter α characterizes the anisotropy. Our analysis has shown that, in this case, the generation regimes are rather similar to those for the case of loss cone losses: stationary generation, relaxation oscillations, and self-sustained oscillations with a small modulation of the wave intensity.

The second type of source function has the form (see Fig. 9.9, right panel)

$$\mathcal{F}_{\rm in} = \begin{cases} C \sin^{\delta}\left(\dfrac{\pi}{2}\dfrac{æ}{æ_0}\right), & æ \le æ_0 \\ C, & æ > æ_0 \end{cases} \tag{9.112}$$

where $æ \equiv \sqrt{\mu}$. This distribution can model, for example, a loss cone distribution if we put $æ_0 \approx \sqrt{\mu_c}$, or any other anisotropic source. The parameter δ characterizes the sharpness of the slope of the distribution function.

A typical wave spectrogram for low source intensity (but still high enough to satisfy (9.110)) is shown in Fig. 9.9. This is a regime with the periodic generation of wave intensity spikes; it is characterized by a rather high modulation of the wave intensity and by the excitation of relatively narrow band waves compared with the regimes discussed earlier. The generation of a spike is accompanied by a shift of the maximum wave amplitude to higher frequencies, and occurs due to the formation of a rather sharp gradient in the energetic electron pitch-angle distribution (Fig. 9.9, right panel) at the boundary between resonant and non-resonant particles; this moves towards higher pitch-angles as the cyclotron instability develops. This feature will be discussed in the next section.

The duration of the spike is determined mainly by the pitch-angle distribution of the source, and is of the order of the inverse initial growth rate $(2|\ln R|/T_g)^{-1}$. The interval between spikes is determined by the source intensity. It is equal to the time needed to supply enough energetic electrons having an anisotropic distribution to satisfy the condition (9.110). Accordingly, the period of spike generation decreases as the source increases. This is seen in Fig. 9.10, where the intensity of the source is four times greater than in Fig. 9.9.

Another interesting feature that appears with an increasing source intensity seen in Fig. 9.10 is that of spikes with two different shapes – one is similar to the case of weak source (Fig. 9.9), and the other has a shape with a much more rapid frequency change but a lower wave amplitude. It is evident that generation of the second type of spike starts immediately after the fading of the previous one, but there is a pronounced gap between the two spikes. The corresponding evolution of the energetic electron distribution function is shown in Fig. 9.10, right panel: there is a sharp gradient during the generation of the first spike which is absent during the second.

The existence of such a generation regime is explained as follows. When the intensity of the source is rather small, its contribution to the generation of a single spike is negligible. With an increase in the source intensity, the source-related modification of the distribution of energetic electrons during spike generation becomes more significant. This leads to an increase in the anisotropy at low ($æ \sim æ_0$) pitch-angles, but pitch-angle diffusion due to waves generated by the sharp gradient at higher æ

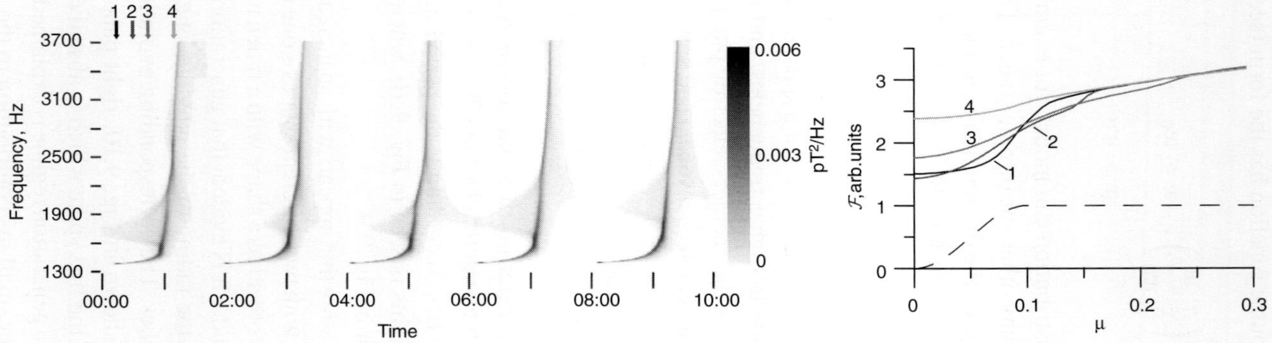

Figure 9.9 Periodic generation of wave intensity spikes in the case of losses due to electron drift and monotonic pitch-angle distribution in the source (Equation 9.112). The parameters in the model $L = 4.4$, $n_{\rm cL} = 55\,{\rm cm}^{-3}$, $W_0 = 45\,{\rm keV}$, $|\ln R| = 2\omega/\omega_0$, the density of drifting energetic (hot) electrons $n_{\rm hL} = 0.04\,{\rm cm}^{-3}$, the duct radius in the equatorial plane is 100 km; $æ_0 = 0.1$, $\delta = 2$. The pitch-angle distribution in the source is shown by the dashed line (right panel). (Taken from Pasmanik *et al.*, 2004a.)

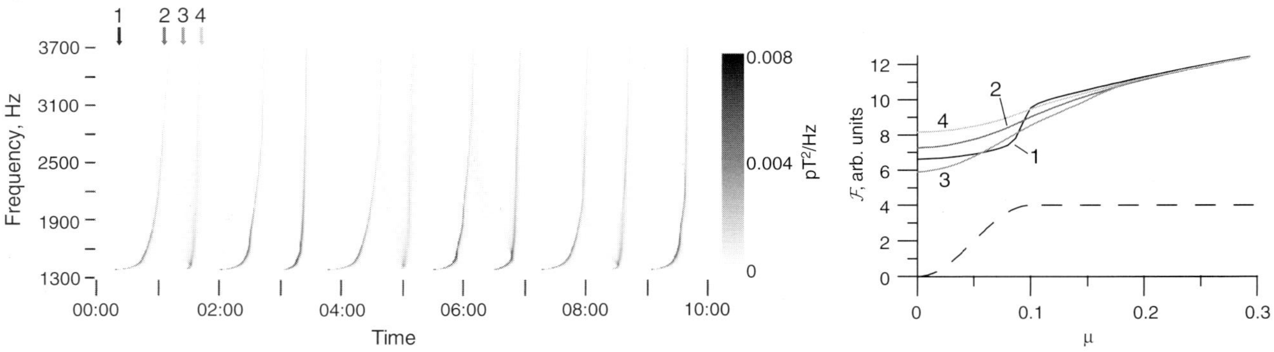

Figure 9.10 Periodic generation of wave intensity spikes in the case of losses due to electron drift, with the appearance of spikes of different shapes. The source of energetic particles is four times greater than in Fig. 9.9 ($n_{\mathrm{hL}} = 0.16\ \mathrm{cm}^{-3}$); all the other parameters are the same. (Taken from Pasmanik *et al.*, 2004a.)

values does not allow the system to reach the instability threshold. Thus, during the generation of the first spike, the source sustains the system near the threshold, so that after the end of the spike the system quickly reaches this threshold, and the second spike is generated. After that, the system comes to its initial state below the threshold.

Variations of other source parameters (i.e. the values of æ and δ) do not change the generation regime qualitatively. An increase in either æ or δ leads to a decrease in the generation period and to an increase of the lowest wave frequency in the spike. This occurs because for the same source intensity (N_0 value) such a change of the source parameters results in an increasing $\mu_{\rm opt}$ and a larger value of the maximum of $\mu\, \partial \mathcal{F}_{\rm in}/\partial \mu$, which actually determine how fast and at what frequency the generation threshold given by (9.110) is exceeded.

The dependence of the generation characteristics on the wave damping frequency profile $R(\omega)$ has been investigated. The actual dependence $R(\omega)$, which is determined by the reflection properties of the ionosphere at the ends of the magnetic flux tube and by the properties of wave propagation along the magnetic field, is not known. To study the role of this parameter, we consider two cases: increasing ($\ln R \sim \omega$) and decreasing ($\ln R \sim \omega^{-1}$) wave losses with increasing frequency. With the former, the period of spike generation increases, because more time is needed to supply enough energetic electrons to reach the instability threshold (9.110). In the latter case, the upper frequency limit of generation increases.

The third type of source pitch-angle distribution is a non-monotonic ('butterfly') distribution:

$$\mathcal{F}_{\rm in} = \begin{cases} C\,[1+\Delta]\sin^{\delta}\left(\dfrac{\pi}{2}\dfrac{æ}{æ_0}\right), & æ \le æ_0, \\ C\left[1+\Delta-\Delta\sin^{\delta_1}\left(\dfrac{\pi}{2}\dfrac{æ-æ_0}{æ_1}\right)\right], & æ_0 < æ \le æ_1+æ_0, \\ C, & æ > æ_1+æ_0 \end{cases} \tag{9.113}$$

which has a maximum at $æ_0$; the parameter Δ characterizes the height of the maximum, while the parameters $æ_1$ and δ_1 characterize the width and steepness of the back slope, respectively. It is clear from (9.103) that a negative derivative of the distribution function ($\partial\Phi/\partial\mu < 0$) leads to wave damping; this may give some interesting effects during wave generation.

The case of a wide negative slope (i.e. large $æ_1$, and not too high values of δ_1) is not very interesting. It may be considered as the case with a monotonic source (9.112) and some additional effective damping term on the right-hand side of (8.41). Small values of Δ change the generation regime insignificantly. Extremely high values of Δ lead to very strong wave damping at high frequencies, so that generation is possible only in the low-frequency band:

$$\omega < \omega_{\max} = \frac{\omega_{BL}/\beta_*}{1 - æ_0} . \tag{9.114}$$

The most interesting case is of moderate Δ values and $æ_1 \sim æ_0$. Three examples for $\Delta = 0.5$ and different values of $æ_1$ are shown in Figs. 9.11–9.13; all the other parameters remain the same as in Fig. 9.9. The first example (Fig. 9.11), corresponding to a narrow peak in the source pitch-angle distribution, is quite similar to the case of a monotonic source (Fig. 9.9). The main difference is in the initial stage of spike generation, when a broader frequency band is excited. The period of spike generation is slightly longer than for the case of Fig. 9.9, which is explained by an effective increase in wave energy losses (see above discussion). With an increase in $æ_1$, the generation of low-frequency whistler-mode waves becomes more intense and takes a longer time, about two minutes (Fig. 9.12). The decrease in the maximum frequency of the generated waves and the increase in the generation period may be explained by an increase in effective wave losses. These generation regimes are qualitatively similar to the case of a monotonic source (9.112). But, for higher values of $æ_1$, the generation regime changes (Fig. 9.13). A persistent low frequency band, with a periodic modulation of its intensity and periodic spikes at higher frequencies, is generated in this case. The period and the maximum frequency of the spike are significantly smaller than for the case of smaller $æ_1$ values.

Now we can summarize the results obtained. Computer modelling of CM dynamics in the framework of the self-consistent quasi-linear (QL) theory reveals several different regimes of wave generation, including stationary generation, relaxation oscillations, and self-sustained oscillations of wave intensity. These regimes appear for both removal mechanisms for the energetic electrons, i.e. due to loss cone precipitation or due to magnetic drift through the generation region. The main parameter determining the type of wave generation in the former case of losses due to precipitation is the intensity of the energetic electron source J_0 (or, more exactly, the dimensionless intensity j_0). Relaxation oscillations of whistler-mode wave intensity take place for small J_0 values. The characteristics of the relaxation oscillations, such as period and quality factor, the constant shape of the wave spectrum generated and the pitch-angle distribution during a spike, are in a good agreement with those expected from an analytical model in the 'two-level' approximation, and considered in Section 9.1. A growth of J_0 results in a transition to self-sustained oscillations of wave intensity, with a notable growth of wave frequency during a spike. The apparent cause of the self-sustained oscillations is the involvement of newly resonant electrons in the instability. This result is in agreement with previous studies (Bespalov and Trakhtengerts, 1986b). In the case of drift losses, the shape of the energetic electron distribution function in the source becomes more significant for the generation regime. A new type of dynamic spectrum was obtained in this computer modelling, which demonstrated alternate spikes with almost constant and rapidly

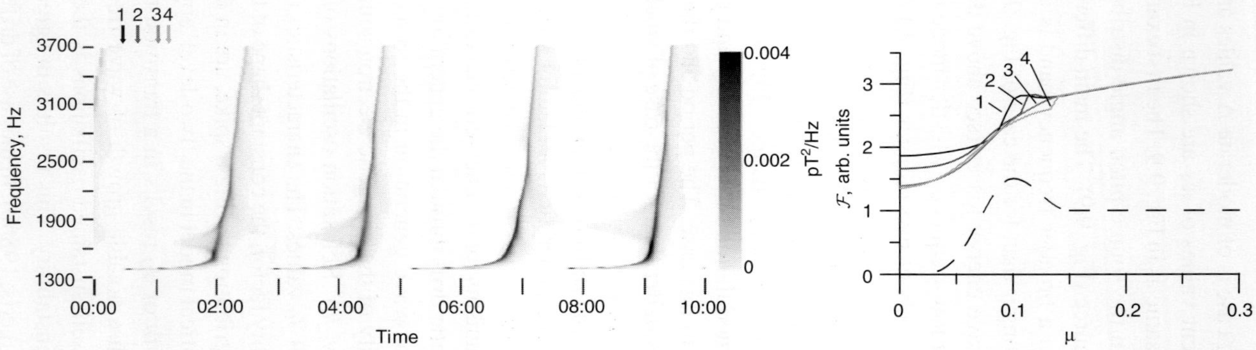

Figure 9.11 Periodic generation of wave intensity spikes in the case of losses due to electron drift and a non-monotonic pitch-angle distribution (Equation 9.113) of energetic electrons in the source; $\Delta = 0.5$, $\delta_1 = 1$, $æ_1 = 0.05$; all the other parameters are as in Fig. 9.9. (Taken from Pasmanik *et al.*, 2004a.)

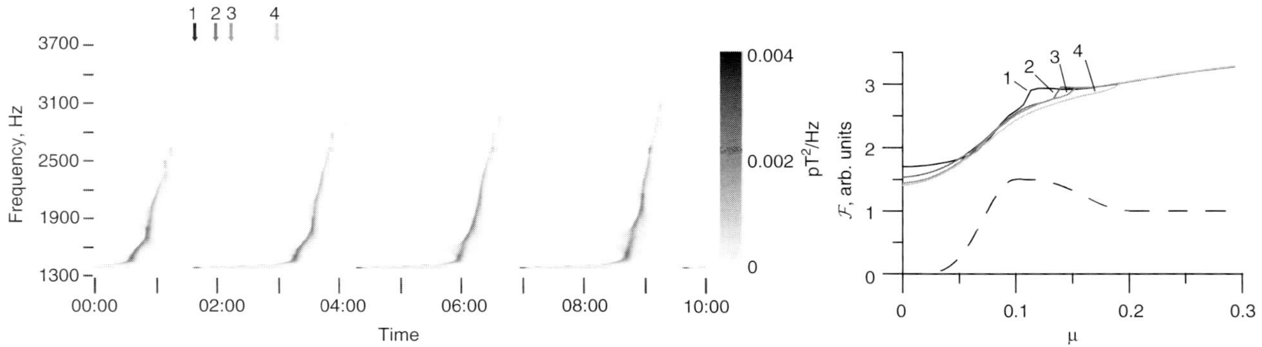

Figure 9.12 The same as Fig. 9.11, but for a wider peak in the source pitch-angle distribution: $æ_1 = 0.1$. More intense low-frequency waves are generated for longer at the beginning of the spikes. (Taken from Pasmanik *et al.*, 2004a.)

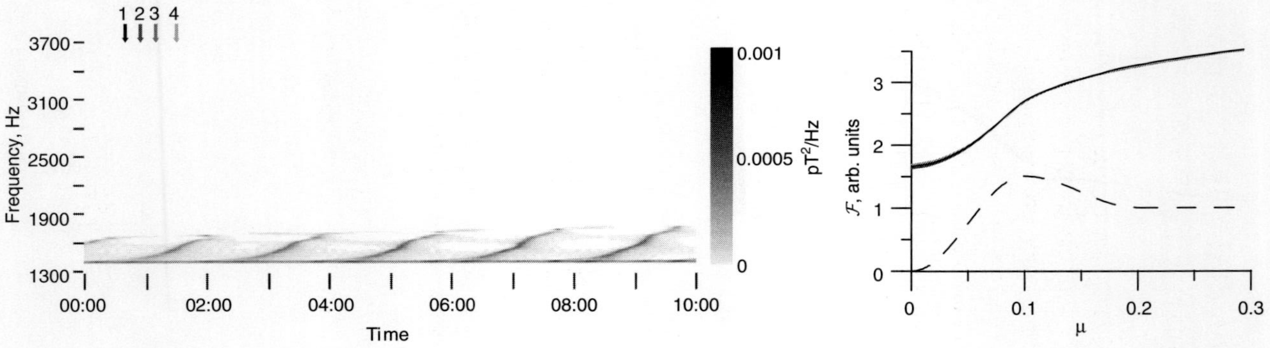

Figure 9.13 The same as in Figs. 9.11 and 9.12, but for an even wider peak in the source pitch-angle distribution: $æ_1 = 0.2$. A continuous low-frequency band is accompanied by periodic spikes generated at higher frequencies. (Taken from Pasmanik *et al.*, 2004a.)

rising frequencies (Fig. 9.10). We shall use the results obtained from these computer simulations in Chapter 10 to explain quantitatively some types of observed ELF/VLF quasi-periodic emissions.

9.4 Deformation of the distribution function during non-stationary QL relaxation

Real world sources of energetic charged particles in the magnetosphere supply the radiation belts (RB) with electrons and protons. As a rule, they have smooth anisotropic distribution functions over energy and pitch-angle. In such a situation as was shown in Chapter 3 and 8, noise-like electromagnetic emissions propagating in the whistler-mode are initially generated by the CI with a rather wide frequency spectrum. According to Fig. 9.14 and the inequalities (8.44) and (8.47), not all energetic electrons (or protons), supplied by a source, interact with whistler (Alfvén) waves via cyclotron resonance. The boundary in phase space between resonant and non-resonant electrons (Fig. 9.14) corresponds to the field-aligned velocity component at the magnetic equator:

$$|v_{\parallel L}|_{\mathrm{st}} = \frac{(\omega_{BL} - \omega_{\mathrm{m}})}{k(\omega_{\mathrm{m}})}, \tag{9.115}$$

where $\vec{k} \parallel \vec{B}_0$, and ω_{m} is the maximum frequency in the spectrum of whistler-mode waves. The region $|v_{\parallel L}| > |v_{\parallel L}|_{\mathrm{st}}$ corresponds to resonant particles, and $|v_{\parallel L}| < |v_{\parallel L}|_{\mathrm{st}}$ to non-resonant particles. It is clear that QL relaxation of cyclotron waves will lead to the isotropization of a distribution function of resonant particles over pitch-angle, and their subsequent precipitation into the loss cone.

As a result of this process, a step-like deformation will appear on the distribution function; this is qualitatively shown in Fig. 9.14, and accounts for the subscript 'st'. According to Chapters 3–5, such a deformation seems to be very important for the generation of discrete electromagnetic emissions, in particular ELF/VLF chorus

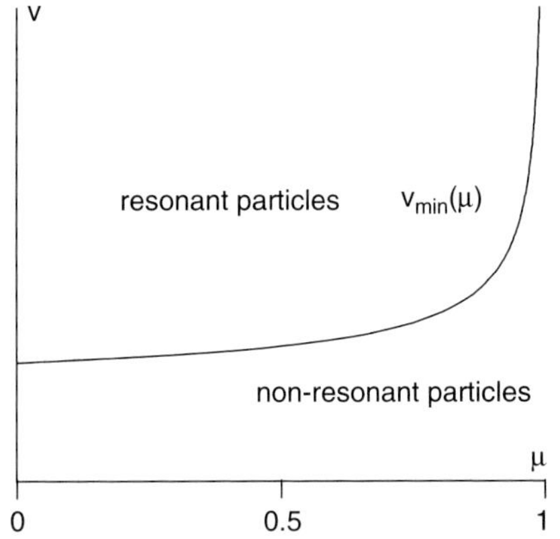

Figure 9.14 Boundary between the resonant and non-resonant region of energetic electrons in phase space.

emissions. It should be mentioned here that a step-like deformation of the distribution function of an electron beam was first discussed by Ivanov and Rudakov (1966) in an application to the beam-plasma instability problem (also see Ivanov, 1977).

Our case of the CI in an inhomogeneous magnetic field differs considerably from the case of this beam-plasma instability. Moreover, for applications to the real magnetospheric phenomenon of chorus generation, a model description of a step-like deformation similar to that of Ivanov (1977) is insufficient. Computer simulations of step formation were undertaken by Bogomolov *et al.* (1991) and Demekhov and Trakhtengerts (1994). More detailed investigations of the fine structure of a step corresponding to magnetospheric conditions were undertaken by Trakhtengerts *et al.* (1996).

The initial system of QL equations for computer simulations has been taken to be the same as in the previous section. These are the equations for the distribution function (9.98) and the wave energy transfer equation (8.41), together with relations (9.100)–(9.104). For the calculations the boundary conditions

$$\left(D \frac{\partial \Phi}{\partial \mu} \right)_{\mu=0;\, 1} = 0 \tag{9.116}$$

and the initial conditions

$$\Phi = 0, \quad \varepsilon = \varepsilon_0 \to 0 \tag{9.117}$$

have been used.

The computational analysis of the system of equations (9.98) and (8.41) for a wide range of values and functional dependencies of ω_0, $J(\mu)$, and $\nu(\xi)$ has demonstrated one very important common property. This is that the distribution function for cyclotron instability conditions evolves towards a function with a step-like deformation, which sharply increases the amplification of the waves. This evolution is somewhat similar to the formation of a shock in physical space, but in velocity space.

It is worth pointing out that the calculations have to be carried out with high accuracy because the step in velocity space is very sharp. This is illustrated in Fig. 9.15 where three examples are given, with different numbers of points (within the μ interval between 0 and 1) being used in the computational programs. Figure 9.15(a) shows the maximum amplification $\Gamma_{\rm m}$ in absolute units for the particular case described in the caption. The amplification $\Gamma_{\rm sm}$ for the smooth distribution in this case is $\Gamma_{\rm sm} \sim 4$ (near the threshold value). Figure 9.15(b) gives the ratio of the maximum averaged growth rate to the spectral width of the amplified line.

Figures 9.16 and 9.17 illustrate the temporal evolution of the cyclotron instability for two cases with different resolutions in the computer code. Electrons are accumulated up to the instability threshold (the linear growth of Γ evident in Fig. 9.15a). After that the CI arises, which leads to whistler wave generation and to the formation of the step-like deformation. This moves toward larger μ values, ‘switching on’ newly resonant particles. At first, a steepening of the deformation causes an increase of the amplification Γ over a narrow frequency range $\Delta\omega$. Accordingly,

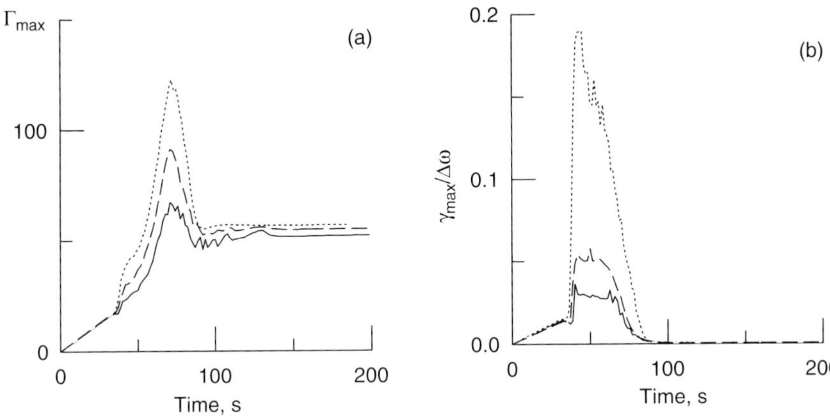

Figure 9.15 (a) The temporal evolution of Γ_m and (b) the ratio of $\gamma_m = \Gamma_m T_g^{-1}$ to the frequency bandwidth at half power (computational results). The following parameters were taken: $L = 6.03$, $\Gamma_{thr} = \nu T_g \simeq 4$, the cold plasma density $N_c \simeq 7\ \text{cm}^{-3}$, and the dimensionless source amplitude $\alpha \simeq 0.03$. The three different curves refer to the different number of points for x used in the computational code, 99, 198, and 592 for the solid, dashed, and dotted curves, respectively. The growth of Γ with increasing numbers of points testifies that the step width is less than the computational step scale. (Trakhtengerts *et al.*, 1996). (Copyright American Geophysical Union, reproduced with permission.)

the parameter $\gamma_{max}/\Delta\omega$, shown in Fig. 9.15(b), grows rapidly ($\gamma = \Gamma/T_g$). Subsequent movement of the step-like deformation towards $\mu \sim 1$ and the excitation of waves with increasing frequency, $\xi \approx \mu$ is accompanied by the decrease of the growth rate ($\gamma \propto (1-\xi)^{1/2}$; see Equation 9.103) and an increase of the frequency width $\Delta\omega = \omega_0 \Delta\xi/(1-\xi)^2$. These processes cause the decrease of the parameter $\gamma_{max}/\Delta\omega$ after $t \sim 50$ s. After $t \sim 10^2$ s, the instability saturates (Fig. 9.15a), the cyclotron amplification Γ approaching the attenuation νT_g. The saturation corresponds to a dynamical balance being established between the energy supply via the electron source J and its loss via wave damping ν.

At this final stage, the steep deformation of Φ is stabilized near $\mu \sim 1$, where the corresponding growth rate is close to the threshold, $\gamma \simeq \nu$. The chosen dimensional parameters correspond to the real situation in the magnetosphere. In particular, we have taken the value for the magnetic field for $L = 6$, and the cold plasma density in the equatorial plane $n_{cL} \simeq 7\ \text{cm}^{-3}$. A typical frequency ω_0 (Equation 9.101) has been taken to be $\omega_0 \simeq 0.25\omega_{BL}$, which corresponds to the mean electron energy $W_0 = mv_0^2/2 \sim 30$ keV. The particle source amplitude can be characterized by a dimensionless value α, which is the ratio of wave damping time ν^{-1} to the accumulation time t_{thr} up to the CI threshold: $t_{thr} \simeq \alpha^{-1}(2t_g/\Gamma_{thr})$. At time $t = t_{thr}$, the CI amplification Γ is equal to Γ_{thr}, which characterizes the wave energy losses in the cyclotron maser.

Under typical conditions, $\alpha \ll 1$ (in our case $\alpha \simeq 0.03$). Thus the results in Figs. 9.15–9.17 can serve as an illustration of the real process of CI development. The increase of Γ_m in the computational results, reaching the relative value $\sim 10^2$, is in good qualitative agreement with the analytical estimates of Section 3.3.5.

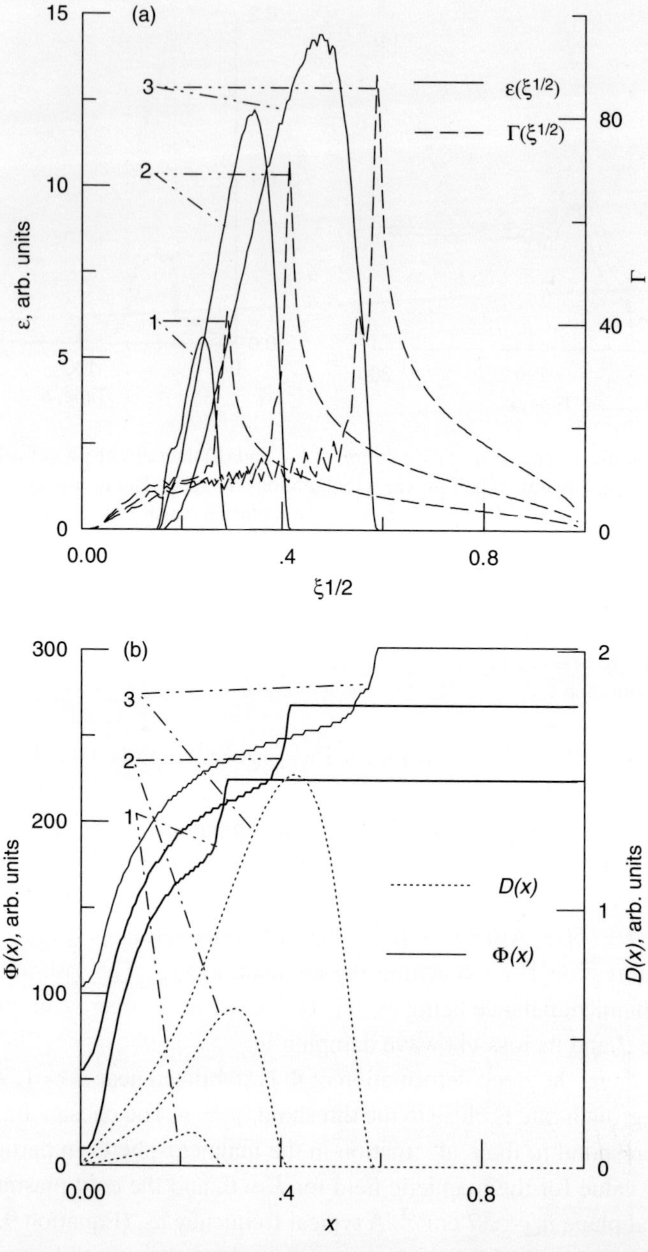

Figure 9.16 The dependencies of the averaged growth rate γ (dashed line) and the whistler wave energy density ε (solid line) on the dimensionless frequency $\xi = 1 - \omega_0/\omega$, and of the distribution function Φ (bold line) and the diffusion coefficient D (dotted line) on the sine of the equatorial pitch-angle, x, at different times; $n_x = 198$. Point 1 denotes $t = 52.5$ s; 2 denotes $t = 63.5$ s; and 3 denotes $t = 70$ s. The formation of the step and wavelet generation are seen, but the quantitative results shown here are not exactly correct, because the step width is less than the computational step (compare with Fig. 9.17). (Taken from Trakhtengerts *et al.*, 1996). (Copyright American Geophysical Union, reproduced with permission.)

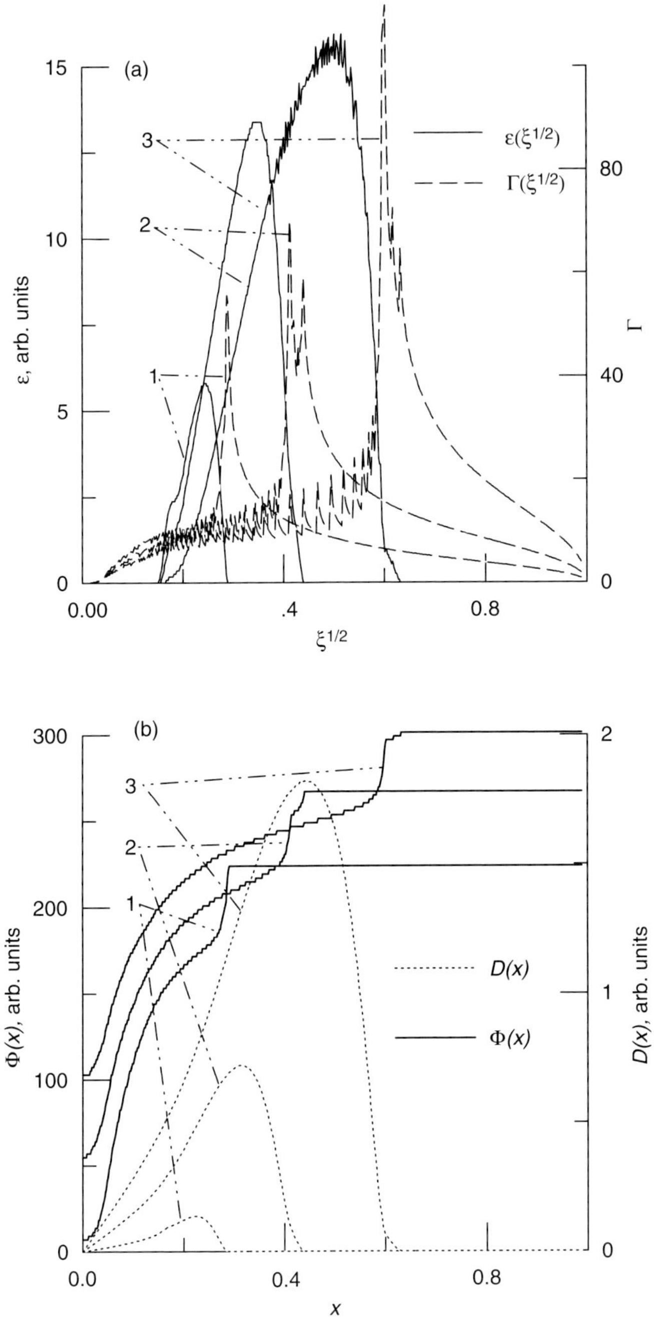

Figure 9.17 The same as Fig. 9.16, but for $n_x = 592$. The results coincide up to $t = 52.5$ s (curves labelled 1 on all plots). After that the difference grows progressively with time (curves 2–3), due to the steepening of the step in the distribution function. (Taken from Trakhtengerts *et al.*, 1996). (Copyright American Geophysical Union, reproduced with permission.)

Unfortunately, the accuracy of the calculations during the maximum phase of the instability is insufficient. This can be seen in Fig. 9.17(a) as the growth of fluctuations.

This self-consistent computational analysis of the cyclotron instability in the Earth's magnetosphere demonstrates a very important property of CI dynamics. This is the formation of a very sharp step in the distribution function of energetic electrons, dividing those electrons which are resonant with cyclotron waves from non-resonant electrons. Both the analytical results and the computations demonstrate the very large increase (by up to two orders of magnitude, ~ 40 dB) of the whistler wave amplification in the presence of such a step-like deformation.

Only the first steps have been taken in the direction of a self-consistent analysis of these effects. It is necessary to improve the accuracy of the calculations carried out in the framework of quasi-linear (QL) theory. However, in this framework, it is impossible to consider phase coherence effects. Therefore new computational codes are needed for a complete solution of this problem. The step-like deformation of the distribution function of energetic electrons is a transient phenomenon, and it will not be simple to observe it in energetic charged particle data from satellites. However, some recommendations can be made.

Usually, *in situ* measurements aboard a satellite are of energetic electron fluxes in some energy and pitch-angle interval. These data can be used for seeking step-like deformations. For that we write the distribution function with a step (3.74) as a function of energy W and pitch-angle θ using the variable μ. The distribution function with a step

$$\mathcal{F}(W, \mu) = \mathcal{F}_{\mathrm{sm}} \cdot \mathrm{He}\left(W_* - W + \frac{WB_L}{B}\, \mu \right) \tag{9.118}$$

depends on the coordinate along the magnetic field line. If the accuracy of particle measurements is sufficiently high (i.e. if the resolution interval over ΔW and $\Delta\mu$ is small), we can select the step to have any height. In Fig. 9.18, the position of the step is shown in the (μ, W) plane for two points in space: the first (1) corresponds to the equatorial plane of the magnetic flux tube where the magnetic field value is B_L, and the second (2) is for some distance away from the equator along the same magnetic flux tube. The shaded areas correspond to non-resonant electrons, the shading with a positive inclination is for a point s below the equatorial plane. At the equator the additional non-resonant electrons appear which are reflected higher than the selected point s (the shading with a negative inclination in Fig. 9.18).

Figures 9.19 and 9.20 present qualitative three-dimensional plots of a step-like feature at the equator and at a finite latitude, respectively. There is some maximum energy value for the second case, given by

$$W_{\mathrm{m}} = W_*(1 - B_L/B)^{-1} \tag{9.119}$$

which selects resonant and non-resonant electrons (Fig. 9.18). Figure 9.21 shows three types of pitch-angle dependence of the electron distribution function, corresponding to different values of the electron energy. In the energy interval

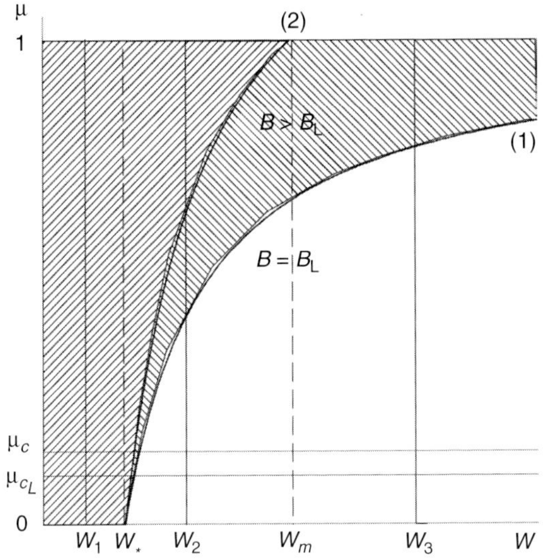

Figure 9.18 Diagram illustrating the boundary between resonant and non-resonant electrons for two points on a magnetic field line, namely, at the equator (1) and between the equator and the ionosphere (2). The shaded areas show corresponding regions in phase space for electrons which cannot resonate with whistler-mode waves. (Taken from Trakhtengerts *et al.*, 1996). (Copyright American Geophysical Union, reproduced with permission.)

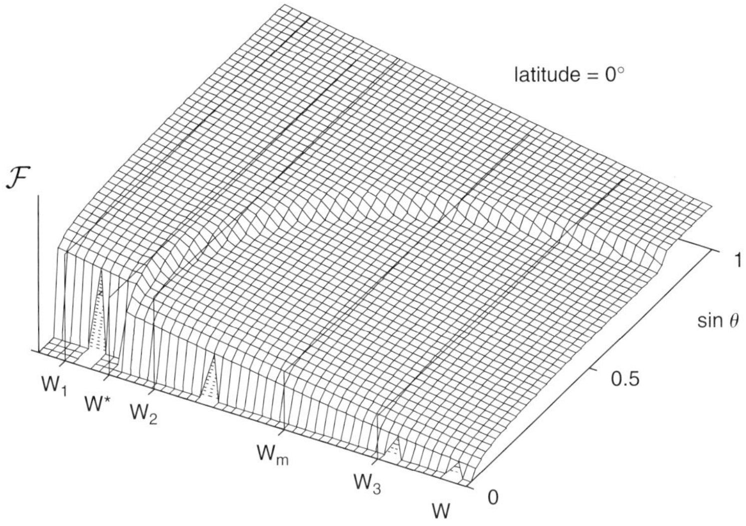

Figure 9.19 Qualitative three-dimensional representation of the distribution function $\mathcal{F}$ (logarithmic scale) with a step, plotted against μ, defined as $\sin^2\theta_L \equiv J_\perp B_L/W$, where θ_L is the pitch-angle at the equator, and $J_\perp$ is the first adiabatic invariant (Equation 2.12). (Taken from Trakhtengerts *et al.*, 1996). (Copyright American Geophysical Union, reproduced with permission.)

$$W_* < W_2 < W_{\mathrm{m}} \tag{9.120}$$

we should observe the step in the $\mathcal{F}$ pitch-angle dependence. According to (9.119) this interval is small for low-altitude satellites ($B_L/B \ll 1$) and goes to infinity on the magnetic equator.

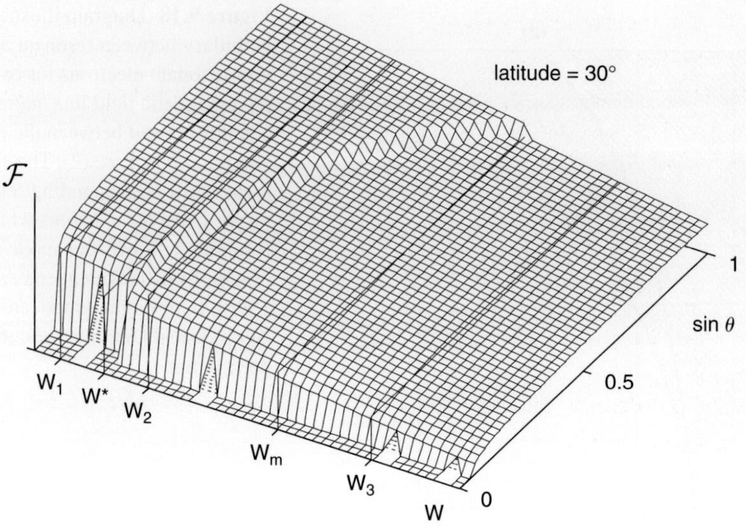

Figure 9.20 The same as Fig. 9.19, but at 30° latitude from the equator. This illustrates that electrons with energy W_2 and W_3 are pushed into the loss cone by interaction with the waves; thus a step is created for these energies at the equatorial plane. A step is not seen for W_3 at the chosen finite latitude. For W_1 no wave–particle interaction takes place. (Taken from Trakhtengerts *et al.*, 1996). (Copyright American Geophysical Union, reproduced with permission.)

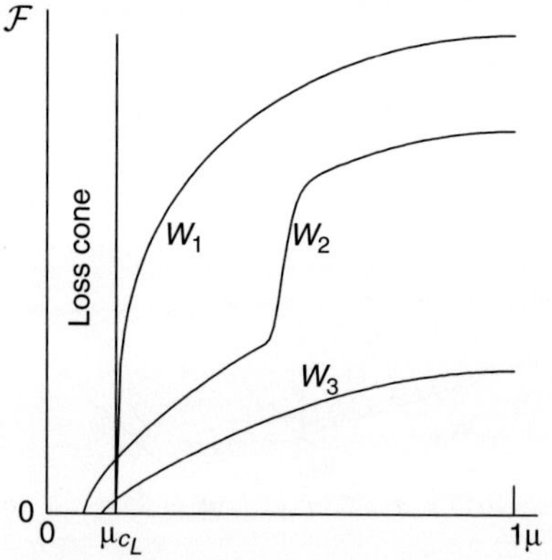

Figure 9.21 Cross-sections of the plot presented in Fig. 9.20 for several energy values. (Taken from Trakhtengerts *et al.*, 1996). (Copyright American Geophysical Union, reproduced with permission.)

A similar step appears in the energy spectrum for a fixed value of μ. It would be worthwhile to select electron (see Chapter 10) data during ELF/VLF hiss and chorus emission events which serve as indicators of CI development. Some new techniques are needed to analyse experimental results to investigate the interrelation between noise-like and discrete ELF/VLF emissions further.

9.5 Passive mode locking in CMs

In previous sections we have considered self-modulation of the wave intensity in CMs with periods $T_M \gg T_g$, the period of group propagation of a wave packet between ionospheric mirrors. Sometimes the nonlinear wave–particle interaction makes the wave emission as a short 'spike', a wave packet with a duration $T_{\rm spike} < T_{\rm g}$, which oscillates between the mirrors. In lasers, this is the well-known regime of passive mode synchronization or passive mode locking. Similar regimes are possible in CMs (see Bespalov, 1984).

We here consider this regime in more detail. We suppose that a short electromagnetic pulse of electric field $E_0(t)$ and duration $T_0 \ll T_{\rm g}$ arrives at the entrance of a CM. This field can be written as a Fourier integral

$$E_0(t) = \frac{1}{2\pi} \int_{-\infty}^{\infty} E_0(\omega)\, e^{-i\omega t}\, d\omega. \tag{9.121}$$

Following to Pereira and Stenflo (1977) and Bespalov (1984) we introduce a transmission coefficient

$$G(\omega) = G_1 + iG_2 \tag{9.122}$$

which characterizes the change of the pulse spectral amplitude after one pass through the magnetospheric resonator:

$$E_1(\omega) = e^G\, E_0(\omega).$$

For two-hop propagation

$$E_2 = e^{2G}\, E_0(\omega),$$

and for n-hop

$$E_n = e^{nG}\, E_0(\omega). \tag{9.123}$$

The total field for $n \to \infty$ can be written as the sum

$$\begin{aligned} E(t) &= \frac{1}{2\pi} \int_{-\infty}^{\infty} E_0(\omega)\,(1 + e^G + e^{2G} + \ldots)\, e^{-i\omega t}\, d\omega \\ &= \frac{1}{2\pi} \int_{-\infty}^{\infty} E_0(\omega)\, \frac{\exp(-i\omega t)}{1 - \exp(G)}\, d\omega. \end{aligned} \tag{9.124}$$

The last equality in (9.124) is obtained (under the condition that the total growth $G_1 < 0$) as the sum of a geometric series. The transmission coefficient G characterizes (in the case of a wave packet oscillating between the ionospheric mirrors) the change of the complex amplitude during the two-hop propagation, which includes the amplitude change due to the CI amplification and to the losses due to reflection from the conjugate ionospheres and the two-hop phase correction.

We consider the case when the one-hop amplification of waves is small, and the phase change is almost a linear function of the frequency. This means that after a long time the spectrum of the signal will be narrow, with a maximum at the frequency ω_0, corresponding to the maximum of $G_1(\omega)$, i.e. $(\partial G_1/\partial\omega)_{\omega_0} = 0$, and

$(\partial^2 G_1/\partial\omega^2)_{\omega_0} < 0$. After that the overall transmission coefficient can be written as

$$G = G_{10} + iG_{20} + iT_{\rm g}\Omega - \delta\Omega^2 \tag{9.125}$$

where

$$T_{\rm g} = \frac{\partial G_{20}}{\partial\omega}, \quad \delta = -\frac{1}{2}\left(\frac{\partial^2 G_{10}}{\partial\omega^2} + i\frac{\partial^2 G_{20}}{\partial\omega^2}\right), \quad \Omega = \omega - \omega_0, \tag{9.126}$$

and all values G_{10}, G_{20}, δ and $T_{\rm g}$ are taken at $\omega = \omega_0$. Using the inequalities

$$|G_{10}| \ll 1 \text{ and } |\delta|\,\Omega^2 \ll 1,$$

we obtain from (9.125) and (9.126) an equation for the envelope $\mathcal{E}$ of the whistler-mode wave, as

$$\begin{aligned}\mathcal{E}(t) &= E(t)\,e^{i\omega_0 t}\\ &= \frac{1}{2\pi}\int_{-\infty}^{\infty} E_0(\omega_0 + \Omega)\,\frac{\exp(-i\Omega t)\,d\Omega}{1 - (1 + G_{10} - \delta\Omega^2)\,e^{i(G_{20}+\Omega T_{\rm g})}}.\end{aligned} \tag{9.127}$$

In cases of our interest G_{10} is a nonlinear function of $\mathcal{E}(t)$, so that (9.127) is a nonlinear integral equation for $\mathcal{E}$. It can be written in the form of a differential equation:

$$\left(1 + \Gamma_0 + \delta\frac{\partial^2}{\partial t^2}\right)\mathcal{E}(t) - e^{-iG_{20}}\mathcal{E}(t + T_{\rm g}) = -e^{-iG_{20}}\,\mathcal{E}_0(t + T_{\rm g}) \tag{9.128}$$

where the envelope $\mathcal{E}_0(t)$ corresponds to the initial field $E_0(t)$. For $t/T_{\rm g} \gg 1$, $\mathcal{E}_0(t) \to 0$; it is possible to neglect the term on the right-hand side of (9.128). We shall be interested in the periodic solution of (9.128), when

$$\mathcal{E}(t + T_{\rm p}) \approx \mathcal{E}(t)\,e^{iG_{20}+i\psi} \tag{9.129}$$

where $T_{\rm p}$ is the period of pulse repetition, and ψ is an arbitrary constant. Further we take into account that in the case of our interest $T_{\rm p}$ is close to the period $T_{\rm g}$ of group propagation of a wave packet between the ionospheres, $T_{\rm g} = T_{\rm p} - \delta_{\rm p}$, where $|\delta_{\rm p}| \ll T_{\rm g}$. In this case

$$\mathcal{E}(t + T_{\rm g}) = \mathcal{E}(t + T_{\rm p}) - \delta_{\rm p}\left(\frac{\partial\mathcal{E}}{\partial t}\right)_{t+T_{\rm p}}. \tag{9.130}$$

Substituting (9.129), (9.130) into Equation (9.128) we obtain the equation describing the envelope form of a separate pulse (Bespalov, 1984):

$$\delta\frac{d^2\mathcal{E}}{dt^2} + \delta_p\ell^{i\psi}\frac{d\mathcal{E}}{dt} + (1 + G_{10} - \ell^{i\psi})\mathcal{E} = 0 \tag{9.131}$$

where δ is determined by (9.126).

We now clarify the meaning of the parameters δ and G_{10} as applied to CMs. The transmission coefficient G for a whistler-mode wave or an Alfvén wave can be formulated as follows:

$$G \equiv G_1 + iG_2 \equiv 2\Gamma + \ln(R_1 R_2) + i2\int_{-\ell/2}^{\ell/2} k\,dz - i2\int_{t_0}^{t} \omega\,dt' \qquad (9.132)$$

where, as earlier, Γ is the nonlinear one-hop cyclotron amplification, $R_{1,2}(\omega)$ are the reflection coefficients from the southern and northern ionospheres, and k is the wave vector. We obtain:

$$\left(\frac{\partial G_1}{\partial \omega}\right)_{\omega_0} \equiv \left.\frac{\partial\big(2\Gamma + \ln(R_1 R_2)\big)}{\partial \omega}\right|_{\omega=\omega_0} = 0, \quad \left(\frac{\partial G_2}{\partial \omega}\right)_{\omega_0} = T_g, \qquad (9.133)$$

$$\delta = \delta_{Re} + i\delta_{Im}, \quad \delta_{Re} = \frac{1}{2}\left(\frac{\partial^2 G_1}{\partial \omega^2}\right)_{\omega_0}, \quad \delta_{Im} = \frac{1}{2}\left(\frac{\partial T_g}{\partial \omega}\right)_{\omega_0}. \qquad (9.134)$$

Without loss of generality, we can take $\Gamma(\omega)$ and $R_{1,2}(\omega)$ in the form:

$$R_1(\omega) = R_2(\omega) = R_0 \exp\left\{-\frac{(\omega - \omega_R)^2}{\Delta_R^2}\right\}, \qquad (9.135)$$

$$\Gamma(\omega) = \Gamma_0 - \left(\frac{\omega - \omega_g}{\Delta_g}\right)^2, \qquad (9.136)$$

suggesting that ω_R and ω_g are close to each other when the conditions for instability are most favourable. In the real ionosphere the reflection coefficient (9.133) for whistler-mode waves has a smooth dependence on ω, thus $\Delta_R \gg \Delta_g$, whereas $\Delta_R \sim \Delta_g$ for Alfvén waves due to the presence of the Ionospheric Alfvén Resonator (IAR, Section 2.5). From (9.135) and (9.136) it is easy to find the frequency ω_0 and the total two-hop amplification $G_{10}(\omega_0)$:

$$\omega_0 = \frac{\omega_g \Delta_R^2 + \omega_R \Delta_g^2}{\Delta_R^2 + \Delta_g^2}, \qquad (9.137)$$

$$G_{10} = 2(\Gamma_0 + \ln R_0) - \frac{(\omega_R - \omega_g)^2}{\Delta_R^2 + \Delta_g^2}, \qquad (9.138)$$

where the subscript means the values for $\omega = \omega_0$, and the case $R_1 = R_2 = R_0$ is taken.

On the basis of relations (9.130)–(9.138), two spike-like generation regimes are analysed in the next subsections. The first regime is due to the nonlinearity of the cyclotron amplification coefficient Γ_0. The second regime is due to the nonlinearity of the ionospheric mirrors (the last term in relation (9.138)).

9.5.1 Mode locking due to the nonlinearity of Γ

According to Chapters 5, 6 and 8, the cyclotron amplification changes as the wave amplitude grows. This change is described, for noise-like emissions, by the effects of the quasi-linear deformation of the distribution function (the system of equations (8.34) and (8.41)) and leads to generation of different dynamic regimes of cyclotron instability (Sections 9.2–9.4). As was shown by Bespalov (1984), in the case of short pulses of whistler waves, $T_0 < T_p$, this dependence can be approximated by the relation

$$G_{10} = g + a\tau - b\tau^2 \tag{9.139}$$

where

$$g = 2\Gamma_0 + 2\ln R_0 \tag{9.139a}$$

is the initial amplification coefficient ($t \to -\infty$), a and b are constants, and τ is determined by the relation

$$\tau = \int_{-\infty}^{t} |\mathcal{E}|^2\, dt. \tag{9.140}$$

It is important that, for $a, b > 0$, Γ_0 depends non-monotonically on time. This is actually due to involving newly resonant electrons in the process of development of the cyclotron instability. As a result VLF waves are generated in the form of a succession of short pulses that reduces the losses in the medium and is going to be energetically advantageous. The solution of Equation (9.131) together with (9.138) and (9.139) for the case of $\Delta_R \gg \Delta_g$ was obtained by Bespalov (1984). It has been shown that the succession of short pulses is generated with a separate pulse in the form of the envelope soliton

$$E = E_p\left[\cosh\left(\frac{t}{t_p}\right)\right]^{i\alpha-1} \exp\left\{i\left(-\omega_0 + \frac{\beta}{t_p}\right)t\right\}. \tag{9.141}$$

The time delay between two neighbouring pulses is close to T_g, and the mean frequency of the electromagnetic radiation varies within a pulse as

$$\langle\omega\rangle = \omega_0 - \frac{\beta}{t_p} - \left(\frac{\alpha}{t_p}\right)\tanh\left(\frac{t}{t_p}\right). \tag{9.142}$$

All characteristics of a pulse sequence, the amplitude E_p, the characteristic time scale t_p, the constants α and β, are known functions of three parameters δ_{Im}/δ_{Re}, g, b/a^2. For example (Bespalov, 1984), in the case of a weak frequency dispersion $\delta_{Im} \to 0$, α and β go to zero, and the approximate relations are valid:

$$E_p \approx \frac{a}{3b}\left(\frac{b}{2\delta_{Re}}\right)^{1/4}, \quad t_p \approx -\frac{a}{g}\left(\frac{2\delta_{Re}}{b}\right)^{1/2}, \quad \alpha \approx \frac{2\delta_{Im}}{3\delta_{Re}} \tag{9.143}$$

where $g < 0$ is determined by (9.139a), and we find from (9.133)–(9.136):

$$\delta_{\mathrm{Re}} = \Delta_{\mathrm{R}}^{-2} + \Delta_{\mathrm{g}}^{-2}. \tag{9.144}$$

It should be borne in mind that the necessary conditions for a pulsed regime

$$a > 0 \text{ and } b > 0 \tag{9.145}$$

are fulfilled for certain types of the distribution functions of energetic particles (see Bespalov (1984) for more details).

9.5.2 Mode locking due to the nonlinearity of the reflection coefficient R

According to Equation (9.138), the nonlinearity of $G_{10}(\omega_0)$ can also be connected with the nonlinear change of the ionospheric reflection coefficient. This can be due to the energetic charged particle precipitation into the ionosphere under the action of the generated cyclotron waves. This case is especially important for Alfvén waves, generated in the proton CM, because the reflection coefficient of these waves from the ionosphere is a strongly non-monotonic function of frequency. This is explained by the presence of the Ionospheric Alfvén Resonator (IAR, see Chapter 2 and Fig. 2.8 for details).

We investigate the regime of weak pitch-angle diffusion when the anisotropy of energetic protons is relatively small and can be considered as constant (the two-level approach). Moreover, the energy of the energetic protons does not vary significantly during the course of wave generation if $\omega \ll \Omega_B$. Under these conditions Γ_0 and ω_0 are not changed in the process of development of the cyclotron instability because the particle losses from the radiation belts are compensated by the action of a weak particle source. At the same time, the precipitating energetic particle flux (connected with pitch-angle diffusion) can modify the ionospheric reflection parameters R_0, Δ_{R} and ω_{R} considerably. According to the numerical calculations of Ostapenko and Polyakov (1990), the most remarkable change due to energetic proton precipitation (i.e. due to small variations of ionospheric electron density and conductivity) during morning and evening hours occurs for the parameter ω_{R}. Thus we take only the change of ω_{R} into account in (9.138), assuming all other parameters (Γ_0, R_0, ω_{g}, Δ_{R}) to be constant. It follows from IAR theory (Belyaev *et al.*, 1990) that ω_{R} depends mainly on the F-layer electron density. A simple formula for ω_{R} for the first IAR eigenmode can be written as:

$$\omega_{\mathrm{R}} = \frac{\pi v_{\mathrm{A}}}{(L+h)}, \tag{9.146}$$

where $v_{\mathrm{A}} \propto n_F^{-1/2}$ is the Alfvén velocity in the F-layer maximum ($F_{\max}$), n_F is the corresponding value of n, h is the F-layer thickness, and L is the characteristic scale of the electron density decrease above $F_{\max}$. The precipitating energetic proton flux modifies both n_F and h; both parameters increase. So we have

$$\Delta\omega_{\mathrm{R}} \approx \frac{\pi v_{\mathrm{A}}}{L+h}\left(\frac{\Delta v_{\mathrm{A}}}{v_{\mathrm{A}}} - \frac{\Delta(L+h)}{L+h}\right) \approx -\omega_{\mathrm{R}}\left(\frac{\Delta n_\Sigma}{n_\Sigma} + \frac{\Delta n_F}{2n_F}\right) < 0, \tag{9.147}$$

where $n_\Sigma = \int n\, dz$ is the integral electron content in the F-layer in a column of unit cross-section.

Now we examine the expression for G_{10} (9.138). We suggest that the main effect of the energetic proton precipitation is to change the total electron content per unit area in the ionosphere: $\Delta\omega_{\mathrm{R}} \simeq -(\Delta n_\Sigma/n_\Sigma)\,\omega_{\mathrm{R0}} \ll \omega_{\mathrm{R0}}$. Here we have neglected the second term on the right-hand side of (9.147). In fact this means that we have assumed that all changes in ω_{R} are due to variations of $(L+h)$. If we had assumed that only n_F varies due to energetic proton precipitation, that would not lead to qualitative changes of the dynamics. For $\Delta n_\Sigma^\pm$ (the index $\pm$ here refers to the ionospheres in the $\pm$ directions, respectively) we have the equation:

$$\frac{d\Delta n_\Sigma^\pm}{dt} = \eta\, S_{\mathrm{A}}^\pm - \Delta n_\Sigma^\pm/\tau_{\mathrm{R}}, \tag{9.148}$$

where $\eta = W_0/\,|e|\,U_0$ is the number of electron–ion pairs created by one energetic proton with characteristic energy W_0, $U_0 \sim 20\,\mathrm{V}$ is the typical ionization potential of atmospheric atoms and molecules, τ_{R} is the recombination life time, and $S_{\mathrm{A}}^\pm$ is the precipitated energetic proton flux density to the ionospheres in the $\pm$ directions, which is proportional to the energy density $\mathcal{E}^\pm$ of the Alfvén waves propagating to the conjugate hemispheres.

According to Polyakov *et al.* (1983),

$$\eta\, S_{\mathrm{A}}^\pm(t) = \frac{\pi e^2}{\Omega_{BL}}\,\frac{N_{\mathrm{h}}\,\eta}{Mc^2}\left(B_{\mathrm{p}}^\pm\right)^2 = q\left(B_{\mathrm{p}}^\mp\right)^2, \tag{9.149}$$

where we have introduced the coefficient $q = (\pi e^2\eta/Mc^2)N_{\mathrm{h}}/\Omega_{BL}$, $N_{\mathrm{h}} \approx \sigma\ell\, n_{\mathrm{hL}}$ is the number of energetic protons in the magnetic flux tube with unit cross-section at the ionospheric level, σ is the mirror ratio, and n_{hL} is the equatorial density of energetic protons. For the case of weak pitch-angle diffusion considered here, the solution of (9.148) can be written in integral form as:

$$\Delta n_\Sigma^\pm(t) = q\exp(-t/\tau_{\mathrm{R}})\int_{t_0}^{t}\left[B_{\mathrm{p}}^\mp(t'-\tau_{\mathrm{D}})\right]^2\exp(t'/\tau_{\mathrm{R}})\,dt', \tag{9.150}$$

where we have taken into account the time delay $\tau_{\mathrm{D}} = \frac{3}{4}T_{\mathrm{g}} - \frac{T_B}{4} \simeq \frac{3}{4}T_{\mathrm{g}}$ between the cyclotron interaction near the equator and the corresponding modification of the reflection coefficient; $T_B/4 \simeq \ell/(2v_\parallel)$ is the time required for precipitated energetic protons to reach the ionosphere from the equatorial region of the magnetic flux tube.

We shall find the solution of the nonlinear equation (9.131) in the particular case when $\delta_p \to 0$ and $\psi \to 0$. In this case Equation (9.131) reduces to the simpler form:

$$(\delta_{Re} + i\delta_{Im})\frac{d^2\mathcal{E}}{dt^2} + G_{10}\mathcal{E} = 0. \tag{9.151}$$

Separating the modulus and the phase of the slowly varying electric field $\mathcal{E} = A\exp(i\Theta)$, Equation (9.151) is transformed into the set of two equations for A and Θ with real coefficients:

$$|\delta|^2 A'' + \delta_{Re} G_{10} A - |\delta|^2 A\left(\Theta'\right)^2 = 0 \tag{9.152}$$

$$\Theta'' + 2\frac{A'}{A}\Theta' - G_{10}\frac{\delta_{Im}}{|\delta|^2} = 0 \tag{9.153}$$

where the prime denotes the time derivative. From the last equation we have:

$$\Theta' = \frac{\delta_{Im}}{A^2}\int_{t_0}^{t}\left(\frac{G_{10}}{|\delta|^2}\right)A^2(t')\,dt'. \tag{9.154}$$

For the wave magnetic field amplitude $B_\sim = B_p\exp\{i\Theta\}$ corresponding to the electric field amplitude $\mathcal{E}$ we have the same Equation (9.151) and Equations (9.152) and (9.153) for B_p and Θ, respectively. To obtain an analytical solution of (9.152) we consider the case when the pulse duration $T_0 \ll \tau_D,\ \tau_R$. If we take into account these inequalities and the periodicity $B_p(t) = B_p(t \pm T_g)$, the relation (9.150) can be rewritten in the form:

$$\Delta n_\Sigma^\pm \simeq q\exp\left[\frac{-(t+\tau_D)}{\tau_R}\right]\int_{-\infty}^{\infty}\left[B_p^\mp(t')\right]^2 dt' \tag{9.155}$$

where τ_R is the recombination lifetime.

In fact, (9.155) means that the pulse is reflected from the ionosphere at the relaxation stage when ionization processes due to the previous pulse have finished. Now we can write the expression for G_{10} in the simple form:

$$G_{10} = 2g - (\tau - \tau_0)^2\frac{2\Delta\omega_{R0}^2}{\Delta_R^2 + \Delta_g^2} \tag{9.156}$$

where $\tau = t/\tau_R$, $g = \Gamma_0 + \ln R_0 > 0$, $\Gamma_0 = \int_{-\ell/2}^{\ell/2}\gamma_L\,dz$,

$$\Delta\omega_{R0} = \frac{q\,\omega_{R0}}{n_{\Sigma 0}}\exp(-\tau_D/\tau_R)\int_{-\infty}^{\infty}\left[B_p^\mp(t')\right]^2 dt' \tag{9.157}$$

where $\Delta\omega_{R0}$ is the maximum nonlinear frequency shift of ω_R due to proton precipitation, and the time corresponding to $G_{10\,max}$ is

$$\tau_0 = \frac{\omega_{R0} - \omega_g - \Delta\omega_{R0}}{\Delta\omega_{R0}}. \tag{9.158}$$

To obtain (9.156), the Taylor series expansion of $\exp(-t/\tau_R)$ has been applied together with the inequality $t/\tau_R \lesssim T_0/\tau_R \ll 1$.

For the case $\Delta\omega_{R0}/\omega_{R0} \ll 1$ which will be considered below it is sufficient to use, in the expression (9.154) for Θ', only the linear term in t. This term is equal to:

$$\Theta' \simeq \frac{\delta_{\mathrm{Im}}\, g\, \tau_R}{|\delta|^2}(\tau - \tau_0) + O(\tau - \tau_0)^3, \tag{9.159}$$

where $\tau = \tau_0$ corresponds to the maximum of B_p^2. After substituting (9.156) and (9.159) into (9.152) we obtain:

$$\frac{d^2 B_p}{du^2} + \left(a - C^2 u^2\right) B_p = 0, \tag{9.160}$$

where $u = \tau - \tau_0$,

$$a = \frac{\tau_R^2\, \delta_{\mathrm{Re}}\, g}{|\delta|^2}, \tag{9.161}$$

$$C^2 = \frac{\tau_R^4\, \delta_{\mathrm{Im}}^2\, g^2}{|\delta|^4} + \frac{\Delta\omega_{R0}^2\, \delta_{\mathrm{Re}}\, \tau_R^2}{|\delta|^2\left(\Delta_R^2 + \Delta_g^2\right)}. \tag{9.162}$$

We discuss some general properties of the solution to (9.160). The pulse-like solution which goes to zero at $|u| \to \infty$ can be written in the form (Kamke, 1959):

$$B_p(u) = B_{pm}\, p(u) \exp(-Cu^2/2), \tag{9.163}$$

where B_{pm} is the maximum value of B_p; $p(u) > 0$ depends on the ratio a/C and determines the internal structure of the wave packet. For example, $p(u) = 1$ in the case that $a/C = 1$.

From (9.163), the pulse duration is equal to

$$T_0 = \tau_R\, C^{-1/2}. \tag{9.164}$$

The frequency drift inside the packet is determined by the function $\omega = \omega_0 - \Theta'$ and, according to (9.159), is equal to:

$$\omega = \omega_0 + \frac{|\delta_{\mathrm{Im}}|\, g}{|\delta|^2}(t - t_0), \quad g > 0. \tag{9.165}$$

The stability condition of the solution (9.163), which follows from (9.160), can be written as

$$\int_{-\infty}^{\infty} B_{\mathrm{p}}^2 u \, du = 0. \tag{9.166}$$

The relations (9.164) and (9.166) together, with (9.157) and (9.161)–(9.163) being accounted for, determine T_0 and B_{m}. In the particular case when $a = C$, (9.166) is satisfied automatically, and we have (if $\delta_{\mathrm{Re}} \gtrsim |\delta_{\mathrm{Im}}|$):

$$T_0 \simeq \left(\frac{|\delta|^2}{\delta_{\mathrm{Re}} \, \Gamma_0} \right)^{1/2}, \tag{9.167}$$

$$B_{\mathrm{pm}} \simeq \frac{\Delta\omega_{\mathrm{R}0}}{\omega_{\mathrm{R}0}} \left(\frac{n_{\Sigma 0}}{q T_0} \right)^{1/2} \exp\left(\frac{\tau_{\mathrm{D}}}{2\tau_{\mathrm{R}}} \right), \tag{9.168}$$

$$\frac{\Delta\omega_{\mathrm{R}0}}{\omega_{\mathrm{R}0}} \simeq \frac{\Delta_{\mathrm{R}} \Delta_{\mathrm{g}}}{\omega_{\mathrm{R}0}} \Gamma_0 \, \tau_{\mathrm{R}} \left(\frac{\delta_{\mathrm{Re}}^2 - \delta_{\mathrm{Im}}^2}{|\delta|^2} \right)^{1/2}. \tag{9.169}$$

We note that this analytical theory provides only a steady periodic solution in the form of a wave pulse oscillating between the ionospheres. There can be pulses with different shapes, depending on the relationship between the parameters a and C. Which of them actually occurs should be the subject of a more detailed study involving numerical computations. We choose here the case $a = C$ because it corresponds to the simplest form of the pulse. In this case, according to (9.167), the pulse duration T_0 does not depend on the wave amplitude.

9.6 Summary

This rather long chapter self-consistently explores some special cases of the feedback between energetic charged particles and waves in the near-Earth space environment. Interactions between both energetic electrons and whistler-mode waves and energetic protons and magnetohydrodynamic waves are considered. Energetic charged particle precipitation from the radiation belts modifies the ionospheric plasma density (see 9.10), which changes the amount of wave energy reflected which, in turn changes the gyroresonant process and the precipitation flux: this constitutes a feedback system having several time constants.

We first investigate a slowly changing situation, with weak pitch-angle diffusion and a narrow spectrum of waves at the frequency of maximum growth rate. This situation is analogous to the two-level approximation approach used in the theory of quantum generators (lasers). A self-consistent treatment shows that, for extremely low-frequency whistler-mode waves, the system of equations is given by (9.12). With a source of charged particles, in the equilibrium state the feedback system exhibits damped (so-called relaxation) oscillations whose frequency is the square root of twice the one-hop whistler-mode growth divided by the product of the one-hop whistler-mode propagation time and the electron lifetime – a fraction of 1 Hz. When the temporal modulation of the ionospheric reflection coefficient is included, these relaxation oscillations can be unstable.

An analysis of (9.19) far from the equilibrium state shows that the nonlinear feedback system exhibits relaxation oscillations with 'spikes' (Fig. 9.2), with the repetition frequency being given by (9.35) and the spike duration by (9.36). If this repetition frequency equals the frequency of an applied fast magnetoacoustic wave, the resulting resonance causes the deep modulation of both the wave intensity and the flux of precipitating energetic particles, which are much deeper than the modulation of the flux of trapped particles (see 9.40).

Section 9.2 considers magnetohydrodynamic (MHD) waves which are generated during this process which causes relaxation oscillations of the radiation belt parameters. The plasma pressure change during a MHD wave propagating along a duct of enhanced cold plasma density adds to that due to the cyclotron instability, see (9.48) and (9.49). Under magnetospheric conditions the most important MHD mode is the Alfvén wave mode whose wavelength along the geomagnetic field is comparable with the magnetic field curvature. For the self-modulational instability due to such Alfvén waves (Section 9.2.2), the theory of the effect of Alfvén waves on the whistler-mode turbulence is treated self-consistently with energetic electrons drifting longitudinally into the interaction region. This positive feedback leads to deep modulation of both the wave intensity and the flux of precipitated electrons. When the frequency of the relaxation oscillations is close to the resonant frequency of Alfvén waves along a magnetospheric flux tube (9.81), the growth rate of the modulational instability is large, and given by (9.85).

Section 9.2.3 discusses the complementary nonlinear situation of an Alfvén wave in gyroresonance with energetic protons when that gyroresonant frequency equals the frequency at which the reflection coefficient from the ionosphere is close to a maximum. If both frequencies change in the same proportion, explosive growth occurs, leading to spike generation (see 9.91).

Computer simulations of these relaxation oscillations for more realistic magnetospheric situations are the subject of Section 9.3. The energetic particle sources may be due to longitudinal drift into the interaction region close to the magnetic equatorial plane or to local acceleration (e.g. due to magnetic compression). Losses may be due to precipitation or to particle drift out of the interaction region. For ducted whistler-mode propagation, energetic electrons having different number densities or different pitch-angle distributions lead to significantly different frequency spectra and different ELF/VLF wave generation regimes (Figs. 9.9–9.13).

Figure 9.14 illustrates the boundary in phase space between resonant and non-resonant particles; quasi-linear relaxation of cyclotron waves leads to the formation of a step-like feature in the distribution function, which increases wave amplification markedly ($\sim$ 40 dB). Figures 9.15–9.17 show the simulation results at $L = 6$

beyond (outside) the plasmapause. Figure 9.19 plots the change of distribution function expected on the magnetic equator, and Fig. 9.20 that at a magnetic latitude of 30°.

Spike generation, dealt with in more detail in Section 9.5, is similar to passive mode locking in a laser (Khanin, 1995). The nonlinear cyclotron amplification (9.139) leads to generation of a quasi-monochromatic pulse, with duration less than T_g, whose envelope is determined by (9.141). Passive mode locking may alternatively be produced by the nonlinear change of the ionospheric reflection coefficient. That is especially important for Alfvén wave–proton interactions because the reflection coefficient varies strongly with frequency, being due to the Ionospheric Alfvén Resonator. A pulse envelope and dynamic frequency spectrum in this case determined by (9.165) and (9.167)–(9.169) can be applied for the quantitative interpretation of hydromagnetic Pc1 pulsations (see Chapter 12).

Chapter 10

ELF/VLF noise-like emissions and electrons in the Earth's radiation belts

10.1 Radiation belt formation for magnetically quiet and weakly disturbed conditions

Satellite data reveal many different types of plasma waves observed in near-Earth space. These waves play important roles in the transport of charged particles across L shells, as well as local acceleration and plasma heating, and energetic particle precipitation from the magnetosphere. Experimental data show that the most important waves, which regulate the population of the electron radiation belts (ERB), are whistler-mode electromagnetic waves in the ELF/VLF range. These waves arise in the magnetosphere from thunderstorm activity and via electron CM generation. The role of ELF/VLF waves in ERB formation has been investigated for a long time; see the review by Bespalov and Trakhtengerts (1986b) and books by Schulz and Lanzerotti (1974), Lyons and Williams (1984), Schulz (1991), and Hultqvist *et al.* (1999), and references therein. Their main contribution is pitch-angle scattering which can be decisive among the loss mechanisms for radiation belt (RB) electrons during magnetically disturbed periods. The most suitable approach for the description of pitch-angle scattering in the case of noise-like ELF/VLF emissions is quasi-linear theory, which has been developed in Chapters 8 and 9; see also Bespalov and Trakhtengerts (1986b) and Schulz (1991). For a comparison with the experimental data on RB particles and waves, the equations of QL theory should include all the real sources and sinks of RB electrons and ELF/VLF waves. The complete solution of this problem for a description of the real RB dynamics is rather complicated.

Two approaches have been used to date. The first is based on the stationary solution of the kinetic equation for RB electrons with pitch-angle scattering by ELF/VLF waves being taken into account, where the spectrum and intensity of waves are taken from satellite data (Lyons and Williams, 1984) and the source of

RB electrons is supplied by the radial diffusion across L shells under the action of ULF electromagnetic waves breaking the third adiabatic invariant. In such a formulation the problem was solved by Kennel *et al.* (1972), Lyons and Thorne (1973) and Bourdarie *et al.* (1996). Together with the basic cyclotron resonance, they took into consideration oblique whistler waves with a wave vector $\vec{k} \nparallel \vec{B}$; they also took account of the resonant interaction of RB electrons with ELF/VLF waves due to the Cherenkov mechanism and higher cyclotron resonances ($s = 2, 3, 4$). The results reproduce reasonably the basic features of the outer zone electron fluxes, including the 'slot' region between the inner and outer radiation belts. Later, pitch-angle diffusion was supplemented by energy diffusion in connection with the problem of the acceleration of RB electrons up to MeV energies. A close correlation was revealed between relativistic electron enhancements in the RB and intense chorus emissions (Meredith *et al.*, 2002, 2003). This stimulated the incorporation of chorus emissions into a scheme for ERB formation on the basis of the diffusion equation, including electron pitch-angle scattering, stochastic acceleration and radial diffusion (Varottsou *et al.*, 2005).

It has to be borne in mind that these results portray electron RB formation in the non-self-consistent approach, with the intensity and the spectrum of ELF/VLF emission being given. Moreover, the motion of the RB electrons has been averaged over their magnetic drift period around the Earth. This means that the treatment is valid only for processes which are changing in time more slowly than over several hours, that is only for magnetically quiet or weakly disturbed conditions. During a magnetic storm, there are many localized (in space and time) wave–particle interaction events, which demand a self-consistent analysis of whistler-mode wave generation and RB dynamics. Just such an approach was developed in Chapters 8 and 9. We shall use this approach further not only for the self-consistent consideration of RB formation but also for the explanation of particular experimental data on ELF/VLF wave generation and electron RB dynamics.

10.2 Radiation belt formation under the joint action of radial diffusion and pitch-angle diffusion

A theory for RB formation for such conditions includes energetic particle transport across L shells under the action of magnetic and electric waves breaking the third adiabatic invariant and different mechanisms of particle losses. Inside the plasmasphere, charged particle loss occurs via pitch-angle scattering due to resonant interactions with waves and to Coulomb collisions. A quantitative description of the time varying distribution function is based on the diffusion equation in phase space, which can be written in the following form (Beutier *et al.*, 1995)

$$\frac{\partial \mathcal{F}}{\partial t} = \sum_{i=1}^{3} \sum_{j=1}^{3} \frac{\partial}{\partial J_i} D_{ij} \frac{\partial \mathcal{F}}{\partial J_j} + \text{sources} \tag{10.1}$$

where the variables J_j refer to three adiabatic invariants. In the non-relativistic approach these invariants (see also Walt, 1994) can be written as

$$\begin{aligned} J_1 &\equiv M = \frac{WL^3}{B_0} && \text{— the magnetic moment,} \\ J_2 &= 2LR_0\,Y(\theta_L)\,mv && \text{— the second (bounce) invariant,} \\ J_3 &= \frac{2\pi R_0^2 B_0}{L} && \text{— the magnetic flux (drift) invariant.} \end{aligned} \tag{10.1a}$$

Here B_0 is the magnetic field strength at the equator on the Earth's surface; R_0 is the Earth's radius.

The diffusion equation is valid when the processes are much slower than the charged particle magnetic drift period T_{MD} (Lyons and Williams, 1984). For electron energies ~ 100 keV and $L \sim 3$, T_{MD} is ~ 2.5 hour. For the case of our interest, with cross L-transport and pitch-angle diffusion being taken into account, the equation is rather difficult to solve because it includes the diffusion terms over all three invariants. Different approximations have been discussed by Kennel *et al.* (1972), Lyons and Thorne (1973), Beutier and Bosher (1995) and Bourdarie *et al.* (1996). In particular the pitch-angle diffusion has been simplified by the loss term $-\mathcal{F}/\tau_L$, where the lifetime τ_L includes the energy density of ELF/VLF waves taken from experimental data. Even such a rather rough approach permitted some important features of RB formation to be explained, in particular the slot in the distribution of RB electrons. Below we consider a self-consistent approach to RB formation which takes account of the possible accumulation of ELF/VLF waves inside the plasmasphere during the development of the cyclotron instability. This approach gives a fuller picture of RB formation.

As has been shown, the cyclotron instability is most effective inside the plasmasphere (Lemaire and Gringauz, 1998) where the majority of RB electrons are involved in cyclotron interactions. The transport of RB particles across L shells promotes CI development because the transverse anisotropy α increases, in particular, for any point in the magnetosphere where the magnetic field near the equatorial plane B is

$$B = B_L\left[1 + (z/a)^{2\nu}\right] \tag{10.2a}$$

and the equatorial magnetic field $B_L \sim L^{-3}$, $a \sim L^{+1}$ we have for the pitch-angle anisotropy factor

$$\alpha \equiv \frac{v_{\perp L}^2}{v_{\parallel L}^2} \sim L^{-\frac{1}{2(\nu+1)}}. \tag{10.2b}$$

In the case of a parabolic approximation of the geomagnetic field with distance z from the magnetic equator $\nu = 1$ and $\alpha \sim L^{-1/2}$. Further, we suggest that the distribution function of RB electrons is given at the plasmapause with some initial anisotropy $\alpha_0 > 1$. We shall then find $\mathcal{F}$ inside the plasmasphere taking into account cross L-transport and the effects of the cyclotron instability. In a self-consistent approach we should add to Equation (10.1) the wave energy transfer equation (8.41) which can be written, in its simplest form, as the balance of wave amplification and losses (for $R_1 = R_2$):

$$\Gamma_{\max} = 2|\ln R|, \tag{10.3}$$

R being the reflection coefficient of whistler-mode waves at the ionosphere. To be interested in qualitative effects we shall neglect the growth of α (10.2b) due to the cross L-transport of RB electrons, suggesting that $v_{\parallel L}$ is changing similarly to $v_{\perp L}$, which is possible when the initial value of α_0 is large.

In this case it is convenient to use new variables $\left(u = \dfrac{v^2}{B_L},\ æ \equiv \mu^{1/2} = \sin\theta_L\right)$ instead of $(M,\ J_2)$ and to write Equation (10.1) in its simplest form:

$$\frac{\partial \mathcal{F}}{\partial t} = L^2 \frac{\partial}{\partial L}\left(L^{-2} D_{LL} \frac{\partial \mathcal{F}}{\partial L}\right) + \frac{\partial}{\partial æ}\left(D_{ææ} \frac{\partial \mathcal{F}}{\partial æ}\right) \tag{10.4}$$

where the operator on L has its traditional form with the diffusion coefficient D_{LL} equal to the sum of magnetic (M) and electric (E) pulse contributions:

$$D_{LL} = D_{LL}^{M} + D_{LL}^{E} \tag{10.5}$$

where, according to Schulz (1991),

$$D_{LL}^{M} \simeq 10^{-13} L^{10} a_M(æ)\ (\mathrm{s}^{-1}) \tag{10.6}$$

and

$$D_{LL}^{E} \approx \frac{10^{-14} L^{10} a_E(æ)}{\tilde{u}^2\left[1 + (\Omega_{\mathrm{d}}\tau_0)^{-2}\right]}\ (\mathrm{s}^{-1}), \tag{10.7}$$

$$\tilde{u} = \left(u/6 \cdot 10^2 c^2\ \mathrm{gauss}\right). \tag{10.8}$$

In (10.6)–(10.8) $a_{M,E}(æ)$ are known functions of æ; further we take $a_{M,E} \approx \mathrm{const}(æ)$, $\Omega_{\mathrm{d}} = 2\pi/T_{\mathrm{d}}$ is the magnetic drift rate, and $\tau_0 \simeq 1200\,\mathrm{s}$ is the characteristic decay time for electrostatic pulses (Lyons and Williams, 1984). Taking into account the general expression for the amplification Γ (8.43), we can write Γ in the following form:

$$\begin{aligned}\Gamma_{\mathrm{m}} &= \frac{(2\pi)^3 e}{c B_L} \int_{\omega_{BL}/k_{\mathrm{m}}}^{\infty} \int_{\mu_{\mathrm{c}}}^{\mu_{\mathrm{m}}} l_{\mathrm{eff}}\, \mu \frac{\partial \mathcal{F}}{\partial \mu} v^3\, dv\, d\mu \\ &= \frac{4\pi^3 e B_L}{c} \int_{u_{\min}}^{\infty} u\, du \int_{\mu_{\mathrm{c}}}^{\mu_{\mathrm{m}}} \mu \frac{\partial \mathcal{F}}{\partial \mu} l_{\mathrm{eff}}\, d\mu.\end{aligned} \tag{10.9}$$

The subscript 'm' indicates the maximum value and the subscript 'c' corresponds to the loss cone boundary. The operator for pitch-angle diffusion can be written according to (8.36) as

$$\frac{1}{T_B}\frac{\partial}{\partial\mu}\mu D\frac{\partial F}{\partial\mu}\approx\left(\frac{2\pi e}{mc}\right)^2\frac{\bar{l}_{\rm eff}}{a\omega_{BL}}\frac{\partial}{\partial æ}\left(æ\mathcal{E}\frac{\partial F}{\partial æ}\right) \tag{10.10}$$

where we put $k_{\rm m}\simeq\omega_{BL}/v$, $T_B\simeq(a/v)æ^{-1}$, $\bar{l}_{\rm eff}$ is taken out of the integral for some middle point, and $\mathcal{E}\simeq\int\mathcal{E}_\omega\,d\omega$. We find from (10.10) in the case that $\bar{l}_{\rm eff}/a\sim 1$:

$$D_{ææ}=æL^3D_0,\quad D_0=\left(\frac{2\pi e}{mc}\right)^2\frac{\mathcal{E}(L)}{\omega_{B0}}, \tag{10.11}$$

where ω_{B0} is the electron gyrofrequency at the equator at the Earth's surface.

In the stationary case, $\partial\mathcal{F}/\partial t=0$, we have the self-consistent system:

$$L^2\frac{\partial}{\partial L}\left(L^{-2}D_{LL}\frac{\partial\mathcal{F}}{\partial L}\right)+D_0L^3\frac{\partial}{\partial æ}\left(æ\frac{\partial\mathcal{F}}{\partial æ}\right)=0 \tag{10.12}$$

$$\Gamma_{\rm m}\equiv\frac{4\pi^3eB_1}{c}\bar{l}_{\rm eff}\int_{u_{\rm min}}^{\infty}u\,du\int_{æ_c}^{æ_m}æ^2\frac{\partial\mathcal{F}}{\partial æ}\,dæ=|\ln R| \tag{10.9a}$$

with the following boundary conditions:

$$\mathcal{F}\Big|_{L=1}=0,\quad \mathcal{F}\Big|_{L=L_0}=\mathcal{F}_0(u,æ) \tag{10.13}$$

$$\mathcal{F}\Big|_{æ=æ_c}=0,\quad \frac{\partial\mathcal{F}}{\partial æ}\Big|_{æ=æ_m=1}=0. \tag{10.14}$$

In (10.13) L_0 is the L-value of the plasmapause ($L_0\sim 6$), where the cyclotron instability is switched on, and RB particles are lost in the atmosphere due to collisions.

The case of weak pitch-angle diffusion is considered, so the distribution function is zero at the edge of the loss cone $æ=æ_c$. In the case of our interest, $æ_c\ll 1$, it is possible to look for the solution to (10.12) as was done in Chapter 9 in the form of the product (see (9.4) and (9.14)):

$$\mathcal{F}=\Phi(L,u)\,Z_1(æ) \tag{10.15}$$

where the pitch-angle dependence of $Z_1(æ)$ is given by the expression (9.17). For simplicity we keep only the term Z_1 with the smallest eigenvalue. Using the solution (9.15)–(9.17) for Z_1 and substituting it into (10.9a) we obtain the following result:

$$\int_0^\infty\Phi u\,du=\Lambda L^2 \tag{10.16}$$

where

$$\Lambda \approx |\ln R|\, \pi^3 R_0 \omega_{B0}. \tag{10.17}$$

We suggest further that the energy (W) spectrum of energetic charged particles at the plasmapause has the following form:

$$\Phi_0(u) = C/(W_0 + u)^n, \quad n > 2 \tag{10.18}$$

with the normalization constant

$$C = \frac{W_0^{2-n}}{(n-1)(n-2)} \tag{10.19}$$

and where W_0 is a characteristic energy of the distribution of energetic electrons.

The next step is to find Φ from Equation (10.12) taking into account (9.14). We have two equations for $\Phi(L)$ and $D_0(L)$, that is (10.16) and (10.12), which can be written, using (10.15), as

$$\frac{1}{L\, D_0(L)} \frac{d}{dL} \left(L^{-2} D_{LL} \frac{d\Phi}{dL} \right) = p_1^2 \Phi \tag{10.20}$$

$$\Phi\Big|_{L=1} = 0, \quad \Phi\Big|_{L=L_0} = \Phi_0(u) \tag{10.21}$$

where, according to (9.16), $p_1 \approx 0.15\pi$, and D_{LL} is determined by (10.5)–(10.8). We consider the case when electric pulses provide the main contribution to D_{LL}. In this case D_{LL} can be written as

$$D_{LL} \simeq D_{LL}^E = \frac{10^{-9} \cdot L^{10}}{10^5 \tilde{u}^2 + L^4} \; (\mathrm{s}^{-1}) \tag{10.22}$$

where in accordance with (10.8)

$$\tilde{u} = \left[\frac{W\,(\mathrm{keV})}{3 \cdot 10^5\ \mathrm{gauss}} \right] L^3. \tag{10.23}$$

It is easy to obtain the analytical solution of (10.20) and (10.16) for two limiting cases, of low and high energy RB electrons. A transition (subscript 'tr') between these cases corresponds to the equality (see (10.22)):

$$10^5 \tilde{u}^2 \sim L^4 \quad \text{or} \quad W_{\mathrm{tr}}(\mathrm{keV}) \sim 10^3 L^{-1}, \tag{10.24}$$

with $\tilde{u}^2$ being given by (10.23). For low energies $W(\mathrm{keV}) < 10^3 L^{-1}$ the diffusion coefficient

$$D_{LL}^{(1)} \sim 10^{-9} L^6 \; (\mathrm{s}^{-1}) \tag{10.25}$$

does not depend on $\tilde{u}$. Taking into account that the number of low-energy particles considerably exceeds the number of high-energy particles, we can find from (10.16) and (10.20) the solutions:

$$D_0(L) \simeq \frac{L}{p_1^2} \cdot 10^{-8} \simeq 5 \cdot 10^{-8} L \ (\mathrm{s}^{-1}) \tag{10.26}$$

and

$$\Phi(u, L) \simeq \Phi_0(u) \frac{L^2 - L^{-5}}{L_0^2 - L_0^{-5}} \approx \Phi_0(u) \frac{L^2}{L_0^2} \tag{10.27}$$

where Φ_0 is determined by the relation (10.18).

For high energies $W(\mathrm{keV}) > 10^3 L^{-1}$, the formation of the electron radiation belt is determined by the same equation (10.20), but the diffusion coefficient D_{LL} differs from (10.25); it is equal to:

$$D_{LL}^{(1)} \simeq 10^{-14} \left(L^{10}/\tilde{u}^2\right) \left(\mathrm{s}^{-1}\right). \tag{10.28}$$

The high-energy electrons contribute only a little to wave generation, so we can use for D_0 in (10.20) the expression (10.26). After substituting (10.26) and (10.28) into (10.20) we obtain for the high-energy electrons:

$$\frac{d}{dL}\left(L^8 \frac{d\Phi}{dL}\right) - 8 \cdot 10^5 \tilde{u}^2 L^2 \Phi = 0. \tag{10.29}$$

This equation should be solved together with the boundary conditions (10.21). The general solution of (10.29) can be written using Bessel functions of imaginary argument in the following form:

$$\Phi = \left[C_1\, I_{7/4}(y) + C_2\, I_{-7/4}(y)\right] L^{-7/2} \tag{10.30}$$

where

$$y \equiv 5 \cdot 10^2 \tilde{u} L^{-2} \gtrsim 1 \tag{10.31}$$

corresponds to the transition from low-energy to high-energy electrons, and coincides with the criterion (10.24). For $y > 1$ we obtain with (10.21) being taken into account:

$$\Phi \approx \Phi_0(u) \left(\frac{L_0}{L}\right)^{5/2} \frac{\exp\{-y\} - \exp\{y - 2y_{\max}\}}{\exp\{-y_0\} - \exp\{y_0 - 2y_{\max}\}} \tag{10.32}$$

where

$$y_0 = 5 \cdot 10^2 \tilde{u} L_0^{-2} \ \text{and} \ y_{\max} = 5 \cdot 10^2 \tilde{u}. \tag{10.33}$$

It is easy to obtain $\Phi(W, L)$ from (10.32), substituting $\tilde{u}$ as a function of W and L according to (10.23).

From (10.32) we observe that $\Phi(W, L)$ has a slot, which is shown in Fig. 10.1. A slot appears in the distribution of high-energy electrons with $W > W_{\mathrm{tr}}$ (10.24) and weakly depends on the initial distribution function, in particular, W_0 and n (see (10.18)). Calculation of the quantitative parameters of a slot demands a more exact treatment, but the qualitative properties of the slot are in a good agreement with experimental data (Lyons and Williams, 1984); $L_{\mathrm{sl}} \sim 3$ to 4, and decreases as the energy increases. In our self-consistent model the formation of the slot is a general feature of the stationary RB, when the radial diffusion and pitch-angle diffusion are determined by (10.7) and (10.11), respectively. It is interesting to note that the energy where the slot appears depends only on the temporal properties of electric pulses (the value of τ_0 in (10.7)) and does not depend on their intensity. In particular, the filling of the slot during a magnetically disturbed period, shown

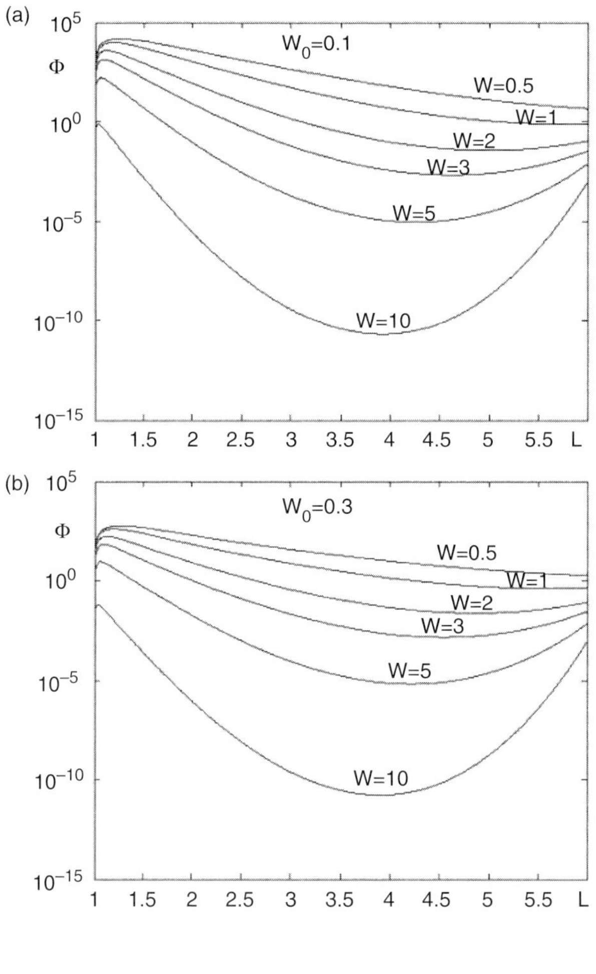

Figure 10.1 Dependence of the distribution function of RB electrons of different energies (in MeV) on L-value: (a) $W_0 = 0.1$ MeV; (b) $W_0 = 0.3$ MeV. A slot is formed in the region $L_{\mathrm{sl}} = 4$.

by Lyons and Williams (1984) may be explained by the decrease of τ_0 during this period.

Using the solution (10.26) it is possible to estimate the intensity and L-dependence of the wave energy flux density $\mathcal{E}$. We have from (10.11) and (10.26):

$$\mathcal{E}(L) \sim L \text{ and } B_{\sim} \approx 10^{-2}\text{nT at } L \simeq 3, \tag{10.34}$$

where $B_{\sim}$ is the wave amplitude. These values are close to those observed experimentally (Lyons and Williams, 1984). Taking into account the wave optimal frequency $\omega \simeq \omega_{BL}/\beta_*$ (3.73), we can estimate for a typical dependence of the cold plasma density on L:

$$\omega \sim L^{-(2 \text{ to } 3)}. \tag{10.35}$$

Using the formula for the integral flux density of RB electrons (see Bespalov and Trakhtengerts, 1986b) on a particular L shell

$$S_L(W, L) \simeq \frac{13.5\, B_0^2}{L^6} \int_{\left(\frac{L}{L_0}\right)^3 \left(\frac{W}{W_0}\right)}^{\infty} \Phi u \, du \tag{10.36}$$

we can find $S_L^*(L)$, corresponding to the RB stationary state. Substituting (10.16) into (10.36) we find (since the lower limit in the integral goes to zero)

$$S_L^*(L) \simeq 13.5 \Lambda B_0^2 / L^4 \tag{10.37}$$

where Λ is determined by (10.17). The L^{-4} dependence (10.37) is in good agreement with experimental data discussed by Kennel and Petschek (1966). We conclude that the self-consistent approach to the problem of radiation belt formation gives very important additional information on the physical processes acting, and is verified using satellite data. Of course, our analytical approach is incomplete as it does not take into account other ELF/VLF waves inside the plasmasphere, such as from lightning discharges and additional resonances (Cherenkov and higher order cyclotron resonances), which can be important at lower energies. We leave considerations of these matters to future investigators.

10.3 A quantitative model for cyclotron wave–particle interactions at the plasmapause

We have previously considered conditions of weak magnetic activity, when energetic electron sources filled the radiation belts over several hours, and the plasmasphere was stationary and quasi-symmetric (i.e. independent of local time). These conditions are broken during a magnetic storm. The beginning of a magnetic storm manifests itself by the pronounced erosion of the plasmasphere and in strong substorm associated energetic charged particle injections which are accompanied by very local and non-stationary precipitation. At this time filaments of detached cold plasma may appear at practically any L shell.

It is clear that under such conditions the problem of cyclotron wave–particle interaction should be formulated as a local injection of energetic electrons, mainly in the night time sector, followed by their magnetic drift through the plasmasphere or into detached cold plasma regions, where the cyclotron interaction is switched on by the increased cold plasma density. Such a formulation of the problem is confirmed by NOAA satellite data, which demonstrate very local RB electron precipitation. Here we apply the theoretical models developed in Chapters 8 and 9 to the quantitative explanation of RB electron parameters, obtained by NOAA satellites.

In the case of time stationary precipitation the self-consistent system of quasi-linear equations can be written in the following form:

$$\Omega_{\mathrm{D}} \frac{\partial \mathcal{F}}{\partial \varphi} = \frac{1}{T_{\mathrm{B}}} \frac{\partial}{\partial \mu} \mu D \frac{\partial \mathcal{F}}{\partial \mu} - \delta \mathcal{F} \tag{10.38}$$

$$\vec{v}_{\mathrm{g}\perp} \frac{\partial \mathcal{E}_\omega}{\partial \vec{r}_\perp} = 2(\gamma - \nu)\, \mathcal{E}_\omega \tag{10.39}$$

where the left-hand term in (10.38) describes the longitudinal magnetic drift, φ is the azimuthal angle, Ω_{D} is the angular drift velocity, T_{B} is the electron bounce period $T_B = \oint dz/v_\parallel$, $v_\parallel$ is the electron velocity component parallel to the Earth's magnetic field, z is the coordinate along a magnetic field line, the quantity $\mu = \sin^2\theta_L$, and θ_L is the pitch-angle in the equatorial plane for a certain L value; the last term in (10.38) characterizes the particle losses via pitch-angle diffusion through the loss cone, with the coefficient δ being equal to

$$\delta = \begin{cases} 0, & \mu \geq \mu_{\mathrm{c}}; \\ \delta_0 = v/l, & 0 \leq \mu \leq \mu_{\mathrm{c}} \end{cases} \tag{10.40}$$

where $\mu_{\mathrm{c}} = L^{-3}(4 - 3L^{-1})^{-1/2}$ indicates the edge of the loss cone for a dipolar magnetic field, v is the electron velocity, and l is the length of the magnetic flux tube (between conjugate ionospheres).

We can see that, in comparison with the previous case in Section 10.2, the particle radial diffusion is changed by the longitudinal magnetic drift; the term $\delta\mathcal{F}$ takes into account the case of weak diffusion (WD) as well as strong diffusion (SD), when the distribution function $\mathcal{F}$ becomes more isotropic, as is seen in the NOAA satellite data. A new term $\vec{v}_{\mathrm{g}\perp} \dfrac{\partial \mathcal{E}_\omega}{\partial \vec{r}_\perp}$ is presented in the wave energy transfer equation. It characterizes the refraction of the whistler-mode waves generated and is important for a description of the geometrical form of the RB electron precipitation pattern. The example of wave ray trajectories taken from Lyons and Williams (1984) is shown

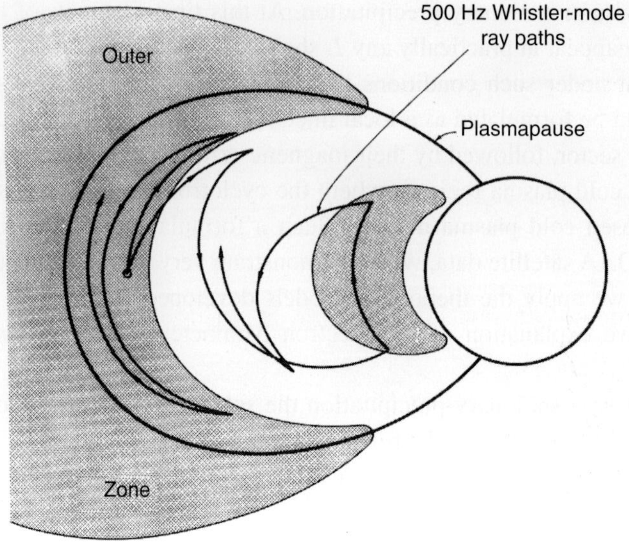

Figure 10.2 Examples of 500 Hz whistler-mode ray paths obtained using a Stanford ray-tracing program (Kimura, 1966), which illustrate how wave energy generated at a point on the magnetic equator in the outer region of the plasmasphere can propagate across field lines so as to fill the plasmasphere with waves (from Lyons *et al.*, 1972). (Copyright American Geophysical Union, reproduced with permission.)

in Fig. 10.2. According to this diagram the component $v_{g\perp}$ can reach a considerable value (tens of per cent) from a full value of group velocity v_g. Further, we consider a one-dimensional problem and put $dr_\perp = R_0 L\, d\varphi$. The electron distribution function is normalized by the relation:

$$N = \pi\mu_c^{-1} \int_0^\infty \int_0^1 T_B \mathcal{F} v^3 \, d\mu \, dv, \tag{10.41}$$

where N is the number of electrons in the magnetic flux tube at a certain L value with unit cross-section in the ionosphere.

The self-consistent system of equations (10.38) and (10.39) must be completed by the initial (on the φ-axis) and boundary (over μ) conditions, which have the form:

$$\begin{aligned} &\varphi = \varphi_0, \qquad \mathcal{E}_\omega = \mathcal{E}_{\omega 0}, \quad F = F_0(\mu), \\ &\mu = 0 \text{ and } 1, \quad \mu \frac{\partial \mathcal{F}}{\partial \mu} = 0. \end{aligned} \tag{10.42}$$

Here φ_0 is the longitude (local time) where eastward drifting (at velocity v_D) energetic electrons meet the sharp cold plasma density enhancement; the boundary conditions over μ are that the flux due to diffusion is equal to zero at the limits of μ.

The solution of (10.38) and (10.39) has many common features with those discussed in Section 10.2. At first we note that the growth (γ) and damping (ν) rates in (10.39) are connected with the amplification coefficient Γ and the effective reflection coefficient $R < 1$ by the relations:

$$2\gamma = \frac{\Gamma}{T_g}, \quad \nu = \frac{|\ln R|}{T_g} \tag{10.43}$$

where the two-hop group period $T_g = 2\int_{-l/2}^{l/2} dz/v_g$. The diffusion coefficient D is determined by the same relation (8.36) as in Section 10.2, and is equal to (10.11) (in the variables $æ = \mu^{1/2}$), when the cold plasma density is sufficiently high. Integrating Equation (10.38) over ω it is possible to obtain the equation for the diffusion coefficient D, which is equivalent to the relation (8.36). Finally we obtain the simplified system of equations for $\mathcal{F}$ and D, which was analysed by Pasmanik *et al.* (1998) to explain the properties of fluxes of trapped and precipitated electrons, measured by a NOAA satellite (see Titova *et al.*, 1998) at a height $\sim 10^3$ km in the evening sector. They found specific localized precipitation at a feature which they termed a 'cliff'. The parameters of the 'cliff' include the fractional increase of trapped electron flux (q), the width of the front of the 'cliff' ($\Delta\varphi_F$), and the characteristic relaxation width ($\Delta\varphi_{rel}$). These parameters are shown in Fig. 10.3 for the cases of strong (S) and weak (W) diffusion. A 'cliff' is formed at the sharp density gradient of the cold plasma region. The increase of trapped flux density S_{tr} is due to pitch-angle isotropization.

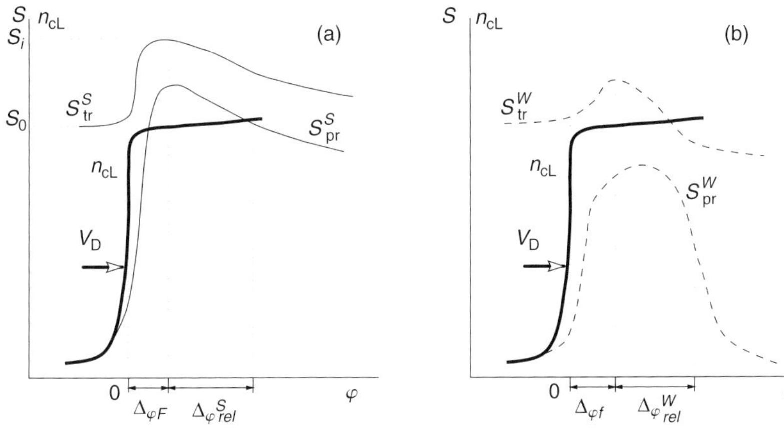

Figure 10.3 Qualitative diagrams of the energetic electron precipitation pattern with longitude φ (or magnetic local time) during a magnetic storm in cases of (a) strong and (b) weak diffusion. S_{tr} and S_{pr} are the fluxes of the trapped and precipitated energetic electrons as measured by a low-altitude satellite; φ is the longitudinal coordinate along the energetic electron drift velocity V_D; n_{cL} is the cold plasma density in the equatorial plane at a certain L-value. Superscripts S and W refer to the strong and weak pitch-angle diffusion regimes, respectively. (Taken from Pasmanik *et al.*, 1998.)

In the case of strong diffusion this increase can be written as

$$q^{\mathrm{s}} = \frac{S^{\mathrm{s}}_{\mathrm{tri}}(z)}{S_{\mathrm{tr0}}(z)} = \left(\frac{1}{\tilde{B}} - \mu_{\mathrm{c}}\right)^{-r}, \quad \tilde{B} = \frac{B(z)}{B_L} \tag{10.44}$$

where the initial distribution has been taken to be of the form:

$$\mathcal{F}_0(\mu) = \begin{cases} (1+r)\, S_{0L}\, (\mu - \mu_{\mathrm{c}})^r, & \mu > \mu_{\mathrm{c}} \\ 0 & \mu < \mu_{\mathrm{c}}. \end{cases} \tag{10.45}$$

The indices 'i' and '0' in the subscripts to S_{tri} and S_{tr0} define the isotropic (i) and initial (o) fluxes; S_{0L} is the initial flux in the equatorial plane. For a weak diffusion regime $1 < q^{\mathrm{w}} < q^{\mathrm{s}}$. According to Pasmanik *et al.* (1998) for strong diffusion

$$\Delta\varphi^{\mathrm{s}}_{\mathrm{F}} \sim \left(\frac{v_{\mathrm{g}\perp}}{L R_0 \gamma_{\mathrm{m}}}\right) \ln\left(\frac{D^{\mathrm{s}}_{\mathrm{m}}}{D_0}\right), \tag{10.46}$$

where γ_{m} is the growth rate at the entrance to the interaction region near the equatorial plane, D_0 and $D^{\mathrm{s}}_{\mathrm{m}}$ are the initial and maximum values of the diffusion coefficient; a typical value for $\ln(D^{\mathrm{s}}_{\mathrm{m}}/D_0)$ is ~ 10. The characteristic relaxation length is

$$\Delta\varphi^{\mathrm{s}}_{\mathrm{rel}} \sim \frac{\Omega_{\mathrm{d}}}{\delta_0 \mu_{\mathrm{c}}} \tag{10.47}$$

where δ_0 is $2/T_B$ (8.35). In the case of weak diffusion, $\Delta\varphi^{\mathrm{W}}_{\mathrm{F}}$ is determined by the same relation (10.46) with the corresponding values γ_{m} and $D^{\mathrm{w}}_{\mathrm{m}}$, but $\Delta\varphi^{\mathrm{w}}_{\mathrm{rel}}$ is more complicated and larger than (10.47) (see Pasmanik *et al.* 1998). Putting, for example, $L = 4.5$, $\gamma_{\mathrm{m}} \sim 15\ \mathrm{s}^{-1}$ (this corresponds to the trapped electron flux $S_{\mathrm{tr}} \sim 10^8\ \mathrm{cm}^{-2}\mathrm{s}^{-1}$ at the equator), and $v_{\mathrm{g}\perp} \sim 10^2$ km/s, we find that $\Delta\varphi^{\mathrm{s}}_{\mathrm{F}} \approx 0.12°$ and $\Delta\varphi^{\mathrm{s}}_{\mathrm{rel}} \sim 0.25°$. These values are in good agreement with the values obtained from the satellite observations (Titova *et al.*, 1998).

Together with the analytical estimates, a computational analysis has been undertaken for the initial distribution function (10.45), with $r = 1$. Examples of the fluxes of trapped and precipitated electrons are given in Figs. 10.4 and 10.5 as functions of magnetic longitude φ, for the strong diffusion (SD) and weak diffusion (WD) regimes, respectively. It is seen that, for the parameters chosen, the characteristic scales of the growth and decay stages are comparable in the WD case, and that the WD decay stage is longer than for the SD case. Also, in the SD case, the trapped and precipitated fluxes are comparable; however, for the WD case, $S_{\mathrm{tr}} \gg S_{\mathrm{pr}}$.

The dependencies of the trapped electron fluxes on $\tilde{B}(z) = B(z)/B_L$ for different values of φ are given in Figs. 10.6 and 10.7. The first curve in these figures ($\varphi_1 = 0$) corresponds to the initial distribution function. The second curve in Fig. 10.6 demonstrates the deformation of the energetic particle distribution during the formation of the cliff ($\varphi_2 \approx 0.07°$), the third curve corresponds to the maximum value of the trapped electron flux at low heights ($\varphi_3 \approx 0.1°$), and the last curve presents the

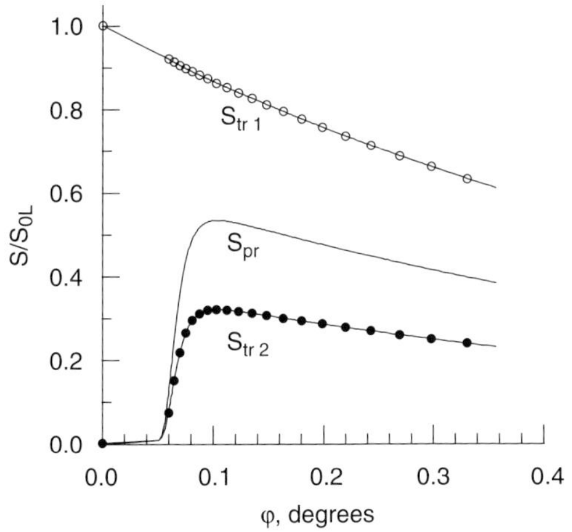

Figure 10.4 The dependence of the fluxes of trapped and precipitated electrons on azimuth φ in the case of strong diffusion. $\mu_c^{-1} = 160$, $D_m \mu_c^{-1} \approx 8$, $v_{g\perp} \approx 10^2$ km/s, $\ln(D_m/D_0) \approx 10$, $S_{tr1} = S_{tr}(\tilde{B} = 1)$, $S_{tr2} = S_{tr}\ (\tilde{B} = \mu_c^{-1}/1.6)$, $\tilde{B} = B(z)/B_L$. (Taken from Pasmanik *et al.*, 1998.)

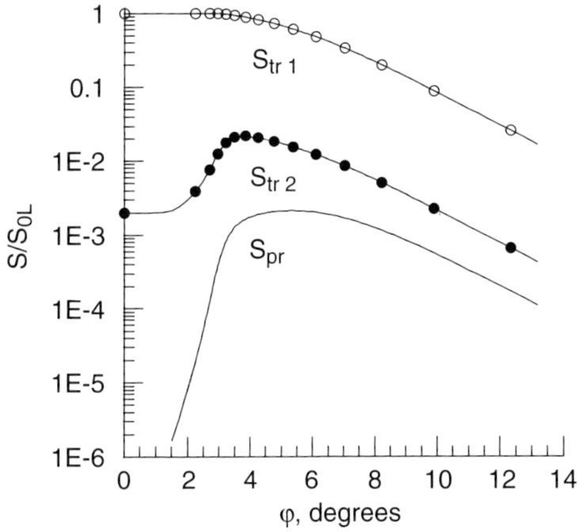

Figure 10.5 The dependence of the fluxes of trapped and precipitated electrons on azimuth φ in the case of weak diffusion. $\mu_c^{-1} = 160$, $D_m \mu_c^{-1} \approx 0.3$, $v_{g\perp} \approx 10^2$ km/s, $\ln(D_m/D_0) \approx 10$, $S_{tr1} = S_{tr}(\tilde{B} = 1)$, $S_{tr2} = S_{tr}\ (\tilde{B} = \mu_c^{-1}/1.6)$, $\tilde{B} = B(z)/B_L$. (Taken from Pasmanik *et al.*, 1998.)

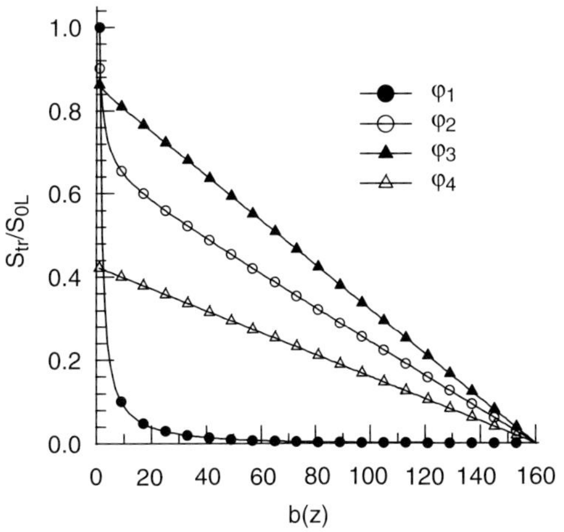

Figure 10.6 The dependence of the fluxes of trapped electrons along a magnetic flux tube on $\tilde{B}(z) = B(z)/B_L$, in the case of strong diffusion. $\mu_c^{-1} = 160$, $D_m \mu_c^{-1} \approx 8$, $v_{g\perp} \approx 10^2$km/s, $\ln(D_m/D_0) \approx 10$, $\varphi_1 = 0$, $\varphi_2 \approx 0.07°$, $\varphi_3 \approx 0.1°$, $\varphi_4 \approx 0.62°$. (Taken from Pasmanik *et al.*, 1998.)

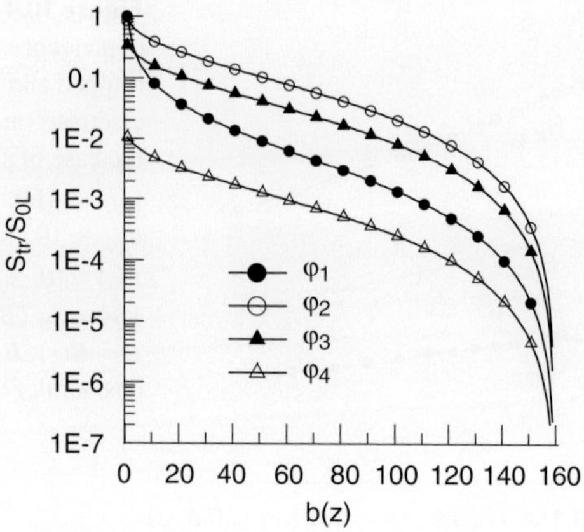

Figure 10.7 The dependence of the fluxes of trapped electrons along a magnetic flux tube on $\tilde{B}(z) = B(z)/B_L$, in the case of weak diffusion. $\mu_c^{-1} = 160$, $D_m\mu_c^{-1} \approx 0.3$, $v_{g\perp} \approx 10^2$ km/s, $\ln(D_m/D_0) \approx 10$, $\varphi_1 = 0$, $\varphi_2 \approx 4.2°$, $\varphi_3 \approx 7°$, $\varphi_4 \approx 14°$. (Taken from Pasmanik *et al.*, 1998.)

relaxation phase ($\varphi_4 \approx 0.62°$). In the SD case the distribution function relaxes to an isotropic distribution, which corresponds to the linear dependence on b (10.44) given in Fig. 10.6, curves 3 and 4. In Fig. 10.7 (the WD regime) the second curve corresponds to the maximum value of the trapped electron flux at low heights ($\varphi_2 \approx 4.2°$), and the two other curves demonstrate the relaxation phase ($\varphi_3 \approx 7°$, $\varphi_4 \approx 14°$). The parallel curves on a logarithmic scale (Fig. 10.7, curves 2, 3 and 4) show that, in the WD case, the distribution approaches a similar dependence on μ with decreasing amplitude, as φ increases.

In both regimes we see a cliff-like deformation of the distribution function at the initial stage of relaxation ($\varphi \to 0$), at sufficiently low heights ($h \sim 10^3$ km), in good accordance with NOAA satellite data (Titova *et al.*, 1998). As a rule, energetic electrons observed by the NOAA satellites show a very sharp cliff and corresponding precipitation, with a typical front width between 0.1° and 0.3°; these values are very similar to the theoretical results for the case of strong diffusion (see Fig. 10.4).

10.4 Quasi-periodic ELF/VLF wave emissions

In Sections 10.1 and 10.2 we limited ourselves to a consideration of the time stationary cyclotron wave–particle interaction. At the same time the analyses of Chapter 9 show that the magnetospheric cyclotron maser can operate in non-stationary regimes with a temporal modulation of the RB electron fluxes and accompanied by more structured ELF/VLF emissions. Among considerable experimental evidence for such non-stationary wave–particle interactions processes the most dramatic are the so-called quasi-periodic (QP) whistler-mode emissions and pulsating aurorae. QP ELF/VLF hiss-type emissions are wideband emissions that are observed inside or near the plasmapause – see, e.g. Helliwell (1965), Sato *et al.* (1974), Kovner *et al.* (1977), Tixier and Cornilleau-Wehrlin (1986), Hayakawa and Sazhin (1992), Sazhin and Hayakawa (1994), and Smith and Nunn (1998). They are characterized by a periodic modulation of the wave intensity with typical periods from several seconds up to a few minutes. The generation of QP emissions is usually accompanied by the precipitation of energetic electrons, also modulated at the same period.

QP emissions can be divided into two classes: QP1 emissions, which are closely associated with geomagnetic pulsations of the same period, and QP2 emissions which do not correlate with geomagnetic pulsations. Usually QP1 emissions have an almost sinusoidal modulation and well defined frequency, while QP2 appear in the form of bursts with regular frequency drifting.

Both types of QP emissions can be explained quantitatively on the basis of the non-stationary CM generation regimes considered in Chapter 9. In particular QP1 events can be associated with the resonant excitation of relaxation oscillations (Section 9) by hydromagnetic waves of the same periods. Actually, according to the results of Section 9, the frequency Ω_J and the quality factor Q_J of the relaxation oscillations are given by (see (9.30)):

$$\Omega_J = \left(\frac{2J_\Sigma \,|\ln R|}{N_0 T_\mathrm{g}}\right)^{1/2} = \left(\frac{4S_\mathrm{A} \,|\ln R|}{N_0 T_\mathrm{g}}\right)^{1/2} \tag{10.48}$$

$$Q_J = \frac{\Omega_J}{2\nu_J} = \left(\frac{N_0 \,|\ln R|}{4S_\mathrm{A} T_\mathrm{g}}\right)^{1/2}. \tag{10.49}$$

Here we have taken into account the relation between the source intensity J_Σ and the precipitation flux density S_A, $J_\Sigma = 2S_\mathrm{A}$, and N_0 is the number of energetic electrons in the magnetic flux tube with unit cross-section at its base in the stationary state. According to Equation (9.58) and relations (9.59), the presence of MHD waves with a field-aligned magnetic component b leads to the excitation of relaxation oscillations of the ELF/VLF wave intensity and of the electron precipitating flux with an amplitude, which is maximum when $\Omega_\mathrm{MHD} \approx \Omega_J$, and is given by:

$$\max(\mathcal{E}_\sim/\mathcal{E}_0) = \max(S_\sim/S_\mathrm{A}) \approx Q_J^2(b/B_L). \tag{10.50}$$

Thus, even an MHD wave with a small amplitude $(b/B_L) \ll 1$ can produce a deep modulation of the wave energy and of the precipitated energetic electron fluxes, if $Q_J \gg 1$.

Let us make some quantitative estimations for $L = 4$, $T_\mathrm{g} \sim 1$ s, and $|\ln R| \sim 0.3$. Suggesting a weak diffusion regime we put the flux density of trapped electrons $S_\mathrm{tr} \sim 10^8$ cm^{-2} s^{-1} and $S_\mathrm{A} \sim 10^6$ cm^{-2} s^{-1} for precipitation. For the energy $W \sim 40$ keV, we find that the energetic electron content of the flux tube $N_0 \sim R_0 L^4 S_0/(2W/m)^{1/2} \sim 2.56 R_0$ (cm), where the Earth's radius R_0 should be taken in cm. Substitution of the above values into the relations (10.48) and (10.49) gives us

$$T_J = 2\pi/\Omega_J \sim 200\,\mathrm{s} \;\text{ and }\; Q_J \approx 10. \tag{10.51}$$

There is qualitative agreement for a decrease of T_J as the source intensity grows, which is observed in experiments (Sato *et al.*, 1974).

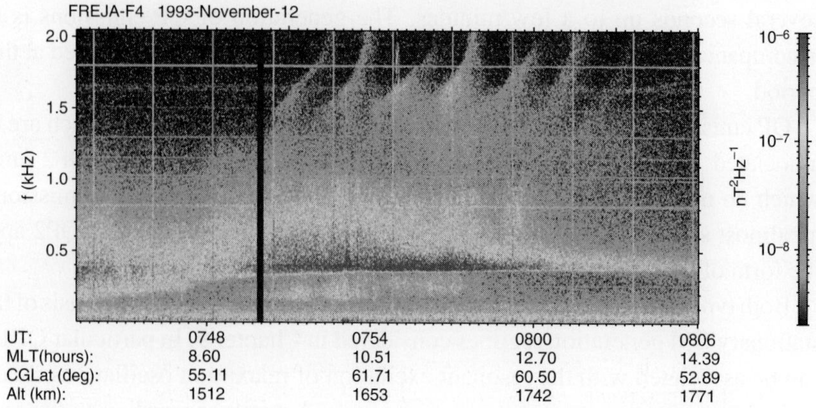

Figure 10.8 Frequency–time spectra of the magnetic component of VLF emissions observed on board the Freja satellite on 12 November 1993. The dynamic spectrum is remarkable in this case by the alternation of QP elements with different frequency drift rates (after Pasmanik *et al.*, 2004b).

Now we demonstrate the application of the rigorous computational theory, developed in Chapter 9, to some particular QP2 events, detected on board the Freja and Magion 5 satellites. Based on these data we choose suitable parameters for the simulation model, such as the source characteristics and cold plasma density, to obtain a good correspondence between the observations and the simulations.

Refering to Pasmanik *et al.* (2004b) for details, we can summarize the modelling of the Freja QP2 events as follows. The most suitable model which quantitatively explains the observed dynamical spectra of ELF/VLF emissions is the flow cyclotron maser (FCM) model with drift losses (Section 9.3.2). The spectral form of the separate elements depends on the source power. An interesting new type of QP2 event was revealed at middle latitudes (57–61°), the QP elements alternating between different frequency drift rates, as shown in Fig. 10.8.

Figure 10.9 demonstrates the results of theoretical modelling, which agree well with the experimental data shown in Fig. 10.8. Such a specific spectrum is expected when the source intensity becomes sufficiently large but, unfortunately, the fluxes of energetic electrons were not measured aboard Freja. However, the Magion 5 included, together with the QP spectrum, cold plasma and energetic electron measurements. A more complicated quantitative model of QP generation explained the frequency range as well as the period and spectral form of the QP elements - see Figs.10.10 (experimental results) and 10.11 (results of the modelling).

An important feature seen when modelling the dynamics of the electron distribution function is the formation of a sharp gradient in the pitch-angle distribution, as discussed already in Section 9.4. An example corresponding to a particular QP event, simulated using the FCM model, is shown in Figs. 10.12 and 10.13. The step feature seems to be very important to explain the relation between QP

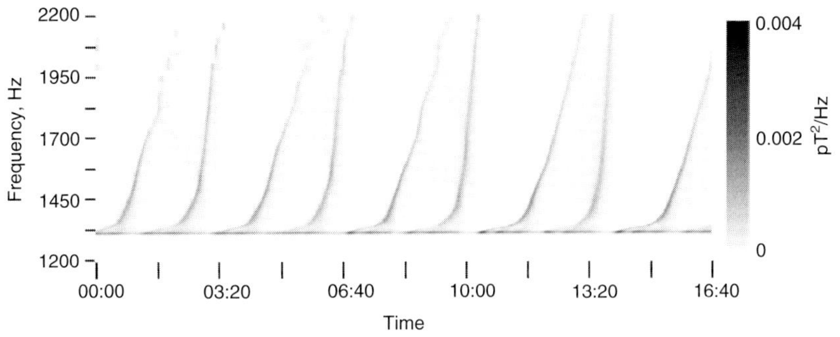

Figure 10.9 Spectrogram of the wave intensity obtained from the FCM model for the experimental case presented in Fig. 10.8. (Taken from Pasmanik *et al.*, 2004b.)

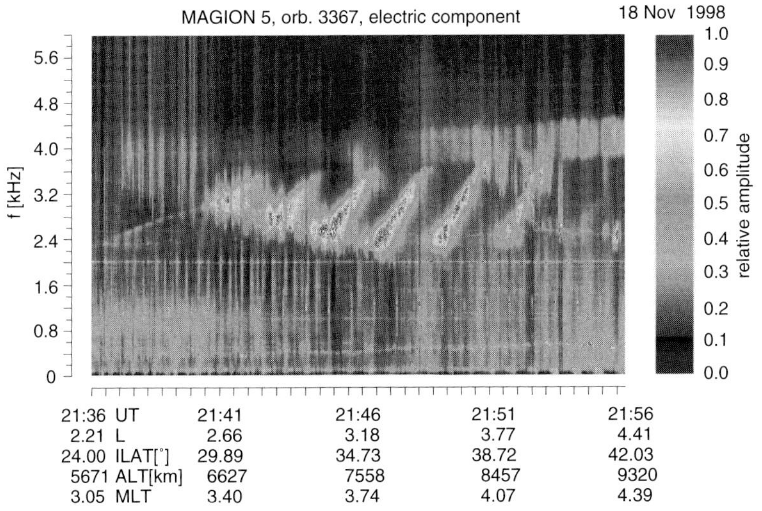

Figure 10.10 Frequency–time spectra of the electric component of VLF emissions observed on-board Magion 5 satellite on 18 November 1998. (Taken from Pasmanik *et al.*, 2004b.)

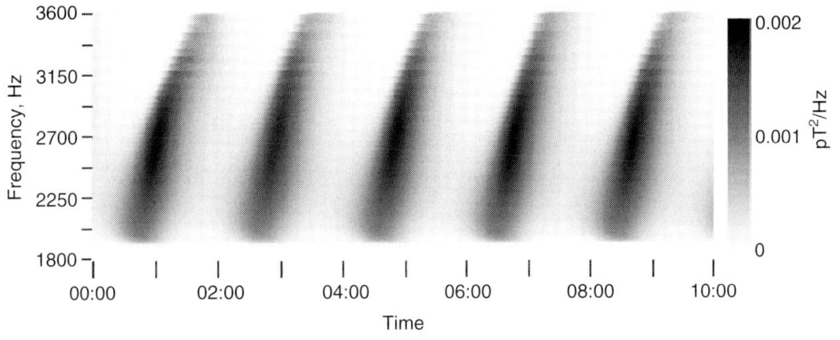

Figure 10.11 Spectrogram of the wave intensity obtained from the FCM model corresponding to the experimental case presented in Fig. 10.10. (Taken from Pasmanik *et al.*, 2004b.)

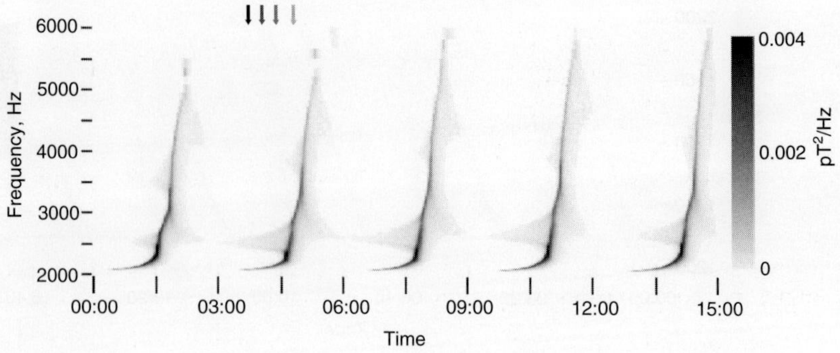

Figure 10.12 Spectrogram of the wave intensity obtained from the FCM model simulation of a particular QP event. (Taken from Pasmanik *et al.*, 2004b.)

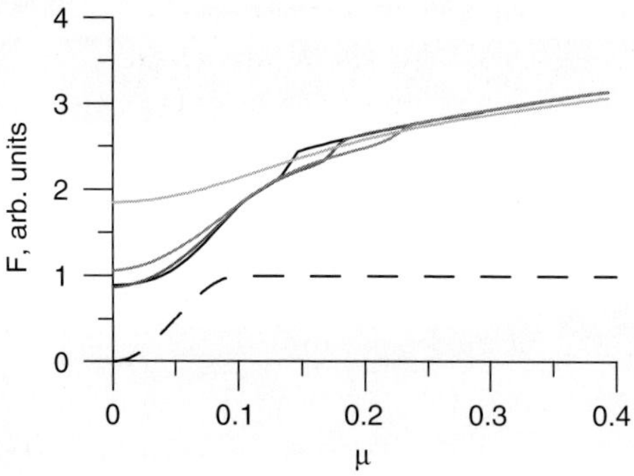

Figure 10.13 Formation of a sharp gradient in the pitch-angle distribution, shown as μ, for energetic electrons at four different times during the generation of an element of the QP emission shown Fig. 10.12; there the arrows show the corresponding times. The pitch-angle distribution in the source is shown by the dashed line. (Taken from Pasmanik *et al.*, 2004b.)

and discrete ELF/VLF emissions (Pasmanik *et al.*, 2004b); it will be discussed more fully in Chapter 11.

10.5 Auroral pulsating patches

A specific type of energetic electron precipitation pulsations and QP ELF emissions occur at auroral and subauroral latitudes during the substorm recovery phase. Then, energetic electrons drift from their injection in or near the midnight sector to the morning side and populate the magnetosphere. Precipitation pulsations are seen in satellite energetic electron data, and in the optical emissions in this night to morning sector are termed auroral pulsating patches.

According to numerous experimental data – see Oguti (1981), Sandahl (1984), Tagirov *et al.* (1986), Yamamoto (1988), Hansen *et al.* (1988), Nakamura *et al.* (1990), Winckler and Nemzek (1993) – these pulsating patches have the following distinctive features:

- typical recurrence of patch shapes from one burst to another;
- general motion of the patches in the local $\vec{E} \times \vec{B}$ drift direction;
- energetic electrons mainly in the energy range 10 keV $\leq W <$ 100 keV;

- characteristic pulsation periods from 2 to 20 s;
- correlation of optical pulsations with bursts of ELF and VLF emissions;
- rather high-frequency modulation in the pulsation maximum, with the frequency $\sim$ 2 to 5 Hz; and
- pitch-angle scattering of energetic electrons in the equatorial plane into the loss cone.

Coroniti and Kennel (1970), Trefall *et al.* (1975), Bespalov and Trakhtengerts (1976b), Davidson (1979, 1986), Trakhtengerts *et al.* (1986), Huang *et al.* (1990) and Davidson and Chiu (1986, 1991) connected the origin of the pulsating patches with the whistler cyclotron instability (CI) occurring on field-aligned ducts having an enhanced plasma density.

Here we present the results of the most advanced modelling of pulsating patches performed by Demekhov and Trakhtengerts (1994), which is based on the flow cyclotron maser (FCM) model. Initially this model was suggested by Trakhtengerts *et al.* (1986), and included drift losses of energetic electrons (see (9.99)) from the generation region, which was a duct with enhanced plasma density. Figure 10.14 shows a diagram of flow cyclotron maser (FCM) operation.

The plasma duct serves as a quasi-optical resonator for ELF whistler-mode waves. Wave reflection from the duct ends is most probably connected with the duct narrowing towards the end and violation of the geometrical optics approximation when the ELF wavelength λ becomes comparable with the waveguide diameter d.

Free energy for the waves is supplied by the initial transverse anisotropy of the distribution function of energetic electron distributions and by the loss cone

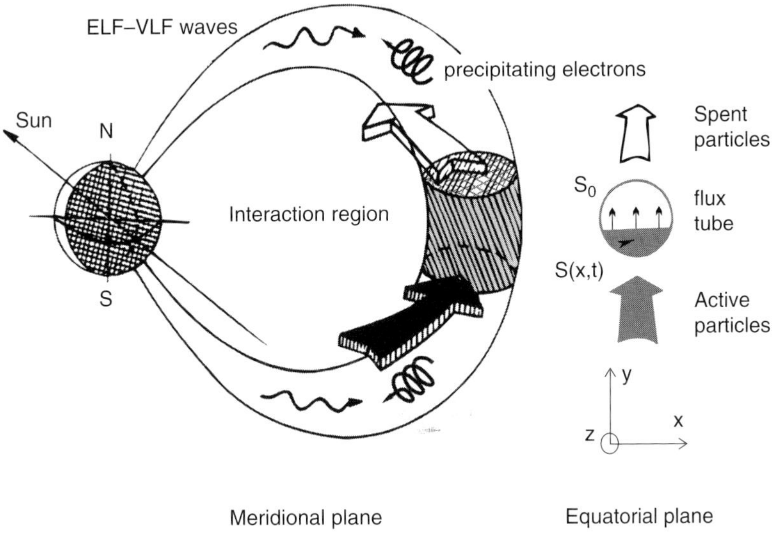

Figure 10.14 A diagram showing flow cyclotron maser (FCM) operation in the equatorial plane. Energetic electrons are the active particles which drift longitudinally into a duct (shaded) of enhanced plasma density; some electrons precipitate into the upper atmosphere and some drift out of the duct (these are termed spent particles). (Based on Demekhov and Trakhtengerts, 1994, and Trakhtengerts and Rycroft, 2000).

that provides this anisotropy all the time. Electrons enter the resonator from the nightside and leave through the dawnside. Initially waves are generated when the resonator stores enough electrons to exceed the generation threshold, determined by the relation (9.110). A mathematical formulation of the FCM model was given in Section 9.3.

We assume that at time $t = 0$ the flow CM begins to be filled with a cloud of energetic electrons with some initial distribution function $\mathcal{F}_0(æ, v)$. For energies $W \geq 10$ keV, $\mathcal{F}_0$ is anisotropic with an anisotropy index

$$\alpha \equiv T_\perp / T_\parallel .$$

Here $T_{\perp,\parallel}$ are the averaged energies of energetic electrons across ($\perp$) and along ($\parallel$) the magnetic field. Transverse anisotropy is caused either by the character of the acceleration mechanism in the midnight magnetosphere, or by the effect of the loss cone. If the threshold (9.110) is not exceeded, the number of energetic particles in the duct grows higher than the threshold value. The CI develops, its characteristic in the initial stage being determined by the amplification Γ_0 calculated by (9.103), with

$$\mathcal{F}_{\text{duct}} = \mathcal{F}_{\text{in}}\, S(x, t)/S_0 \simeq \mathcal{F}_{\text{in}}\, t/t_{\text{D}} \tag{10.52}$$

where $S(x, t)$ is the area of duct filled at time t (Fig. 10.8), and $t_{\text{D}} = d/v_{\text{D}}$ is the drift time through the duct. Formula (10.52) determines the threshold value of the duct-averaged number of energetic electrons $N_{\text{h}}(x, t) = N_0 S(x, t)/S_0$; here N_0 is the number of energetic electrons outside the duct. A spike of emission is generated, with duration t_i, of the order of the inverse initial growth rate $\gamma^{-1} = T_{\text{g}}/2\,|\ln R|$. In the case $|\ln R| > 1$ we can thus use the value

$$t_i \sim T_{\text{g}} . \tag{10.53}$$

As mentioned earlier, the formation of the spike-like regime of CI is closely connected with involving newly resonant particles in the instability and the expansion of the resonance region for growing waves into a larger æ domain. It provides an increase of the growth rate at the initial stages of the instability. At the same time the pitch-angle diffusion by whistler-mode waves excited by newly resonant particles prevents the restoration of free energy in the region of æ responsible for the instability switching on. Owing to this a ‘flash’ ends with a low free energy state, so that after a flash a pause occurs during which it is necessary to store a new portion of free energy of electrons in the duct. From this it follows that a pulsating regime of the CI can exist when the particle source power is sufficiently weak; otherwise damped oscillations are formed. Damping of the pulsations can be thought of as a consequence of the overlapping of subsequent flashes, so a rough estimate for the maximum source power can be obtained from the condition that a pause between the pulses is longer than a flash:

$$T_{\text{p}} > t_i . \tag{10.54}$$

For the case of well-distinguished pulses the duration of the pause is approximately equal to the time of accumulation of free energy from zero to the threshold value:

$$T_{\rm p} \approx t_J \approx t_{\rm D}\, \Gamma_{\rm m0}/|\ln R| \tag{10.55}$$

where the maximum ELF wave amplification $\Gamma_{\rm m0} \equiv \Gamma_{\rm m}(\mathcal{F}_{\rm duct})$, see Equations (9.110) and (10.52). According to Equations (10.54) and (10.55) there is an upper value of the energetic electron number N_0, which corresponds to the transition from a pulsating condition to a quasi-stationary CI regime in a duct. At the same time the condition for whistler-mode wave growth (9.110) forms a lower limit of N_0, so that the duration of the pause between spikes in the flow CM cannot be larger than the electron drift time across the duct:

$$T_{\rm p} \leq t_{\rm D} = d/v_{\rm D}. \tag{10.56}$$

Numerical calculations of the flow CM dynamics have been performed assuming that the energetic electrons have a narrow energy spectrum. The pitch-angle distribution of electrons at the front of the duct, $\mathcal{F}_{\rm in}$, was chosen to be of the form (9.112), with the normalization condition (9.106).

We present the result of calculations for the equations averaged over the duct cross-section – Equations (9.98)–(9.106). For Equation (9.98) to be solved it is necessary to know both $\mathcal{F}_{\rm in}$ and $\mathcal{F}_{\rm out}$. The first is determined mainly by transverse acceleration processes in the nightside magnetosphere and by the loss cone feature; here it is assumed to be given (see Equation 9.112). The function $\mathcal{F}_{\rm out}$ results from the evolution of $\mathcal{F}_{\rm in}$ during particle drift across the duct, so that it is dependent on the CI dynamics in the duct. According to Demekhov (1991), the influence of the loss cone on the pitch-angle distribution inside a duct results in damping of the pulsations. At the same time, the corresponding variations of the distribution of electrons escaping through the side of the duct lead to partial compensation of this effect because the anisotropy of the outgoing electrons decreases with increasing wave amplitude. Hence, the anisotropy of the effective source J in the kinetic equation (9.98) becomes larger.

In our calculations we neglect both of these effects and concentrate on the role of pitch-angle diffusion outside the loss cone. We restrict ourselves to the situation when the initial anisotropy is due to some transverse acceleration process, and is considerably larger than that due to the loss cone feature. In this case, the instability dynamics is determined by particles with pitch-angles $\mu \gg \mu_{\rm c}$, and the loss cone plays a passive role, only providing electron precipitation into the ionosphere. In the region outside the loss cone, we can assume the outgoing distribution $\mathcal{F}_{\rm out}$ to be isotropic, its amplitude being determined by the approximate particle conservation law in the duct:

$$\int \mathcal{F}_{\rm in}\, T_B\, d\mu \simeq \int \mathcal{F}_{\rm out}\, T_B\, d\mu. \tag{10.57}$$

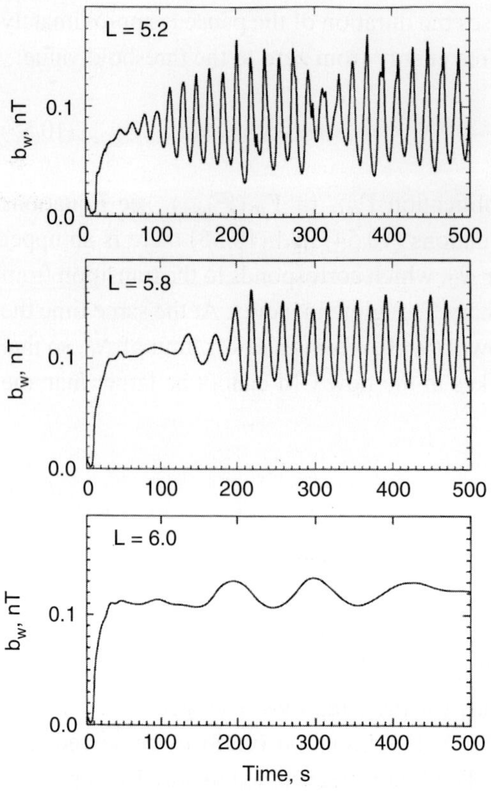

Figure 10.15 Whistler-mode wave magnetic field amplitude b_W in pulsating aurorae (results of numerical solution of the CM equations). Dimensional parameters are chosen as follows: $W = 15$ keV, $|\ln R| = 20$, $n_{hL} = 0.3\ \text{cm}^{-3}$, L values are specified on the plots. The source pitch-angle distribution was chosen to be of the form (9.112), with $æ_0 = 0.2$, $\delta = 5$. (Taken form Demekhov and Trakhtengerts, 1994). (Copyright American Geographical Union, reproduced with permission.)

For simplicity we assume that $\mathcal{F}_{\text{out}}$ is isotropic at all μ. This leads to only small errors in the pitch-angle region for the instability dynamics, provided that $\mu \gg \mu_c$.

Some results of calculations are presented in Figs. 10.15 and 10.16; parameter values are indicated in the figure captions. The value of $æ_0$ was chosen to be $æ_0 = 0.2$, so that $æ_0 \gg \sqrt{\mu_c}$, and the relative contribution of the loss cone anisotropy to the total growth rate is small.

Figure 10.15 shows oscillograms of whistler-mode wave magnetic field amplitude b_W computed from the equality

$$\frac{b_W^2}{8\pi} = \int \frac{\mathcal{E}_\omega}{v_g}\, d\omega$$

for different L values, with the other parameters being fixed. For smaller L, clearly distinguished pulsations are obtained, whereas for larger L we see a quasi-stationary time pattern with only a small modulation. The calculation result is consistent with the observed north/south transition from a quasi-stationary to a spike-like pulsation pattern (Yamamoto, 1988).

Detailed information on the evolution of the wave spectrum and electron pitch-angle distribution during a spike is presented in Fig. 10.16. From Fig. 10.16(b)

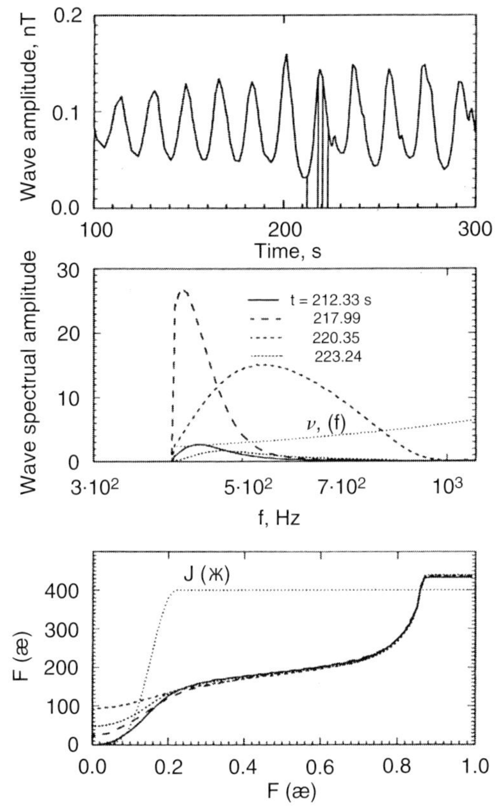

Figure 10.16 Results of numerical calculations of the flow CM equations (8.34) and (8.41) together with (9.102), (9.103). The parameters are as follows: $L = 5.2$, $W = 15$ keV, $n_{hL} = 0.3$ cm^{-3}, and $n_{cL} = 10^2$ cm^{-3}. The upper panel shows the wave magnetic field amplitude proportional to the precipitating electron flux. The middle and bottom panels show the wave spectrum and pitch-angle distribution function at the times marked by vertical lines in the upper panel. The source function $J(æ)$ and frequency dependent wave damping rate $\nu = 2\,|\ln R|/T_g$ are shown (arbitrary scale) by dotted lines in the bottom and middle panels, respectively. (Taken form Demekhov and Trakhtengerts, 1994). (Copyright American Geographical Union, reproduced with permission.)

it is seen that higher-frequency components are excited at the equator during later stages of the flash, in agreement with experimental data (see Tsuruda *et al.*, 1981).

The important feature of the dynamics of the pitch-angle distribution presented in Fig. 10.16(c) is that it does not change so much in the region of æ responsible for the changes of wave growth rate ($æ \geq æ_0 = 0.2$). The pause phase is not very pronounced, and accordingly free energy is partially restored during a flash; the rather small modulation of the distribution function gives a moderate (50–70%) modulation of the whistler-mode wave amplitude.

For electrons with a given energy, the cold plasma density in a duct is limited from below by the requirement that the gyroresonance condition is met for waves with frequencies $\omega < \alpha\omega_{BL}$. In particular, for $W \sim 20$ keV, $L = 6.5$, and $\alpha \sim 0.3$, we find $n_{cL} > 6$ cm^{-3}. As values of $n_c = 4$ cm^{-3} are typical for the plasma trough region, even at the very beginning of the storm recovery phase, we can assume larger densities in the plasma ducts, especially during quieter periods.

Taking into account that the maximum CI growth rate is

$$\gamma_m \sim \alpha \frac{n_{hL}}{n_{cL}} \omega_{BL} \tag{10.58}$$

we find from (10.3) that

$$\frac{n_0}{n_{cL}} \geq \frac{v_g \,|\ln R|}{\alpha \omega_{BL}\, a}\,. \tag{10.59}$$

For $L = 6.5$, $W \sim 20$ keV, $\alpha \sim 0.5$ and $|\ln R| \sim 10$, the ratio of the energetic electron density to the cold electron density in the equatorial plane is

$$\frac{N_0}{N_{cL}} \geq 10^{-3}\,. \tag{10.60}$$

It is evident that conditions for the CI to develop in a plasma duct depend strongly on the effective reflection coefficient determined by the duct properties as a whistler-mode resonator. Having a sufficiently high coefficient of reflection ($R > 0.1$) from the ends, the relation $\lambda \sim d$ (the duct diameter) must be fulfilled in some cross-section of the duct. Under the conditions where $n_c(z) \propto B(z)$, variations of the wavelength λ are small up to heights $h \sim 10^3$ km, and d decreases with height as $z^{-3/2}$. Thus, for d_{min} we can take the diameter found from optical auroral measurements: $d_{min} \sim 50$ km. Assuming $n_{cL} \sim 5$ cm^{-3} and $L \sim 6.5$, we have: $\lambda > d_{min}$, if frequency $f < 0.8$ kHz. The spectrum of the observed ELF waves in the region conjugate to the pulsating aurorae is close to this frequency.

We now discuss the relation between the characteristic time scale of the pulsations as predicted by the model and as derived from the observations. The period of pulsations is $T = T_p + t_i$, where $t_i \gtrsim T_g$, $T_p \leq d/v_D$. Specifically, for $d \sim 50$ km and $W \sim 20$ keV, we have $T_p \leq 15$ s, and $T_g \simeq 4$ s for $n_c \simeq 5$ cm^{-3}. These estimations are in agreement with the observed data, as well as being supported by numerical calculations.

Analytical and numerical investigations confirm that the flow cyclotron maser (FCM) model proposed by Trakhtengerts *et al.* (1986*b*), can account for the observed characteristics of morningside pulsating aurorae. The model explains the formation of the spike-like regime of whistler-mode wave generation on the basis of self-consistent quasi-linear equations describing variations of the wave spectrum and the electron pitch-angle distribution in a duct of enhanced background plasma density. Within the framework of the model, the pulsation 'on' time is determined by the nonlinear development of the CI in a duct and is of order of the wave travel time between ionospheres, $t_i \gtrsim T_g/2$. The 'off' time is controlled by the process of restoring free energy in the CM via the supply of energetic electrons into the duct by longitudinal drift. This is approximately equal to the time of energetic electron accumulation in a duct from zero to the threshold concentration, and depends on the background plasma and energetic electron parameters as well as on the wave damping rate.

Estimations and numerical calculations give reasonable values for the pulsation parameters such as the repetition period, $T \sim 5$–30 s, flash (spike) duration,

$t_i \sim 1$–5 s, and wave amplitude $b_w \sim 0.01$–0.1 nT. The model also provides a consistent explanation for the experimentally observed longitudinal and latitudinal dependence of the pulsation time pattern. Rapid intensity modulations can be attributed in this model to the formation of a sharp gradient in the pitch-angle distribution, in which case nonlinear phase effects begin to play an important role.

10.6 Summary

In this chapter earlier cyclotron instability results have been applied to a self-consistent treatment of:

- energetic electrons trapped in the Earth's magnetosphere and constituting the radiation belts, including both their inward radial (cross-L) diffusion and pitch-angle diffusion, which leads to the formation of the well-known 'slot' at $L = 3$ to 4, under quiet or steady conditions of geomagnetic activity;
- electron precipitation at L-values just inside the plasmapause, or in detached plasma regions at higher L-values, under geomagnetically disturbed conditions; both weak and strong diffusion conditions are considered;
- the formation of quasi-periodic (QP) whistler-mode emissions; and
- features of the accompanying pulsating aurorae which exhibit the same periodicity.

The key results obtained are, respectively, that:

- the integral flux density S_L of trapped energetic electrons under quiet and steady conditions is proportional to the inverse fourth power of the L-value (Equation 10.37) and the wave energy flux density $\mathcal{E}$, generated, is proportional to the first power of the L-value (relation 10.34);
- a sharp 'cliff' in the flux of both trapped and precipitated electrons, is predicted at a certain longitude away from the night-time injection point, where they drift into a region of large cold plasma density, which is in good agreement with NOAA satellite data sets;
- two types of QP emissions find quantitative explanation in the framework of two physical mechanisms, where one is due to the CI growth rate modulation by external MHD waves; even an MHD wave with a small amplitude can initiate a deep modulation of the whistler-mode wave intensity and of the flux density of precipitated energetic electrons, if the frequency Ω_M of MHD wave is close to the frequency Ω_J of the relaxation oscillations;
- the second mechanism, which explains QP2 burst-like emissions whose frequency drifts with time, is due to self-modulation effects as the cyclotron instability develops; the computer simulations of this case are in good agreement with observations; also, QP2 emissions alternately exhibiting very

different frequency sweep rates (Fig. 10.9) are well explained by the flow cyclotron maser theoretical model, including drift-losses; and

- the flow cyclotron maser (shown diagramatically in Fig. 10.14) explains the properties of the precipitating electron fluxes observed in pulsating auroral events on the morningside, and predicts a whistler-mode wave amplitude ~ 0.1 nT.

Chapter 11

Generation of discrete ELF/VLF whistler-mode emissions

Discrete ELF/VLF signals, i.e. those whose strength exhibits a well-defined signature in the frequency–time domain (see Fig. 1.3), abound in the Earth's magnetosphere. Whistlers generated by lightning discharges give important information on the distribution of plasma above the ionosphere. They are an effective diagnostics tool for near-Earth plasmas, and their study led to the discovery of the plasmapause and the plasmasphere. We shall be interested here in other types of discrete ELF/VLF signals, which include artificial signals termed triggered emissions excited by external quasi-monochromatic waves from ground-based transmitters, and so-called chorus radiation. Both of these discrete emissions are generated by gyroresonant interactions of whistler-mode waves with radiation belt electrons, and are described here on the base of the nonlinear theory presented in Chapters 3–7.

In this chapter we mainly consider quantitative descriptions of the generation of these signals, referring for details of the history and for other theoretical details to the above mentioned chapters and to the reviews by Helliwell (1969, 1988, 1993), Matsumoto (1979), Omura *et al.* (1991), Trakhtengerts and Rycroft (2000), and Demekhov and Trakhtengerts (2001).

11.1 Overview of experimental data on triggered VLF emissions

The typical frequency of a triggered VLF emission is below half the electron gyrofrequency in the equatorial plane of the magnetic flux tube where the emission is generated. The triggering signals are pulses of length from 0.05 to 10 s produced by terrestrial VLF (3–30 kHz) radio transmitters with bandwidths normally of the order of, or less than, 20 Hz.

VLF emissions, triggered by the high-power US military transmitters NAA and NPG, were first reported by Helliwell *et al.* (1964) and Helliwell (1965). It was soon discovered that triggering was not confined to high-power VLF transmitters. Kimura (1968) noted triggering by the 100 W Omega transmitter at a frequency of 10.2 kHz, located near $L = 3.5$. Extremely interesting results on triggering have been obtained using the purpose-built VLF transmitter at Siple on the Antarctic plateau ($L = 4.2$, $f = 2.5$–7 kHz, radiated power 400 W). The results obtained have been reviewed by Helliwell and Katsufrakis (1974), Stiles and Helliwell (1975) and Helliwell (1983, 1993). The triggering process is very complicated and we are far from understanding it fully. The most general features of triggered VLF emissions can be now quantitatively explained by modern sophisticated computer simulations, developed by Nunn (1990, 1993), Nunn and Smith (1996), Nunn *et al.* (1997), Smith and Nunn (1998), and Nunn *et al.* (2005). At the same time, based on the analytical theory developed in Chapters 5–7, it is possible to suggest simplified models describing the formation of the dynamical spectrum of triggered signals, and to compare them with the results obtained from the computer simulations. In this connection see as well Helliwell (2000), Demekhov and Trakhtengerts (2001), Trakhtengerts *et al.* (2001), and Trakhtengerts *et al.* (2006).

Some examples of triggered VLF emissions are shown in Fig. 11.1 and 11.2. The following features of triggered emissions form the focus of our attention

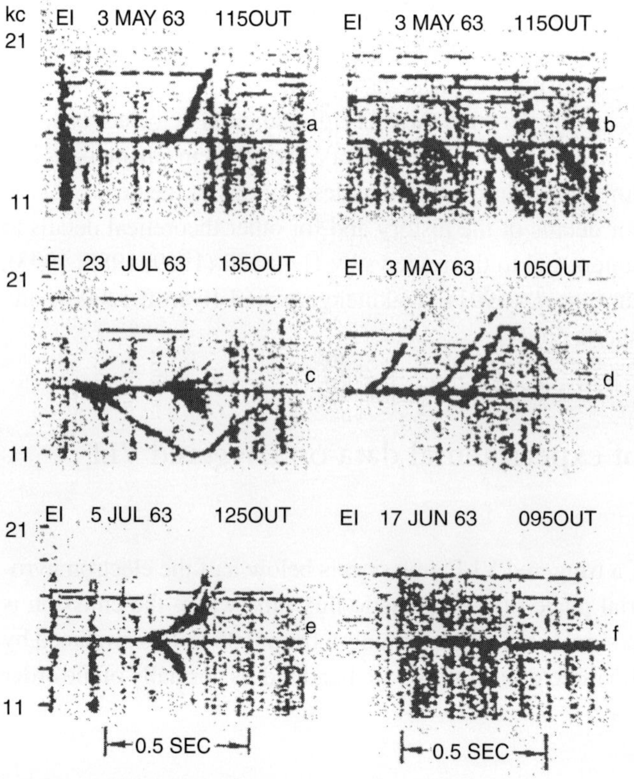

Figure 11.1 Examples of the dynamic spectra of VLF emissions triggered by Morse code signals observed at Eights Station, Antarctica, after one-hop propagation in the whistler-mode (after Kimura, 1968). (Copyright American Geophysical Union, reproduced with permission.)

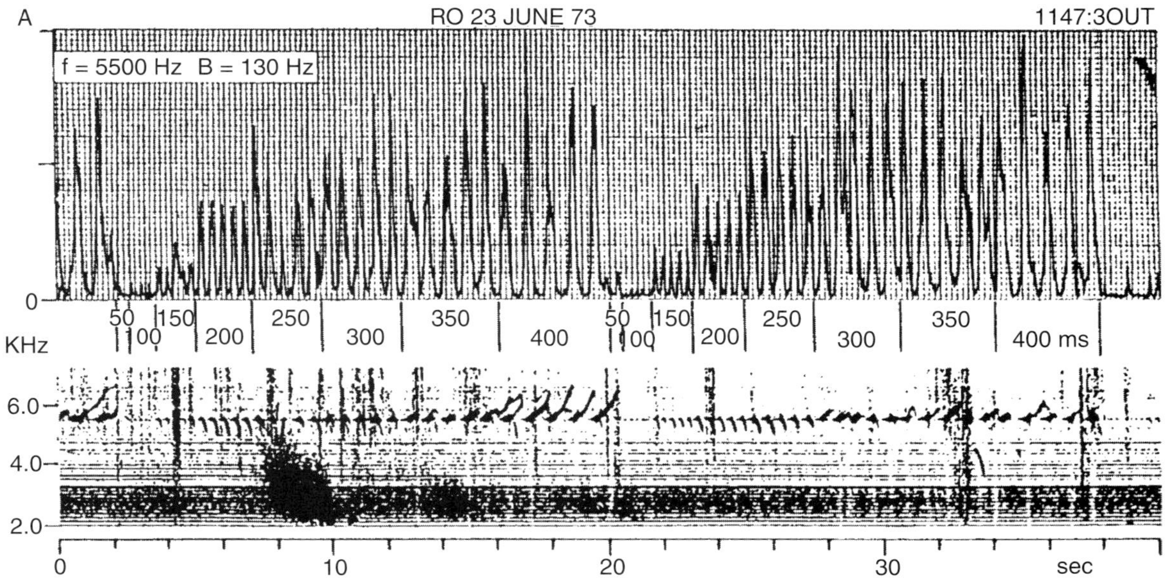

Figure 11.2 Examples of VLF emissions triggered by two identical sequences of coherent pulses from the Siple transmitter varying in duration from 50 to 400 ms, and observed at Roberval, Canada, at the northern end of the geomagnetic flux tube of the Siple transmitter. The dynamic spectra in the lower panel show fallers triggered at the ends of shorter ($\leq$ 250 ms) pulses and risers triggered prior to the termination of longer ($\geq$ 300 ms) pulses. The upper panel shows the relative amplitude of signals received in a 130 Hz band centred on the carrier at 5500 Hz, from Helliwell and Katsufrakis (1974). (Copyright American Geophysical Union, reproduced with permission.)

(see Matsumoto, 1979, Omura *et al.*, 1991):

- The emissions have a narrow bandwidth, usually less than 100 Hz and sometimes as little as 10–20 Hz.
- They are long enduring, sometimes having a duration approaching 1 s, many times that of the triggering pulse.
- They have an amplitude which saturates at a level which is typically 30 dB above that of the triggering signal.
- Triggered emissions usually have a continuously sweeping frequency, typically at a rate of the order of 1.0 kHz/s. Most commonly the frequency will rise, giving a so-called 'riser'. However falling tones or 'fallers' are occasionally observed. The total frequency change may be considerable, up to 30% of the starting frequency. More complex emission forms are sometimes noted. For example, risers sometimes have downward hooks, and fallers may turn into risers.
- VLF emissions can be triggered from ground-based radio transmitters with L-values ranging from 2 to 5.

In the next section we present some results on numerical simulations of triggered VLF emissions.

11.2 Numerical simulations of triggered VLF emissions

Strictly quantitative results on triggered VLF emissions can be obtained using numerical simulations based upon the nonlinear equations (5.62) given in Chapter 5. Here we consider the results of a numerical parametric study of triggered VLF signals, presented by Nunn *et al.* (2005), which gives the fullest picture of this phenomenon to date. The simulation code assumes that the VLF waves are ducted through the magnetosphere in ducts of enhanced cold plasma density so that their wave vectors are approximately parallel to the ambient geomagnetic field. The ambient geomagnetic field amplitude is, to a good approximation, a parabolic function of the distance z from the equator. The simulation box which straddles the equator is of the order of 6000 km in length. As the frequency of the simulated emission rises or falls, the simulation centre frequency moves at every time step to be at the centre of gravity of the simulation wavefield spectrum. The time-varying phase space box and grid are defined in z, $v_{\|}$, ψ space for each value of pitch-angle, where $v_{\|}$ is the energetic electron's velocity parallel to the geomagnetic field, z the distance from the equator and ψ the gyrophase. Simulation electrons fill phase space with a given value of density, and are followed forwards continuously in time. Electrons leaving the phase space box are removed from the simulation and new electrons are continuously introduced at the phase box boundaries. The integrated energy change dW is calculated for each electron; applying Liouville's theorem, the distribution function $\mathcal{F}$ is then known at all the phase space points occupied by electrons.

Hence the dynamic spectrum of the emission and the resonant current of electrons are calculated.

The aim of the simulations was to investigate the dependence of the properties of triggered signals on the most important parameters; these are the cold plasma density n_c, linear equatorial growth rate γ_{lin}, triggering pulse length T_t and triggering pulse input amplitude. The data from the Siple experiments were taken for comparison.

The typical situation chosen as the 'base case' was with the L-value $L = 4.4$, the transmitter pulse frequency $f = 3600$ Hz and $n_c = 400$ cm^{-3} (inside the plasmapause). The triggering pulse length was 277 ms, the initial amplitude $B_{\text{in}} = 0.1$ pT, and the linear growth rate $\gamma_{\text{lin}} = 60$ dB s^{-1}.

A stable riser was generated with a sweep rate $\sim$ 1 kHz s^{-1}, as shown in Fig. 11.3. Figure 11.4 shows a snapshot (at $t = 1.525$ s) of the wave field amplitude (in pT), the component of resonant electron current in phase with the wave electric field (J_{Re}) and at 90° to that (J_{Im}), all as functions of z, the distance from the equator along the field line, positive z being towards the receiver in the northen hemisphere. It represents a view of the structure of the generating region for a triggered riser. When γ_{lin} was reduced, the riser was replaced by a short faller. Triggering was terminated when γ_{lin} decreased to 30 dB s^{-1}. Triggered signals also disappeared when the duration of the initial pulse was less than some critical value ($\sim 10^2$ ms). Figure 11.5 shows the change of dynamic spectrum of a triggered signal for a higher linear growth rate of 90 dB s^{-1}. A long stable faller is generated with a sweep rate of -400 Hz s^{-1}; Figure 11.6 shows the structure of the generation region in this

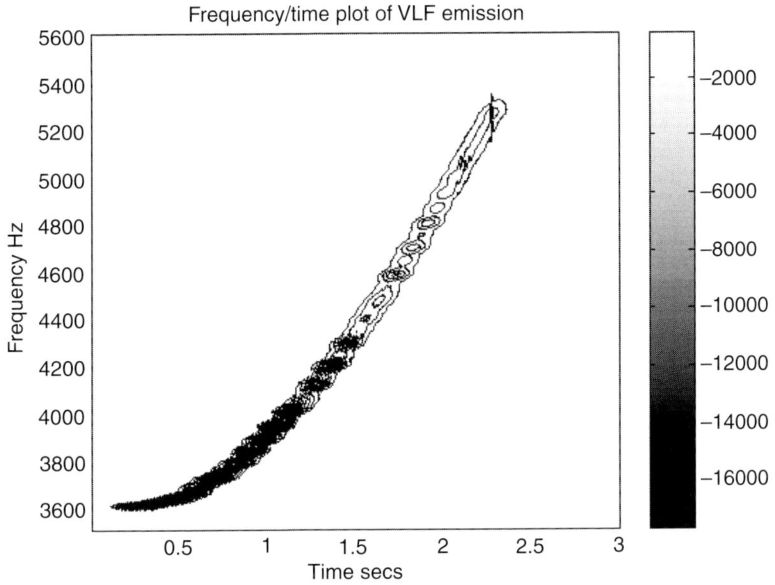

Figure 11.3 Frequency–time spectrogram of the stable riser at the exit of the simulation box in the equatorial plane corresponding to the 'base case'. The units of spectral power are arbitrary in all the spectrograms shown (from Nunn *et al.*, 2005).

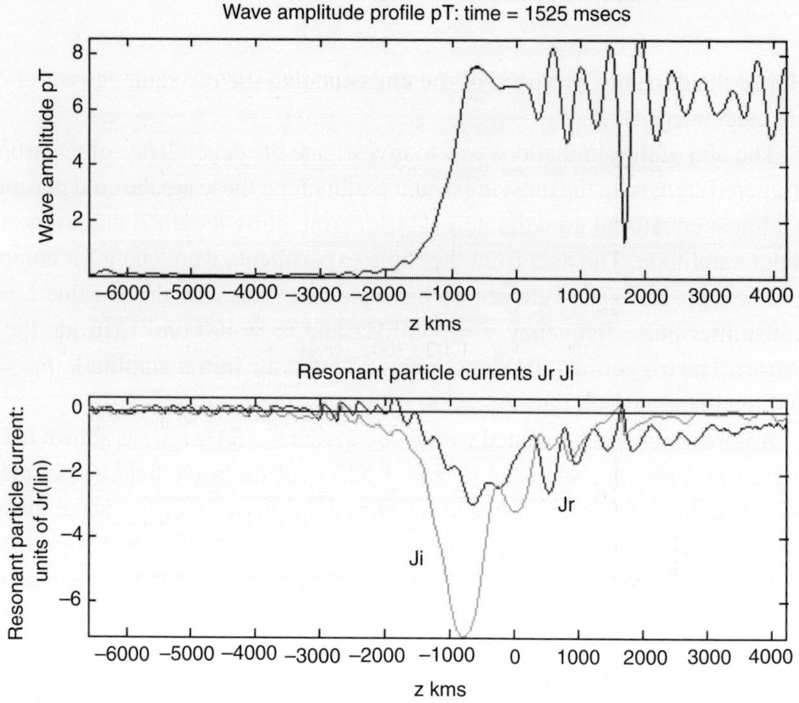

Figure 11.4 Snapshot of wave amplitude and real and imaginary resonant electron currents in the riser generation region in the equatorial zone at $L = 4.4$, for the base case at a time $t = 1525$ ms after the imput of the pulse into the simulation box (from Nunn *et al.*, 2005).

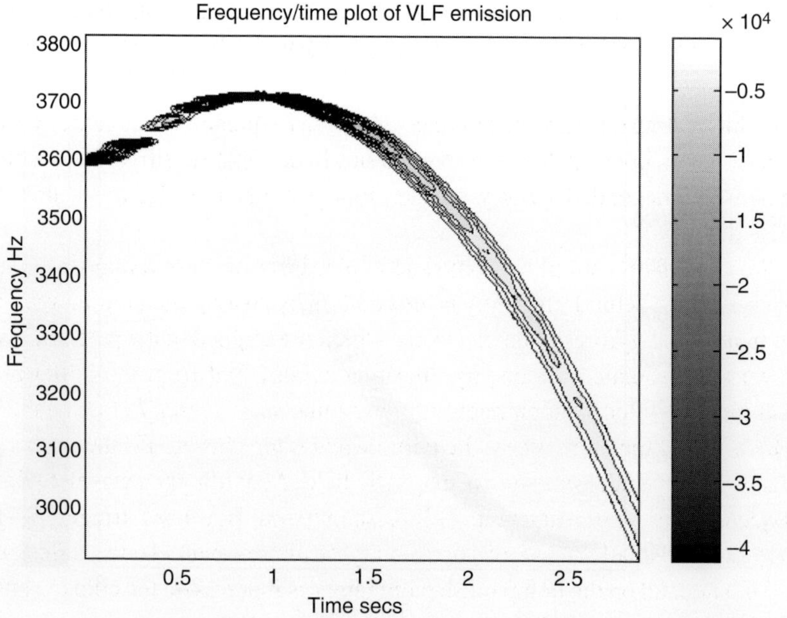

Figure 11.5 Spectrogram of the strong stable faller produced with a linear growth rate of 90 dB s^{-1}. The spectral form is a downward, or falling, hook with an offset frequency of +20 Hz (from Nunn *et al.*, 2005).

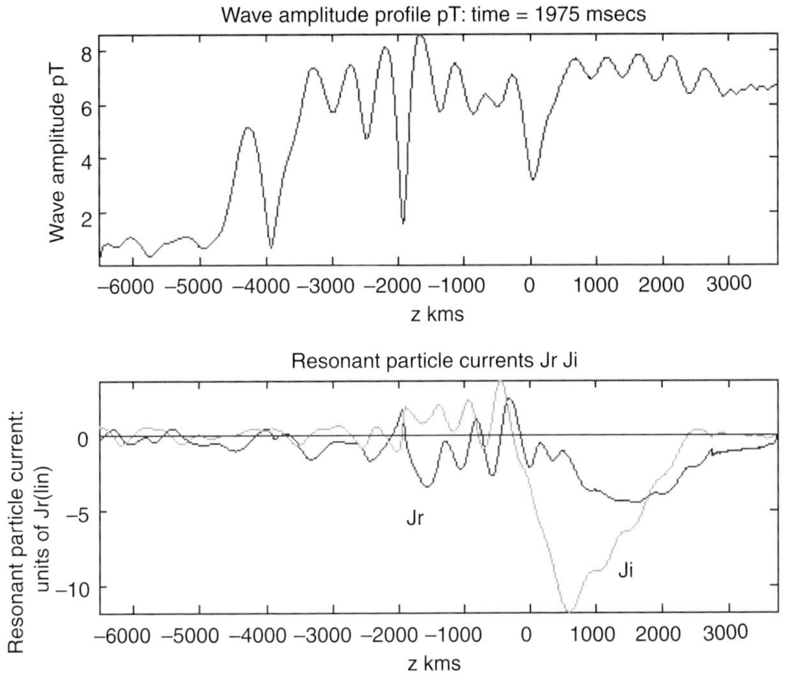

Figure 11.6 Snapshot of the wave amplitude and resonant electron currents (real and imaginary) in the faller generation region of the downward (falling) hook shown in Fig. 11.5, at $t = 1975$ ms. Note that the wave profile extends well upstream of the equator (i.e. on the transmitter, southern, side) and that consequently electron trapping by the wave occurs in this region; this results in a positive value for J_i, the resonant particle current in the wave magnetic field direction (from Nunn *et al.*, 2005).

case. Interesting results were obtained when the frequency sweep rate of the triggered emission was investigated. This turned out to be most sensitive to the cold plasma density and depended weakly on the other parameters ($\gamma_{\rm lin}$, input amplitude and pulse duration).

We now give a simplified physical explanation of these numerical simulations. The basic idea behind triggering is the creation by an amplified initial wave of coherent bunches of gyroresonant electrons, which are trapped in the potential well of the wave; these radiate a secondary emission as they exit from the initial pulse. The field-aligned velocity component of these bunches is initially close to the resonant velocity $v_{\rm R}$ at the point where the pump wave is terminated. Further, this velocity is changing in the inhomogeneous magnetic field. Accordingly a wavelet with another frequency will be in resonance with these bunches. Because a triggered signal produces a very small change of the velocity of the resonant electrons the sweep rate will not depend on the pump pulse parameters as much as on the cold plasma density and the geomagnetic field parameters.

In this scenario, the dynamic spectrum of a triggered emission will be determined in the main by the position of the generation region relative the equator. This is seen clearly from Figs. 11.4 and 11.6. A riser is generated by the resonant current moving away from the equator (towards the northen hemisphere), while the hook includes the equatorial region within the generation region. However, the full dynamics of the triggered signals and of the distribution function of the resonant electrons are much more complicated and include many fine scale features, some of which will be discussed in the next section.

11.3 Formation of the dynamic frequency spectrum of triggered VLF emissions: an analytical approach

These numerical simulations show that only sufficiently large values of $\gamma_{\text{lin}} \sim 60$–$90$ dB s^{-1} trigger emissions. Such large growth rates can only occur for specific energetic electron distribution functions such as a beam or a step-like distribution in the geomagnetic field-aligned velocity component. According to the discussions given in Chapters 8 and 10, the most likely condition to occur in nature is the formation of a step-like distribution function. Radiation belt electrons with such a distribution function can exist in a metastable state for a long time until an external whistler wave, which is in cyclotron resonance with the step, modulates the velocity distribution, initiating a strong (nonlinear) instability and generating a secondary emission. In such a situation two possibilities for ELF/VLF triggering exist. Either the source of energy is an external strong wave, which produces a secondary beam of electrons with a well-defined velocity, or a step, modulated by an external VLF pumping wave, is the source of the secondary VLF emission.

We first consider the first possibility. An analytical self-consistent approach in this case is based on the second-order cyclotron resonance effect (Chapter 7). This self-consistency includes:

- calculation of the cyclotron amplification of an initial quasi-monochromatic whistler wave packet;
- formation of an energetic electron beam, produced by this packet; and
- analysis of the secondary whistler wave generated by the beam, with the geomagnetic field inhomogeneity being taken into account.

A schematic picture of the generation of a triggered VLF emission by the secondary beam is given in Fig. 11.7. Figure 11.8 demonstrates advantages of the initial wave amplification Γ_{step}, by a step, in comparison with the amplification Γ_{sm} by a smooth distribution function: $\Gamma_{\text{step}}/\Gamma_{\text{sm}} \sim 10$.

The next question concerns the formation of the electron beam, which was considered in Chapter 6. Accordingly, the velocity of escaping electrons is equal to the resonance velocity of the pump wave at the trailing edge of the pulse, and is pushed further along the magnetic field line by the inhomogeneity of the geomagnetic field. This beam can generate the secondary wavelet. The maximum amplification of this wavelet occurs when the second-order cyclotron resonance condition (7.45) is fulfilled, which gives us the dynamic frequency spectrum of the triggered signal.

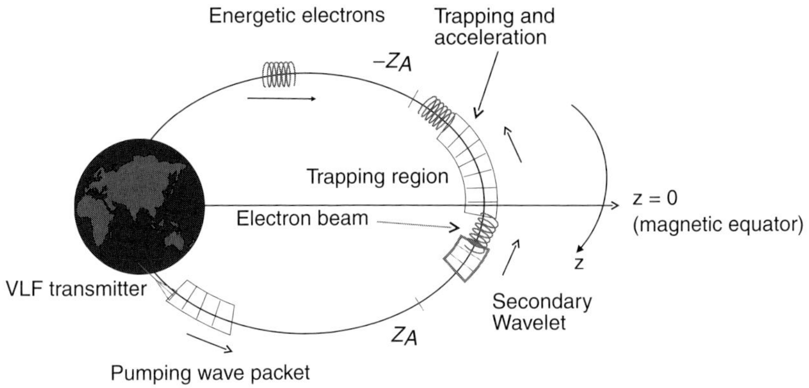

Figure 11.7 Diagram illustrating the generation of a triggered VLF emission. (based upon Trakhtengerts *et al.*, 2001).

Trakhtengerts *et al.* (2001) gave results for the particular parameters: $L = 4$, $n_c = 300\ \text{cm}^{-3}$, $\omega/\omega_{BL} = 0.5$, the final amplitude of the pumping wave after amplification $B_\sim \sim 16$ pT, pump wave duration $T_t = 0.5$–1.3 s. The results of these calculations are shown in Fig. 11.9. The interaction regions are located in the region of negative z in the case of shorter T_t ($s = -0.15$ to -0.3, where $s = z/a$ is the normalized coordinate), becoming closer to the equator as T_t increases. On the other hand, the interaction regions for longer T_t are located in the region of positive s, ranging from 0.2 to 0.3, the region for riser generation. In contrast with the case of small T_t (faller emissions), the position of the interaction region for large T_t (risers) does not depend strongly on T_t.

These results agree qualitatively with the numerical simulations, but the times at which the triggered emissions begin correspond poorly with the experimental data. A more exact solution should include the bunching of a beam at the exit from the pump wave together with the second-order cyclotron resonance effects.

We now consider the second possibility, namely the generation of triggered emissions by a step modulated by a VLF whistler-mode wave. Numerical simulations of triggered emissions show very complicated changes of the distribution function of the resonant electrons. Together with phase-modulated beams some regions of velocity space appear with very sharp velocity gradients at their edges. When we have the initial distribution function as a step over the field-aligned velocity component such a modulated deformation, which appears under the interaction with a pump wave, is shown schematically in Fig. 11.10. According to the nonlinear treatment of cyclotron interactions (Chapters 5 and 6) a pump wave traps resonant electrons near the resonant velocity (6.11) $v_R = (\omega - \omega_B)/k$ over a small range of velocities

$$\Delta v_R \sim \Omega_{tr}/k \tag{11.1}$$

where the trapping frequency Ω_{tr} is determined by the relation (6.10). Now we have two sources for secondary emission: the beam of trapped electrons considered earlier

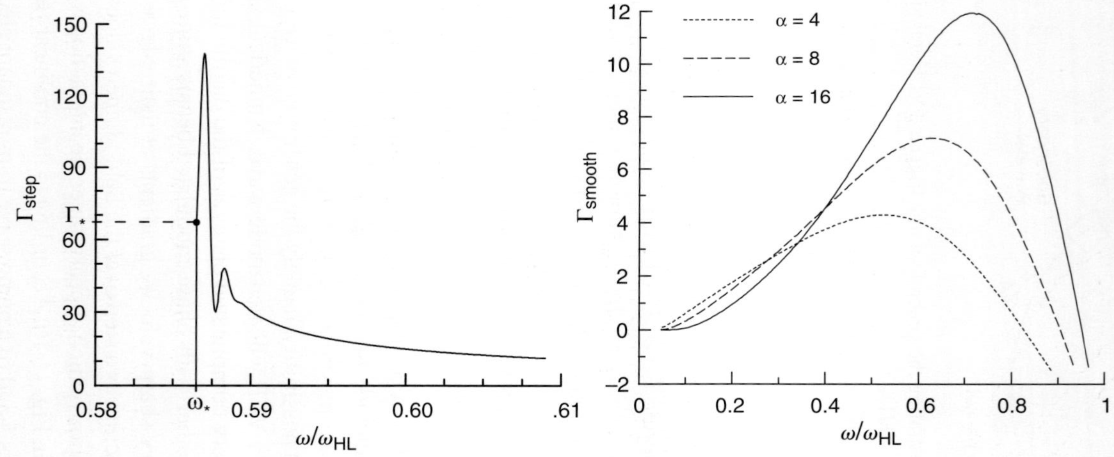

Figure 11.8 Amplification as a function of frequency ω, normalized by the electron cyclotron frequency at the equator at a particular L-value, ω_{BL} (shown here as ω_{HL}), in the case of a smooth (left) and step-like (right) distribution function. The parameters correspond to the experimental data presented by Bell *et al.* (2000). These are: $L = 3.4$, $\omega_{BL}/2 \sim 22$ kHz, $n_{\text{h}} = 10^{-2}$ cm^{-3}, $\omega_{pL} = 9 \cdot 10^5$ Hz, and $v_0 = 8 \cdot 10^4$ km s^{-1} ($W_0 \sim 18$ keV); α is the anisotropy parameter. According to both experimental data and numerical simulations (Nunn *et al.*, 1997, 2003), the necessary value of Γ should be ≥ 20 (≥ 60 dB). (Taken from Pasmanik *et al.*, 2002). (Copyright American Geophysical Union, reproduced with permission.)

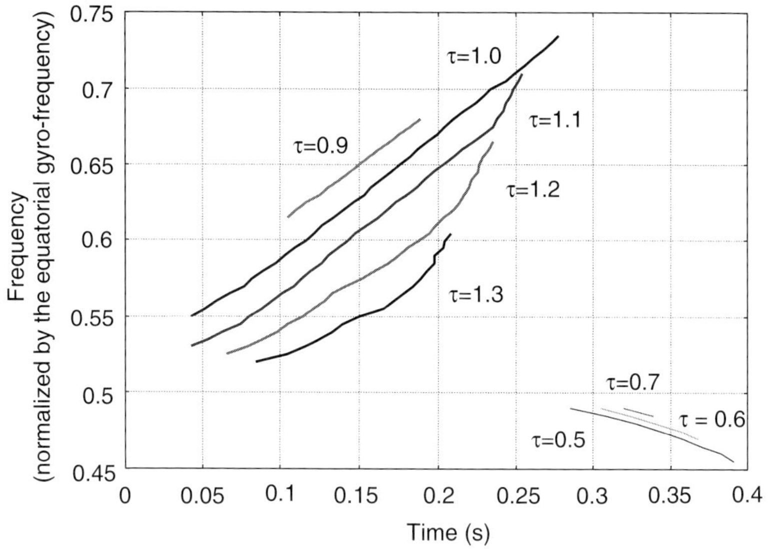

Figure 11.9 The frequency–time spectrogram of calculated triggered VLF emissions for different pulse durations ($T_t = 0.5, 0.6, 0.7, 0.9, 1.0, 1.1, 1.2$, and 1.3 s). Time t is reckoned from the time when electrons begin to escape from the quasi-monochromatic pumping wave packet; the emission frequency is normalized to the equatorial gyrofrequency. (Taken from Trakhtengerts *et al.*, 2001). (Copyright American Geophysical Union, reproduced with permission.)

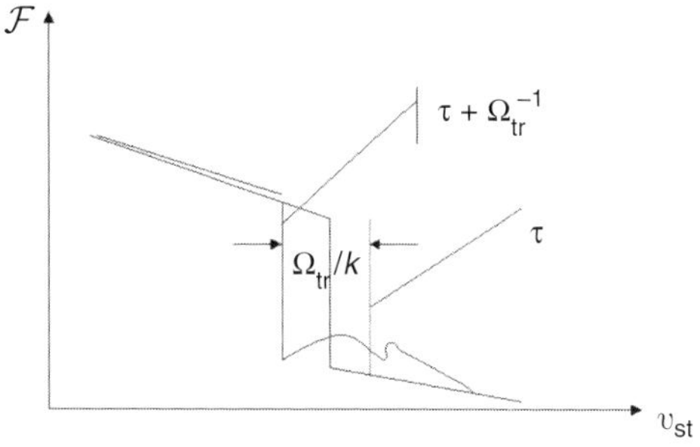

Figure 11.10 Diagram showing the evolution of the distribution function $\mathcal{F}$ of resonant electrons. Erosion of the step (subscript 'st') occurs subsequently in time over a small range of velocities $\Delta v_R \sim \Omega_{tr}/k$ for the time interval $\Delta t \sim 2\pi/\Omega_{tr}$, v_R being the resonance velocity (11.2), and Ω_{tr} the trapping frequency (6.10).

in the first part of this section, and a new modulated step, which corresponds to a sharp edge in velocity space (Fig. 11.11).

VLF triggering by a new modulated step can be understood as follows. As it propagates along the magnetic flux tube, the pump wave creates a strong modulation of the step at points s_{01} and s_{02} where the step velocity v_{st} coincides with the resonance velocity v_R. The deepest modulation will be at the trailing edge of an amplified pump packet. Due to the antenna effect (Trakhtengerts *et al.*, 2003b) in the inhomogeneous magnetic field, the modulated step will generate a wavelet with changing frequency, which is in local cyclotron resonance with the step. Actually this model coincides with the first phenomenological model of triggered VLF emissions, proposed by Helliwell (1967). We now know the place and the time of wavelet generation, and can find its frequency dependence on the coordinate z and time t

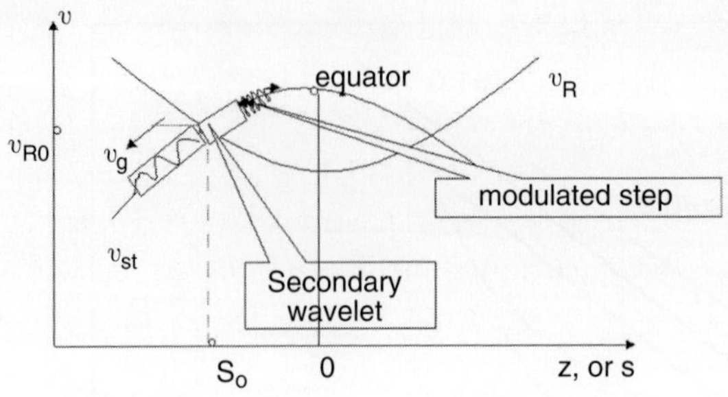

Figure 11.11 Diagram illustrating the generation of a secondary emission by a wave-modulated step in the distribution function, occurring within a few thousand km of the equator. The modulation appears under the interaction with a pump wave at the point S_o, where the initial step velocity v_{st0} coincides with the pump wave resonant velocity v_{R0}.

from the relation

$$v_R \equiv (\omega_B - \omega)/k \approx v_{st}. \tag{11.2}$$

To use this we should know the spatial and temporal dependence of the step velocity, i.e. $v_{st}(s, t)$. In the case of our interest, when $\Delta v_R \sim \Omega_{tr}/k \ll v_{st}$, the equation of motion of the step, averaged over the oscillations of electrons in the potential well of the wave, can be written as

$$\frac{dv_{st}}{dt} = -\frac{1}{2}\mu v^2 \frac{dB}{dz} - \frac{\Omega_{tr}^2}{2\pi k}, \quad \frac{dz}{dt} = v_{st} \tag{11.3}$$

where the first term on the right-hand side of Equation (11.3) describes the mirror force in the drift approximation, and the second nonlinear term takes into account the erosion of the step (11.1), which occurs during the time interval $\Delta t = 2\pi/\Omega_{tr}$ (Fig. 11.10).

Equations (11.2), (11.3) yield the coordinate z_g and time t_g of the generation of every frequency $f = \omega/2\pi$ in the dynamical spectrum of the secondary signal. The values of z_g and t_g are related by the relation:

$$t_g = t_0 + \int_{z_0}^{z_g} \frac{d\xi}{v_{st}(\xi)} \tag{11.4}$$

where $t = t_0$ corresponds to the time when the step meets the termination (trailing edge) of the pump wave ($z = z_0 < 0$, Fig. 11.11). At any arbitrary point z along the wave propagation path, the dynamic spectrum of a triggered emission can be found from the relation:

$$t(f, z) = t_g + \int_{z}^{z_g} \frac{d\xi}{v_g(f, \xi)} \tag{11.5}$$

where v_g is the group velocity, and the coordinate ξ is the variable of integration.

According to the results of Chapter 4 the most effective amplification of a quasi-monochromatic whistler-mode wave by a step-like distribution function takes place near the equatorial plane of a magnetic flux tube, where the geomagnetic field can be approximated by a parabolic expression (2.8)

$$B = B_L(1 + a^{-2}z^2), \quad a \simeq \left(\sqrt{2}/3\right) LR_0. \tag{11.6}$$

Suggesting further that the frequency change in a triggered emission $f - f_0$ is not too large ($\Delta f_{\text{trig}}/f_0 < 0.3$) and that the wave amplitude is constant, we can write the solution of (11.3) as

$$v_{\text{st}}^2 = v_{\text{st0}}^2 + v_{\perp L}^2(s_0^2 - s^2) + rsv_{\text{st0}}^2, \quad \int_{z_0}^{z} \frac{d\xi}{v_{\text{st}}(\xi)} = t \tag{11.7}$$

where

$$r = \frac{\Omega_{\text{tr}}^2 a}{2\pi k v_{\text{st0}}^2}, \tag{11.7a}$$

$s = z/a$, $v_{\perp L}$ is the energetic electron velocity component across the magnetic field at the equator at the L-value considered and Ω_{tr} is determined by the relation (6.10). For the case of low frequencies $(f/f_{BL}) \ll 1$ it is easy to find from (11.2) that

$$\frac{f}{f_0} = \frac{(1+s^2)^2}{(1+s_0^2)^2 + 2r(s_0 - s) + \dfrac{v_{\perp L}^2}{v_{\text{st}}^2}(s^2 - s_0^2)} \tag{11.8}$$

where f_0 is the frequency of the pump wave, s_0 is the coordinate of the crossing of the step velocity's curve (11.7) when $r = 0$ and of a curve corresponding to the cyclotron resonance condition for the pump wave (Fig. 11.11).

We can estimate from (11.8) the wave amplitude, when the nonlinear effect (the term $2r(s_0 - s)$) can be neglected in comparison with the magnetic field inhomogeneity term. If the magnetic field inhomogeneity's effect prevails during the formation of a triggered emission's spectrum, we can find from (11.8) the maximum frequency deviation (for $s_0^2 \ll 1$, $s = 0$) as

$$\delta f_{\text{m}} = \frac{\Delta f_{\text{m}}}{f_0} \approx \left(2 + \frac{v_{\perp L}^2}{v_{\text{st}}^2}\right) s_0^2. \tag{11.9}$$

The nonlinear effects can be neglected if

$$2r < \left(\frac{v_{\perp L}^2}{v_{\text{st}}^2}\right) s_0. \tag{11.10}$$

Substituting s_0 from (11.9) and r from (11.7a), we find (for $\delta f_{\text{m}} \sim 0.3$, $v_{\perp L} \sim v_{\text{st}}$)

$$\omega_{\text{b}} \equiv \frac{eB_\sim}{mc} < \frac{v_{\perp L}}{a} \tag{11.11}$$

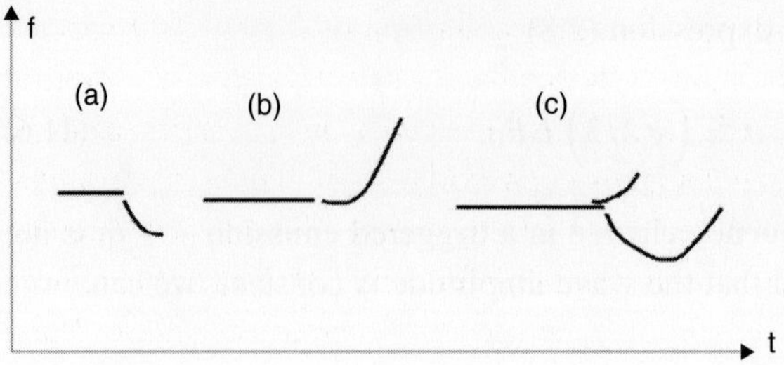

Figure 11.12 Three types of spectrum for triggered emissions.

where $B_\sim$ is the wave magnetic field amplitude (after amplification). It gives, for $L = 4$ and $v_{\perp L} \sim 10^{10}$ cm s^{-1}, $B_\sim < 50$ pT. The observed amplitude of the triggered emission is, as a rule, less than 50 pT.

Now we can find the spectrum of the triggered emission from the simplified formula, which follows from (11.8) under the condition that $r \to 0$, and s^2, $s_0^2 \ll 1$:

$$\frac{f}{f_0} \approx \left[1 + \frac{v_{\perp L}^2}{v_{s0}^2}(s^2 - s_0^2)\right]^{-1}. \tag{11.12}$$

Figure 11.12 shows three types of spectrum which can be produced. The spectral form 11.12(a) is produced by a weak and short pump pulse after double amplification at the points s_{01} and s_{02} (Fig. 11.11). The spectral form 11.12(b) is generated when the modulated step moves from s_{02} into a region of increasing geomagnetic field. The third spectral form 11.12(c) is generated when the interaction is effective at both points s_{01} and s_{02}. It is clear that the spectra shown in Fig. 11.12 are close to those observed in the Siple experiments (Helliwell, 1965, 1983).

If the wave amplitude $B_\sim$ is sufficiently large ($B_\sim \gtrsim 50$ pT), nonlinear effects should be taken into account. Such a situation occurs for chorus, considered in the next section.

11.4 Generation mechanism for ELF/VLF chorus

Chorus emissions are the most intriguing signals amongst naturally occurring ELF/VLF radiation. They consist of a succession of discrete elements with rising frequency, with a repetition period of $T \sim 0.1$–1 s. According to ground-based observations, the typical duration of a chorus event is 0.5 to 1 h or more, in the morning hours. Recent Cluster satellite measurements (Cornilleau-Wehrlin *et al.*, 2003, Parrot *et al.*, 2003) have shown that chorus is generated close to the equatorial plane of a certain geomagnetic flux tube, as whistler-mode waves whose wave vectors $\vec{k}$ are close to the direction of the magnetic field line $\vec{B}$. As a rule, chorus is accompanied by hiss emissions which serve as a lower frequency background out of which the discrete elements appear. For more detailed information about chorus emissions we refer to Helliwell (1965, 1969) and Sazhin and Hayakawa (1992).

It is generally accepted that the generation mechanism of chorus is connected with the cyclotron instability of radiation belt electrons. The similarity of the spectral forms of chorus elements and of triggered ELF/VLF emissions has stimulated the application of the theory of triggered signals to an explanation of chorus. However, we meet the same difficulties as in the case of VLF triggered emissions. In particular, a computational analysis of the cyclotron instability with the distribution function of energetic electrons as measured in satellite experiments (Nagano *et al.*, 1996, Nunn *et al.* 1997) showed that a chorus-like element appeared from weak initial whistler-mode waves, but the CI growth rate had to be one or two orders of magnitude greater than followed from estimates based on the smooth distribution functions of the energetic electrons observed.

A second problem arises for chorus generation: that is how to explain the appearance of a succession of discrete elements. Bespalov and Trakhtengerts (1978a) suggested a mechanism based on a resonance between the oscillations of a wave packet reflected between conjugate ionospheres and the bounce oscillations of energetic electrons. However, experiments have revealed that the repetition period is often smaller than both the bounce period of electrons between the magnetic mirror points and the one-hop wave propagation time between conjugate ionospheres.

The solution of these two problems, namely strong amplification and a succession of discrete elements, seems to be linked to the character of the quasi-linear relaxation of the smooth distribution function of energetic electrons in the process of CI development. This leads to the formation of a specific step-like deformation of the distribution function, which drastically changes all further development of the wave–particle interactions. We suggest that chorus generation begins in regions of rather dense cold plasma, which correspond to ducts outside the plasmapause or at the plasmapause. Energetic electrons come into the generation region during their magnetic drift (in longitude, primarily) from an injection point which is often near midnight local time. The initial distribution function of injected electrons is taken to be anisotropic but smooth in velocity space with $\alpha = v_{\perp 0}^2/v_{\parallel 0}^2 > 1$, where $v_{\perp 0}$ and $v_{\parallel 0}$ are the electron's characteristic velocities across and along the magnetic field, respectively; noise-like ELF/VLF emissions are excited in the process of the CI development.

At this stage a step-like deformation of the distribution function is formed, which divides velocity space into two regions, corresponding to electrons in resonance with waves and to electrons for which the gyroresonant condition cannot be fulfilled (Chapters 8 and 9). The characteristic time for this preliminary quasi-linear stage T_{QL} is equal to (see (8.34) and (8.36)):

$$T_{\text{QL}} = T_B(\Delta\theta_L)^2/D, \qquad D \approx \left(\frac{4\pi e}{mc}\right)^2 \left(\frac{a\mathcal{E}_{\text{w}}}{kW_0}\right), \tag{11.13}$$

where θ_L is the electron's pitch-angle in the equatorial plane, T_B is the energetic electron bounce period, $W_0 = \dfrac{m}{2}\,(v_{\perp 0}^2 + v_{\parallel 0}^2)$, e and m are the charge and mass of an electron, c is the velocity of light, a is the characteristic scale of the magnetic field inhomogeneity (11.6), k is the whistler-mode wave vector, ε is the whistler wave energy density, and D is the pitch-angle diffusion coefficient.

After the formation of the step on the distribution function, the preliminary stage is finished, and a new very fast hydrodynamic (HD) stage of the CI starts, with the growth rate (Trakhtengerts, 1995) being given by

$$\gamma_{\mathrm{HD}} \approx \omega_{BL} \left(\frac{\Delta n_{\mathrm{h}}}{n_{\mathrm{c}}} \right)^{1/2} \gg \gamma_{\mathrm{sm}} \tag{11.14}$$

Here ω_{BL} is the electron gyrofrequency at the equator, n_{h} and n_{c} are hot (energetic electron) and cold plasma densities, respectively, Δn_{h} is the step height, and γ_{sm} is the growth rate for a smooth distribution function with the same density of energetic electrons. This stage begins, when the step width $k\,\Delta v_{\mathrm{st}}$ becomes less than the growth rate γ_{HD}

$$|k\,\Delta v_{\mathrm{st}}| < \gamma_{\mathrm{HD}}. \tag{11.15}$$

The hydrodynamic stage of the CI manifests itself as the transition to a new, backward wave oscillator (BWO) generation regime, when the absolute CI develops in a narrow region which is symmetrical about the equatorial plane. A schematic diagram of a BWO generator is shown in Fig. 11.13. Usually, the absolute instability appears in generators of electromagnetic emissions due to positive feedback which is organized by mirrors. These mirrors reflect waves back into the generation volume, and the wave intensity grows in time up to the saturation level which is determined by nonlinear effects. In the case of BWO generation, a whistler-mode wave packet meets, at the entrance of the generation region near the equator, the phase-bunched electron beam (the beam wave mode) formed by the previous wave packet; this amplifies the next packet more strongly. Actually this is the positive feedback connection through a beam wave mode. The BWO generation regime was considered in detail in Chapters 4, 5 and 6.

Two necessary conditions have to be satisfied for a BWO to operate: (1) the wave has to interact with a beam moving in the opposite direction, and (2) the velocity

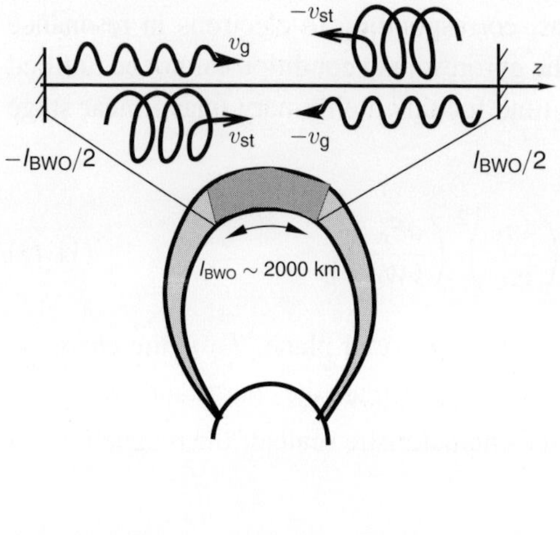

Figure 11.13 Schematic diagram of a magnetospheric BWO. Electrons having a step-like distribution move in the ($\pm z$)-directions with the velocity $\pm v_{\mathrm{st}}$ and independently generate chorus waves propagating in the opposite directions (i.e. with group velocities $\mp v_{\mathrm{g}}$, respectively). (Taken from Trakhtengerts *et al.*, 2004). (Copyright American Institute of Physics, reproduced with permission.)

spread of the beam (step) must be as small as defined in (11.15), for the hydrodynamic stage of the CI to be realized. In the opposite case to Equation (11.15), the thermal motion of the beam electrons suppresses the beam wave mode.

Both of these conditions are fulfilled for the whistler-mode wave if the distribution function $\mathcal{F}$ of the electron beam is a delta-function

$$\mathcal{F} \sim \delta(v_{\|m} - v_{\|})\,\delta(v_{\perp m} - v_{\perp}) \tag{11.16}$$

or, as a function with a step-like deformation,

$$\mathcal{F} \sim \Phi(v_{\perp}, v_{\|})\,\mathrm{He}(v_{\mathrm{st}} - v_{\|}) \tag{11.17}$$

where $\Phi(v_{\perp}, v_{\|})$ is a smooth function of $v_{\perp}$ and $v_{\|}$. $\mathrm{He}(x)$ is equal to unity, if $x > 0$, and to zero if $x < 0$. Here v_{st} is the velocity corresponding to the step.

The BWO regime is excited when the beam density exceeds some threshold value. With an increase of the beam density the BWO regime changes from a continuous regime (when the wave amplitude is constant) to a periodic regime, which is accompanied by a periodic modulation, of wave intensity, and then to stochastic modulation (Table 5.1). These bifurcations are characterized by a single dimensionless parameter, an effective length l_{eff}, which is given by

$$l_{\mathrm{eff}} = \gamma_{\mathrm{HD}}\, l (v_{\mathrm{st}} v_{\mathrm{g}})^{-1/2} \tag{11.18}$$

where γ_{HD} is the hydrodynamic growth rate in the case of the distribution function as a step function, l is the 'working' length (the geometrical length of a device along the magnetic field, if it is homogeneous), v_{st} is the step velocity, and v_{g} is the group velocity. In this case the threshold value of l_{eff} for BWO generation (the continuous regime) is equal to two. The first bifurcation, the transition to the periodic regime, takes place, when $l_{\mathrm{eff}} \approx 3$, and the stochastic behaviour begins when $l_{\mathrm{eff}} \geq 6$. The periodic regime is characterized by the modulation period

$$T_{\mathrm{BWO}} \simeq 1.5(1/v_{\mathrm{g}} + 1/|v_{\mathrm{st}}|). \tag{11.19}$$

Now we discuss the length l of the magnetospheric generator, which determines l_{eff} and T_{BWO}. This length depends strongly on the inhomogeneity of the geomagnetic field. At first we estimate the minimum value of l, termed l_{BWO}, neglecting second-order cyclotron resonance effects. In this case the wave frequency is fixed, and l_{BWO} can be found from the condition of limited phase mismatching, relation (4.54). So, we shall have:

$$l_{\mathrm{BWO}} = \left(\frac{R_0^2 L^2}{k} \right)^{1/3} \tag{11.20}$$

where R_0 is the Earth's radius, and L is the McIlwain parameter of the geomagnetic flux tube where the BWO is operating. For real magnetospheric conditions, the second-order cyclotron resonance is important: then the cyclotron resonance is supported along the step trajectory due to a suitable change of the wave frequency $\omega(z)$ and to a nonlinear change of the electron velocity component $v_{\rm st}$.

Indeed, after the transition to the periodic BWO regime, we deal with the particular case of triggered emissions, when a succession of quasi-monochromatic wavelets starts from the equatorial region of the magnetic flux tube and triggers signals with rising frequency. The parameters of these signals can be estimated from a theoretical model of triggered emission, developed in Section 11.3. This model considered VLF triggered emissions generated by a modulated step, when the step parameters were below the threshold (4.27) for BWO generation. In this case the dynamical spectrum was described by (11.12).

ELF/VLF chorus signals are usually more intense than VLF triggered emissions. According to the satellite data (Hattori *et al.*, 1991, Nagano *et al.*, 1996, Santolik *et al.*, 2003) the wave magnetic field amplitude of chorus can reach $B_\sim \gtrsim 100$ pT and, occasionally, 300 pT. For these values and $L \geq 4$ the main contribution to the dynamic spectrum of chorus, the expression (11.8), is determined by the nonlinear term $2r(s_0 - s)$. As a result we have, inside the generation region:

$$\frac{f}{f_{\min}} \approx \frac{1}{1 + 2r(s_0 - s)} \approx \frac{1}{1 - 2r \cdot \tau_{\rm st}}, \qquad s_0 - s < 0 \tag{11.21}$$

where $\tau_{\rm st} = v_{\rm st0}t/a$, and the other definitions are as in (11.7a), (11.8); $f_{\min}$ is the minimum frequency which is generated at the entrance of the energetic electrons to the BWO, with an initial step velocity $v_{\rm st0}$ (Fig. 11.11).

We estimate the wave amplitude $B_\sim$ coming into the expression for r (11.7a), from the relation, which is valid in the case of an absolute instability (see (5.60)),

$$(\Omega_{\rm tr}/\gamma_0) \simeq (16/3\pi)^2 \simeq 3.2 \tag{11.22}$$

where γ_0 is the initial growth rate. In our case this is the growth rate for BWO generation after the transition to the periodic regime. When $r\tau_{\rm st} \ll 1$, the frequency sweep rate, according to (11.21) and (11.22), is equal to:

$$\frac{df}{dt} \sim \left(\frac{2rv_{\rm st}}{a}\right) f \sim \frac{f}{2f_{BL}}\gamma^2_{\rm BWO}. \tag{11.23}$$

We can summarize our theoretical model as follows. Energetic electrons enter the generation region (a detached cold plasma cloud, or the plasmapause) during their magnetic drift in longitude. After that the CI switches on and QL relaxation begins, which prepares the transition to the BWO generation regime. We can estimate the duration of this stage, taking into account equation (11.13) and using GEOS-1 data (Hattori *et al.*, 1991). For $L = 6.6$ and $f_{BL} \simeq 3$ kHz, the intensity of hiss emission, accompanied by chorus generation, $\varepsilon_f \simeq 600\,({\rm pT})^2/{\rm Hz}$, the energy of energetic electrons $W_{\rm e} \sim 20$ keV, their bounce period $T_B \sim 1.4$ to 1.8 s, and the chorus frequencies $f \sim 1$ to 1.5 kHz. Substituting the corresponding values into

Equation (11.13) we obtain (for $L = 6.6$)

$$T_{\mathrm{QL}} \sim 3T_B \sim 5 \text{ s}.$$

This is in accordance with the supermodulation period of chorus, which is seen in experiments (Sazhin and Hayakawa, 1992). From our point of view these modulations are interpreted as spatial-temporal features of chorus generation, which are due to local, different stages of QL relaxation before the BWO regime starts. The spatial scale of such localizations across the magnetic field near the equatorial plane is

$$d \sim v_{\mathrm{D}}\, T_{\mathrm{QL}} \sim (1 \text{ to } 3) \cdot 10^2 \text{ km}$$

where v_{D} is the electron's magnetic drift velocity in the equatorial plane.

After the formation of a step-like deformation to the distribution function the fast stage of chorus generation occurs, with a total duration $T_{\mathrm{ch}} \geq T_B$. The succession period of chorus elements, T_{BWO}, is determined by Equation (11.19) and, for the particular case of GEOS-1 data, occupies the interval $0.2 < T_{\mathrm{BWO}} < T_B \sim 1.6$ s, which is in good agreement with the GEOS-1 data.

Recent measurements of chorus emissions by the four closely spaced Cluster spacecraft at $L \sim 4.4$ provide important information concerning the chorus generation mechanism. They confirm our assumed properties of the wave source such as its strong localization near the equator of a particular geomagnetic flux tube, an almost parallel average wave-vector direction with respect to the geomagnetic field, and an energy flux direction pointing away from the generation region (Parrot *et al.*, 2003, Santolik *et al.*, 2003). Inside this region, Cluster discovered strong temporal and spatial variations in the amplitude, with correlation scale lengths of the order of only 100 km across the magnetic flux tube (Santolik *et al.*, 2004). The wave electric field reached 30 mV m^{-1}, corresponding to a wave magnetic field of up to 300 pT for the conditions of the experiment, and the maximum growth and damping rates are very large, of the order of a few hundreds s^{-1}. Both these and other properties of the detected chorus emissions are explained on the basis of the backward wave oscillator mechanism. Table 11.1 summarizes a comparison of BWO model results with Cluster data.

The typical value taken for the relative height of a step $b = \Delta n_{\mathrm{h}}/n_{\mathrm{h}}$ follows from computer simulations (see Chapters 9 and 10).

Evidence confirming the BWO model as the mechanism responsible for the formation of the chorus spectrum (relations (11.2)–(11.5) and (11.21)) was obtained from analyses of Cluster data (Trakhtengerts *et al.*, 2007). The theory yields the spectral form of the dependence of a separate chorus element on the wave energy flux direction and on the position of observation point inside the generation region. In particular, it predicts that only part of a chorus element is visible inside the source region (Fig. 11.14). For rising frequency chorus elements, the lower frequencies are generated downstream and, hence, becomes invisible as a receiver moves upstream within the source region (Fig. 11.14). These spectral features were verified using

Table 11.1 Summary of chorus parameters in the BWO model. Numerical estimates are obtained for the event studied experimentally by Santolik *et al.* (2003). The basic parameters are $L = 4.4$, cold plasma density $n_c \sim 2\ \mathrm{cm}^{-3}$, and $\omega/\omega_{BL} \approx 0.45$. From that, we obtain the resonant parallel energy $W_{\mathrm{res}} = (m/2)((\omega_{BL} - \omega)/k)^2 \simeq 62$ keV and the wavelength in the plasma medium $\lambda \sim 26$ km. The flux density of energetic electrons is assumed to be $S \sim 4 \cdot 10^8\ \mathrm{cm}^{-2}\mathrm{s}^{-1}$, and the threshold parameter q is determined by (4.28).

	Length of generation region, l_{BWO}	Characteristic period between successive elements, $T > T_{\mathrm{BWO}}$	Growth rate, γ_{BWO}	Frequency drift, df/dt	Wave amplitude, $\lvert B_{\sim}\rvert$
Theory	$1.76\left(\frac{R_E^2 L^2}{k}\right)^{1/3}$	$> l_{\mathrm{BWO}}\left(\frac{1}{v_{\mathrm{g}}} + \frac{1}{v_{\mathrm{st}}}\right)$	$\frac{\pi}{T_{\mathrm{BWO}}}\left(q - \frac{\pi}{2}\right)$ $\sim 1.2 \cdot 10^2\ \mathrm{s}^{-1}$	$0.5\gamma_{\mathrm{BWO}}^2$	$\frac{10mc}{ek\,v_{\perp 0}}\gamma_{\mathrm{BWO}}^2$
	2600 km	> 0.025 s	Note that $b = \frac{\Delta n_{\mathrm{h}}}{n_{\mathrm{h}}} = 0.17$	$0.7 \cdot 10^4/\mathrm{Hz\ s}^{-1}$	~ 100 pT
Observation	≈ 2000 km	0.02–0.5 s	34–420 s^{-1}	$1.5 \cdot 10^4/\mathrm{Hz\ s}^{-1}$	100–300 pT

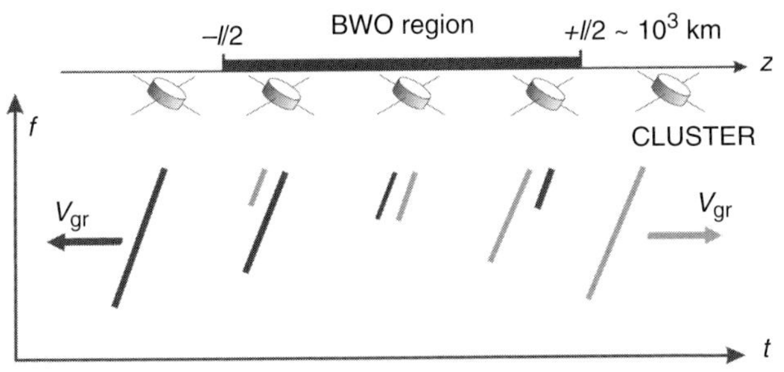

Figure 11.14
Diagram showing the formation of chorus elements. Black and grey lines on the dynamic spectrogram show elements propagating to the left and right, respectively, with group velocity v_g. (Taken from Trakhtengerts *et al.*, 2007). (Copyright American Geophysical Union, reproduced with permission.)

Cluster data. The chorus source was localized using multicomponent measurements of the wave electric and magnetic fields. The distributions of normalized lower and upper frequencies of chorus elements for different Cluster 1 locations, with respect to the chorus source, are shown in Fig. 11.15. The analysis confirmed that, in agreement with the theoretical model, the spectrum of chorus elements detected lacked the lower frequencies at the centre of the source region, at the magnetic equator.

Chorus is a most effective mechanism for accelerating radiation belt electrons. Its efficiency is due to its discrete spectral form as a quasi-monochromatic wave packet with varying frequency, which traps electrons and moves them through velocity space. Triggered emissions, discussed in chapter 6, act similarly. The non-relativistic analysis of Trakhtengerts *et al.* (2003) showed that the energy gain made by an electron interacting with just a single wave packet whose frequency varies by $\sim$ (0.1–0.2) ω_B, e.g. from a lighting-generated whistler, can be several k_eV. Demekhov *et al.* (2006) generalized this analysis for the relativistic case and for chorus emissions. They showed that electrons whose perpendicular energies are greater than those which generate the chorus are accelerated effectively. An electron interacting with one chorus element just beyond the plasmapause can increase its energy by up to 30 keV.

11.5 Summary

After a brief review of the characteristics of triggered whistler-mode emissions observed at $L \sim 4$, particularly their narrow bandwidth, duration, saturation amplitude and frequency sweep rate, we discuss in this chapter how these features may be modelled and explained.

We first discuss the comprehensive numerical simulations of risers, fallers and hooks presented by Nunn *et al.* (2005) in a parametric study. The most important parameter determining the properties of the triggered emission in the case of weak

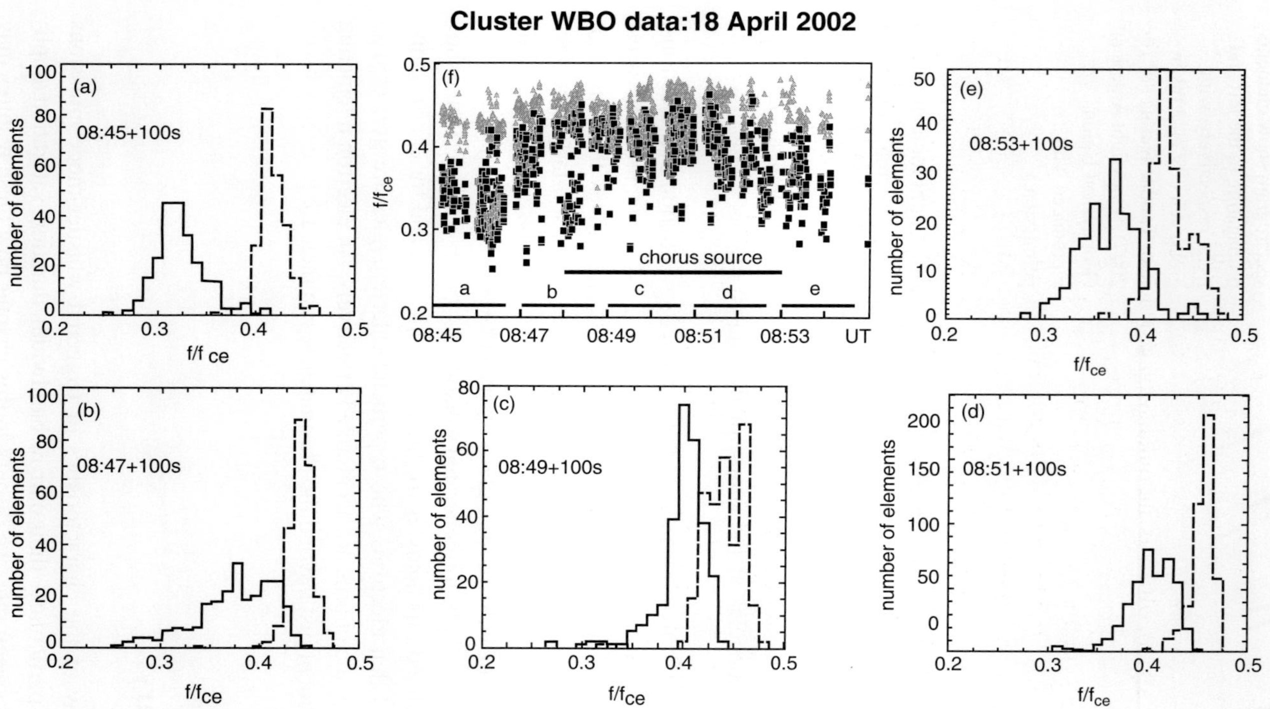

Figure 11.15 Top central panel shows the normalized lower (black) and upper (grey) frequencies of chorus elements detected by the Cluster 1 spacecraft at $L \simeq 4$ on 18 April 2002. The chorus source region obtained from Poynting flux measurements is marked by a horizontal line. The distributions of normalized lower and upper frequencies of the chorus elements, obtained for 100-s intervals for different Cluster 1 locations with respect to the chorus source, are shown in the other panels by solid and dashed lines, respectively; the electron cyclotron frequency in the equatorial plane (normally termed f_{BL}) is shown here as f_{ce}. The Universal Time of the start of each panel's data is indicated. (Taken from Trakhtengerts *et al.*, 2007). (Copyright American Geophysical Union, reproduced with permission.)

signals is the cold plasma density in the equatorial plane of the geomagnetic flux tube where the wave–particle interactions take place.

Next we consider an analytical and self-consistent model of a wave, amplified via a step-like distribution function, which produces a secondary beam of electrons. At the trailing edge of the pulse this generates a secondary wavelet which achieves maximum amplification when it obeys the second-order cyclotron resonance condition. Input pulses of varying durations produce fallers or risers (Fig. 11.9).

We then discuss a step which is modulated by a whistler-mode wave. In the spatially varying geomagnetic field along a flux tube, this generates a wavelet of changing frequency, predicted by Equation (11.8) or (11.12).

Finally, the important problem of chorus generation is tackled. Recent Cluster satellite observations within the source region (scale $\sim$ 1000–3000 km only) near the equatorial plane at $L \sim 4$ are well explained by the Backward Wave Oscillator (BWO) model. The characteristic interval between chorus elements, $\sim$ 0.02–0.5 s, is given by Equation (11.19). A nonlinear treatment of chorus generation predicts the spectral shapes observed in the source region (Equation (11.21) and Fig. 11.14), and also the sweep rate (Equation 11.23). Further, the supermodulation period of chorus, observed at $L \sim 6.6$, $\sim$ 5 s, is explained.

Chorus is generated when energetic electrons drift into a region of higher cold plasma density, e.g. at the plasmapause. The growth rate can reach 120 s^{-1} (Table 11.1); the BWO model predicts an amplitude of $\sim$ 100 pT in the source region, as observed by Cluster (Table 11.1).

Chapter 12

Cyclotron instability of the proton radiation belts

In previous chapters we have illustrated cyclotron wave–particle interactions using, as examples, the electron radiation belts. However, all the physical effects discussed are valid for the proton radiation belts too. In particular, a sufficiently dense cold plasma is needed for the proton cyclotron instability to be switched on. Thus the plasmapause and detached plasma filaments outside the plasmasphere are the most likely regions where the proton cyclotron instability can develop, as is the case for the electron radiation belt (Section 11.3).

As for electrons, the inward transport of protons across L shells serves as the main source, and the losses are determined by wave–particle interactions (which are most important during the main phase of a geomagnetic storm) and by other mechanisms of energetic ion loss; see Lyons and Williams (1984) and Watt and Voss (2004) for more details. However, when considering the temporal and spatial properties of an Alfvén cyclotron maser compared with a whistler-mode cyclotron maser, we find that the quantitative differences are rather large. Both the characteristic frequency and the growth rate of the proton CI are less by the ratio of their masses ($M/m \sim 2000$) in comparison with the electron CI. But we have to remember that the density of energetic protons is larger than the density of energetic electrons, so that the actual difference in growth is not so big. The ratio of the Alfvén and whistler wavelengths $\lambda_A/\lambda_w \gtrsim (M/m)^{1/2} \sim 40 \gg 1$. The wavelength of the Alfvén waves generated is comparable with the characteristic scale of the ionospheric F-layer, hundreds of km, which can serve as a resonator for the Alfvén waves (Belyaev *et al.*, 1990; *JASTP*, special issue, 2000). This leads to the non-monotonic dependence of the reflection coefficient $R(\omega)$ of the wave by the ionosphere (Fig. 9.5). We shall see in Section 12.3 that such a dependence of $R(\omega)$ can be responsible for a burst-like generation regime of Alfvén waves, with a frequency drift inside a burst.

Another peculiarity of proton cyclotron masers in comparison with electron CMs is the influence of ions, helium and oxygen, on the development of the CI. Even small concentrations of these ions drastically changes the dependence of the refractive index on ω near the gyrofrequencies of these ions, which often occur in the middle of the unstable frequency range. Some effects of heavy ions are discussed in Section 12.5.

12.1 Precipitation pattern of energetic protons due to cyclotron wave–particle interactions

The precipitation pattern of energetic charged particles is directly related to the ionospheric projection of geomagnetic flux tubes favourable for the operation of cyclotron masers. These tubes have to contain a sufficiently dense cold plasma, such as at the plasmapause or in detached plasma clouds outside the plasmasphere which appear during the main phase of a geomagnetic storm. Independently of the particular source of energetic protons (substorm injections, transport across L shells, or magnetic compression) precipitation of energetic protons will be related to these regions.

During a substorm new energetic protons appear in the magnetospheric plasma sheet at local midnight (Fig. 12.1) and participate in magnetospheric convection. The ring current formed at this stage is composed of many ions ($W_i \gtrsim 10$ to 200 keV), and accounts for a considerable fraction (up to 0.5) of the energy accumulated in the magnetosphere during a geomagnetic storm. The trajectories of these energetic protons are their oscillations between magnetic mirror points superposed on drifts in the electric and magnetic fields, as illustrated in the equatorial plane of the magnetosphere in Fig. 12.2. The grey area in Fig. 12.2 represents the plasmasphere during the main phase of the storm; plasma is eroded causing its radius to decrease and a tail of detached cold plasma to form in the evening-noon sector. The cyclotron instability starts when energetic protons cross the plasmapause, so the geometric form of the precipitation pattern is determined by the contact line between the ion drift trajectories and the plasmaspheric boundary, as shown schematically at ionospheric heights in Fig. 12.3.

A strict consideration of the precipitation pattern is based on the system of self-consistent equations for the distribution function $\mathcal{F}$ of energetic charged particles and for the wave energy flux density $\mathcal{E}_\omega$, formulated in Chapter 8 (see (8.34) and (8.41)). We first consider the case of strong pitch-angle diffusion when the distribution function is almost isotropic and the loss cone is filled. In this case the precipitated proton flux density S_{pr} at the footprint of the magnetic flux tube is equal to

$$S_{\mathrm{pr}} = \frac{1}{2} \int (\mathcal{F} v_\parallel)_{s=l/2}\, d^3\vec{v} \tag{12.1}$$

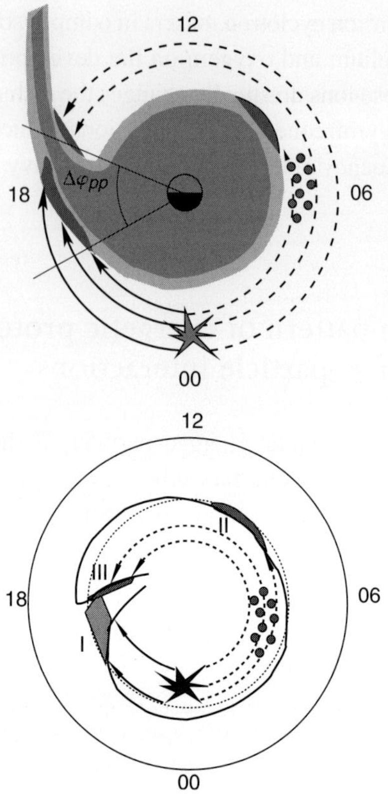

Figure 12.1 Schematic illustration of the equatorial plane the interaction of westward drifting (clockwise) ring current protons with the plasmapause bulge near 18LT (shaded); energetic electrons drift eastwards (anticlockwise) and interact with plasma filaments outside the plasma-pause near 06LT. (Taken from Trakhtengerts and Demekhov, 2005).

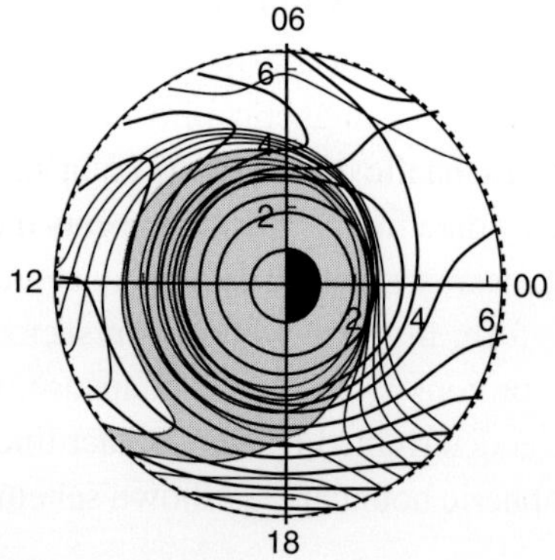

Figure 12.2 Crossing of convection trajectories of energetic protons and cold plasma particles (from Liemohn *et al.*, 2001). Thin and thick lines show the trajectories in the equatorial plane of cold plasma and protons with energy $W_p = 12$ keV, respectively, and the grey region represents the plasmasphere. (Taken from Trakhtengerts and Demekhov, 2005).

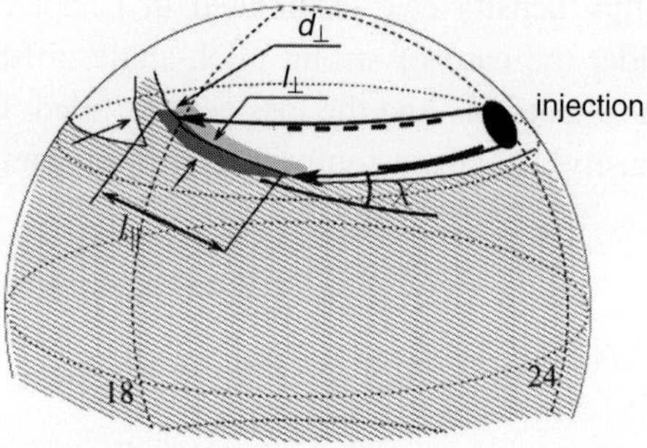

Figure 12.3 Diagram showing the ion precipitation region near 18LT at ionospheric heights.

where $s = l/2$ is the coordinate of the flux tube footprint (l is the total length of the magnetic flux tube), and $v_\parallel = v\sqrt{1 - \mu B/B_L}$ is, as earlier, the charged particle's field-aligned velocity component. For strong diffusion the distribution function is constant along a magnetic field line specified by its L-value, so that

$$\mathcal{F}(s = \pm l/2) \approx \mathcal{F}(s = 0) \equiv \mathcal{F}_L. \tag{12.2}$$

To describe the precipitation pattern it is enough to use only the kinetic equation for the distribution function $\mathcal{F}$ in the following form:

$$\frac{\partial \mathcal{F}}{\partial t} + B\frac{\partial(\mathcal{F} v_\parallel B^{-1})}{\partial s} + \frac{v_\mathrm{D}}{r \sin\xi}\frac{\partial \mathcal{F}}{\partial \varphi} = \hat{D}_{\mu\mu}\mathcal{F} \tag{12.3}$$

where (r, φ, ξ) are the geomagnetic polar coordinates (ξ is the colatitude and φ is the azimuthal angle), $s = (z/a)$ is the dimensionless distance along the magnetic field line from the equatorial plane, (2.8), v_D is the drift velocity, and $\hat{D}_{\mu\mu}$ is the pitch-angle diffusion coefficient (see (8.23)). For simplicity we consider rather energetic protons in the magnetosphere at $L \leq 6$, whose energy lies in the instability region $W_\mathrm{p} \gtrsim 30$ keV. In this case the electric drift in (12.3) can be neglected, so that $v_\mathrm{D} \simeq v_\mathrm{DM}$ (2.23).

Integrating (12.3) over phase space and the volume of a magnetic flux tube with unit cross-section at the ionosphere, we obtain for the two symmetric hemispheres:

$$\int_{-l/2}^{l/2} \frac{B_I}{B}\, ds \int d^3\vec{v}\, \frac{v_\mathrm{DM}}{r \sin\xi}\frac{\partial \mathcal{F}}{\partial \varphi} = 2S_\mathrm{pr} \tag{12.4}$$

where S_pr is determined by the relation (12.1), and B_I is the magnetic field at the ionospheric footprint of the magnetic flux tube. Taking into account the expression (2.21) for $\vec{v}_\mathrm{DM}$, we can obtain from (12.4) the following relation between S_pr and kinetic pressure p of the energetic protons (see Trakhtengerts *et al.* (1997) for details)

$$S_\mathrm{pr} = \frac{cB_I L^2}{B_0 R_0^2 e}\frac{\partial p}{\partial \varphi}\frac{\partial w}{\partial L} \tag{12.5}$$

where $B_0 = B_I\left(\xi = \pi/2\right.$, at the geomagnetic equator$\left.\right)$, R_0 is the Earth's radius, e is the electron (or proton) charge, the volume of the magnetic flux tube with unit cross-section at the ionosphere w is (for $L \gg 1$) equal to

$$w = \int_{-l/2}^{l/2} \frac{ds}{B} \approx \frac{32}{35}\frac{R_0 L^4}{B_0}. \tag{12.6}$$

Taking the kinetic pressure p to be isotropic

$$p = p_\parallel = p_\perp = 2\pi \int_{-\infty}^{\infty}\int_0^{\infty} \mathcal{F}_L v_\parallel^2 v_\perp\, dv_\perp\, dv_\parallel, \tag{12.7}$$

it is related to S_{pr} by the relation (M being the proton mass):

$$p = M\bar{v}S_{pr}, \quad \bar{v} = \frac{4}{3}\frac{\int_0^\infty v^4 \mathcal{F}\, dv}{\int_0^\infty v^3 \mathcal{F}\, dv}. \tag{12.8}$$

Note that (12.5) represents the particular case of the well-known formula (Vasyliunas, 1970) for the field-aligned current $j_\parallel$ at the ionospheric level (subscript I):

$$j_\parallel = \frac{c}{2}\left(\vec{s}_0 \left[\nabla w \times \nabla p\right]\right)_I \tag{12.9}$$

when this current is determined by precipitating energetic protons; here c is the velocity of light, and $\vec{s}_0$ is the unit vector along the magnetic field line.

Actually the expression (12.5) together with (12.7) serves as the equation for finding p as a function of φ and L. This equation can be written in the form:

$$\frac{\partial p}{\partial \varphi} = -qp, \quad q = \frac{35\, R_0}{128\, r_{B0}} \cdot \frac{L^{-5}}{(4 - 3/L)^{1/2}} \tag{12.10}$$

where the proton gyroradius $r_{B0} = \dfrac{\bar{v}}{\Omega_{B0}}$, $\Omega_{B0} = \dfrac{eB_0}{Mc}$, with B_0 being the magnetic field at ionospheric heights at the geomagnetic equator.

The solution of (12.10) is written as

$$p(L, \varphi) = p_0(L) \exp\left[-q\left(\varphi - \varphi_0(L)\right)\right] \tag{12.11}$$

where $\varphi_0(L)$ is the plasmapause bulge boundary (see Fig. 12.1) and $p_0(L)$ is the ring current pressure at this boundary in the equatorial plane. In the case of a sharp plasmapause, the precipitation is switched on at this boundary and decreases along φ similarly to the pressure:

$$S_{pr} = \frac{p_0(L)}{M\bar{v}} \exp\left[-q\left(\varphi - \varphi_0(L)\right)\right]. \tag{12.12}$$

The relaxation of the ring current and the precipitation of energetic protons are described by (12.11) and (12.12) in the interval

$$\varphi_0(L) \le \varphi \le \varphi_1(L) \tag{12.13}$$

where φ_1 is the boundary at which the CI is switched off due to a decrease in the ring current flux density or to a drop in the cold plasma density.

An important question is the formation of the partial ring current, remaining in the night-evening sector of the magnetosphere (Tverskoy, 1968). It is clear that pitch-angle diffusion affects this if the precipitation leads to a significant variation of the pressure p over the local time interval (or longitude) $\Delta\varphi \equiv \varphi - \varphi_0 = q^{-1} \lesssim \pi$.

This gives us an estimate for the maximum L value, $L = L_{max}$, at which the precipitation is important for the formation of the partial partial ring current:

$$L_{max} \simeq \left(\frac{R_0}{2r_{B0}}\right)^{1/5}. \quad (12.14)$$

For energetic protons with $W_p \sim 20$ keV, $L_{max} \approx 5$. It follows from (12.14) that $L_{max} \propto W_p^{0.1}$, i.e. L_{max} depends very weakly on the energy of the ring current particles.

The geometrical shape and the strength of the precipitation region are determined by the position and the shape of the plasmapause in the region of its contact with drifting protons (Figs. 12.1–12.3). In the simplest case, the ionospheric projection of this region can be approximated by a strip forming an angle χ with the energetic ion drift velocity (Fig. 12.3); its dimensions can be determined using (12.12) and (12.13). Taking into account the variation of the flux tube cross-sectional area along the field line, we obtain for the projections onto the Earth's surface:

$$\begin{aligned} l_\perp &= R_0 \sin\theta_I \, |\Delta\varphi| \sin\chi \\ l_\parallel &= R_0 \, |\Delta\theta_I| \sin^{-1}\chi \end{aligned} \quad (12.15)$$

where, for a dipole model of the geomagnetic field, $\sin^2\theta_I = L^{-1}$. Finally, we have for $\Delta\varphi \sim q^{-1}$ and $\Delta L/L \ll 1$:

$$\begin{aligned} l_\perp &\simeq 3.4\, r_{B0} L^{4.5} (4 - 3/L)^{1/2} \sin\chi \\ l_\parallel &\simeq \frac{R_0\, \Delta L}{2L(L-1)^{1/2} \sin\chi}. \end{aligned} \quad (12.16)$$

Putting $W_p \sim 20$ keV, $L \sim 4$, and $\Delta L \sim 1$, we obtain

$$\begin{aligned} &\text{for } \chi = \frac{\pi}{2}: \quad && l_\parallel = 460 \text{ km}, \quad && l_\perp = 2000 \text{ km} \\ &\text{for } \chi = 0.1: \quad && l_\parallel = 4600 \text{ km}, \quad && l_\perp = 200 \text{ km} \end{aligned} \quad (12.17)$$

which are both quite sizeable sectors.

The regime of strong pitch-angle diffusion, which was assumed in these calculations, corresponds to the maximum phase of a rather large geomagnetic storm. However, quantitative corrections to the above estimates can be introduced to account for deviations from the strong diffusion limit. The main correction concerns a decrease in the azimuthal decay rate q of the ring current in (12.10). This decrease is proportional to the ratio of the trapped flux S_{tr} to the precipitated flux S_{pr}. The azimuthal extent $\Delta\varphi \sim q^{-1}$ and linear scales of the precipitation region increase correspondingly (Trakhtengerts and Demekhov, 2005).

12.2 Dynamical regimes of an Alfvén cyclotron maser. Generation of IPDP and Pc1 pearl pulsations

During a geomagnetic storm, short-period geomagnetic pulsations are generated in the frequency range $f = 0.1$–10 Hz, which are typical frequencies of the cyclotron instability for the radiation belt protons. Numerous types of pulsations in this frequency range appear during different phases of a magnetic storm – see Jacobs (1970), Guglielmi and Troitskaya (1973), and Kangas *et al.* (1998). We select for discussion here two well-known types of pulsations, which demonstrate the most important dynamical regimes of an Alfvén CM in the magnetosphere. These are non-structured, or somewhat structured IPDP pulsations of increasing frequency (IPDP=irregular pulsations of diminishing periods) and highly structured Pc1 'pearl' pulsations. Examples of the dynamic frequency spectra of these pulsations are given in Fig. 12.4.

IPDP accompany, as a rule, the formation of the ring current with protons having softer energy spectra ($W_\mathrm{p} \gtrsim 10$–300 keV); Pc1 pearls appear during the recovery phase of a magnetic storm and are connected with harder energy protons ($W_\mathrm{p} > 10^2$ keV). Most IPDP events occur in the evening sector and correlate with new injections of energetic protons drifting from the night toward the evening sector. We suppose that these protons meet detached plasma clouds and are caused the proton cyclotron instability. The situation is similar to the generation of ELF/VLF emissions under the cyclotron interaction of newly injected energetic electrons with detached plasma regions, considered in Chapters 9 and 10. The dynamic spectra of ELF/VLF emissions produced by such an interaction are shown in Figs. 10.8 and 10.10. The frequency inside every noise-like element increases due to the evolution of the distribution function of the energetic electrons, which is accompanied by the formation of a step-like deformation in the distribution function (Fig. 10.13).

The dynamics of the proton distribution function is similar as the proton cyclotron instability develops, and this permits us to explain the rising frequency of IPDP signals. According to the analyses presented in Chapters 9 and 10, the temporal characteristics of noise-like emissions in CMs (ELF/VLF or IPDP emissions) are determined by the frequency of the relaxation oscillations Ω_J (9.30). This frequency also determines the frequency drift rate (see Figs. 9.9–9.11). The difference between whistler and Alfvén masers is that the temporal and spatial scales for the Alfvén CM are increased by a factor $\sim (M/m)^{1/2}$ over those for the whistler CM. Actually, taking into account relations (10.38), (10.39) we can write $T_J = 2\pi/\Omega_J$ and the quality factor Q_J in the following form:

$$T_J = \frac{2\pi}{\Omega_J} = \pi \left(\frac{S_\mathrm{tr} T_\mathrm{g} l}{S_\mathrm{pr} \, |\ln R| \, v_{0\alpha} \mu_\mathrm{c}} \right)^{1/2}, \qquad (12.18)$$

$$Q_J = \left(\frac{S_\mathrm{tr} \, |\ln R| \, l}{4 S_\mathrm{pr} T_\mathrm{g} v_{0\alpha} \, \mu_\mathrm{c}} \right)^{1/2} \qquad (12.19)$$

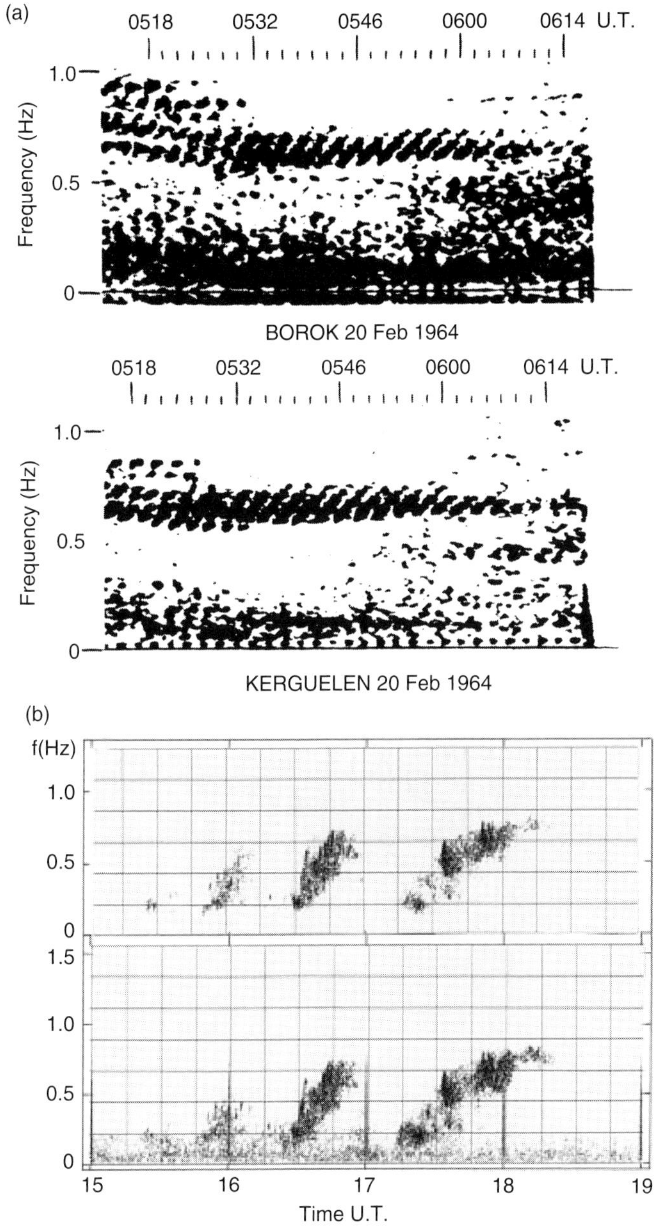

Figure 12.4 Spectrogram illustrating (a) non-structured (IPDP) (from Kangas *et al.*, 1998) and (b) highly structured (Pc1) pulsations (from Gendrin and Troitskaya, 1965).

where we use the relation $N_0 = l\, n_{hL}\, \mu_c^{-1}$, l is the length of the magnetic field line, $\mu_c \simeq L^{-3}$ is the loss cone boundary, n_{hL} is the number density of energetic particles on that L shell, $S_{tr} = n_{hL}\, v_0$ and S_{pr} are the flux densities for trapped and precipitated particles, respectively, v_0 is the typical particle velocity, R is the reflection coefficient, and T_g is the one-hop propagation time. The duration of an entire IPDP event is $T_{\rm IPDP} \sim 0.5\, T_J$ and the frequency change $\Delta\omega/\omega \sim 1$ during this interval. Putting, for example, $L \sim 5$, $S_{\rm tr}/S_{\rm pr} \sim 10$, $|\ln R| \sim 3$, the particle energy $W_{\rm p} = Mv_0^2/2 \sim 50$ keV, and $T_{\rm g} \sim 10^2$ s, we find that $\tau_{\rm IPDP} \sim 0.5$ hour, in good agreement with the experimental data. The quality factor $Q_J \sim 15$ for the chosen values of the parameters, and is a consequence of the auto-oscillation regime (Chapter 9).

During the recovery phase of a magnetic storm when magnetic conditions are quieter, Pc1 (continuous) 'pearl' pulsations are produced as a result of isolated pulses having a positive frequency drift inside each individual pulse. The repetition period between individual pearls is consistent with the group delay of Alfvén waves between the geomagnetically conjugate ionospheres. Pc1 pearls are observed during rather quiet periods when the proton injection rate and wave amplitudes are fairly small. Only one wave packet oscillates between the conjugate ionospheres in pearl events.

We shall show that the properties of Pc1 pearls can be explained by accounting for special properties of the ionospheric mirrors for Alfvén waves. These mirrors also create a Fabry–Perot resonator, which is termed the ionospheric Alfvén resonator (IAR) (see Chapter 9 and Fig. 9.5). The IAR determines the resonance structure of the ionospheric reflection coefficient R as function of ω (Fig. 9.5). Such a structure of R leads to the selection of a definite range of frequencies for which the conditions for CI development are most favourable. Moreover, the precipitated flux of energetic protons, which appears as a result of the CI instability which develops, can change this structure and lead to an auto-oscillation regime of the Alfvén maser being excited. All these peculiarities were taken into account in the model of an Alfvén sweep maser, which was developed in Chapter 9 (subsection 9.5.2). We discuss some quantitative results of this model in order to compare theory with the experimental data on Pc1 pearl pulsations.

The important result of the Alfvén sweep maser model is that only one pulse is excited, which oscillates between conjugate ionospheres. According to subsection 9.5.2, the particular solution for the wave magnetic field amplitude of the pulse at the exit from the ionosphere $b(t) = B_{\rm p} \exp\{i\Theta\}$ can be written in the following form (see Fig. 12.5):

$$B_{\rm p} = B_{\rm pm} \exp\{-t^2/T_{\rm p}^2\}, \tag{12.20}$$

$$T_{\rm p} = \left(\frac{|\delta|^2}{\delta_r\, g}\right)^{1/2} < T_{\rm g}, \tag{12.21}$$

$$\Theta'_t \equiv \omega = \omega_0 + \frac{|\delta_I|\, g}{|\delta|^2}\, t \tag{12.22}$$

where $g = \Gamma_{00} + \ln R$, Γ_{00} being the maximum value of the one-hop cyclotron amplification, (9.136), ω_0 is the corresponding frequency, R is the reflection

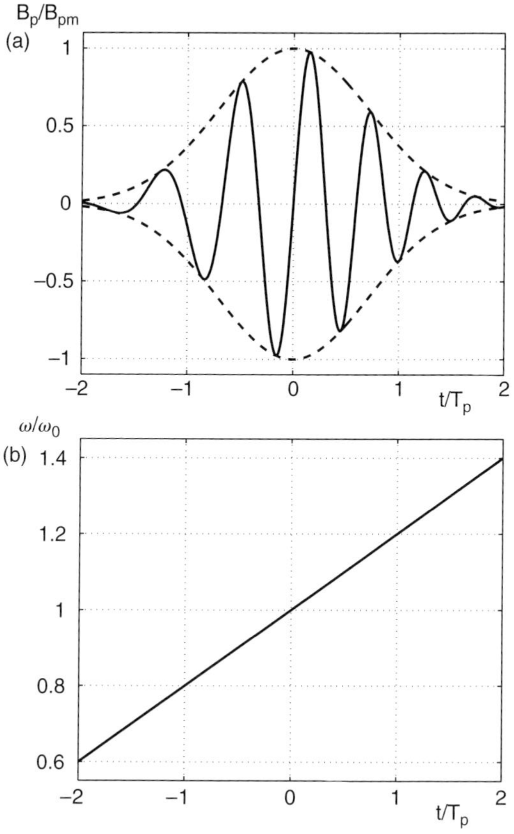

Figure 12.5 Diagram illustrating the envelope of a pulse (a) and the temporal dependence of a frequency inside a pulse (b) which follow from the nonlinear theory of Pc1 pulsations.

coefficient of the undisturbed ionosphere, $\delta = \delta_{\mathrm{Re}} + i\delta_{\mathrm{Im}}$, where δ_{Re} and δ_{Im} are determined by the relations (9.134); T_{p} is the pearl pulse duration, T_{g} is the one-hop Alfvén wave propagation time, and $\Theta' \equiv \omega$ is the instantaneous frequency emitted at time t. Using the formula $\delta_{\mathrm{Im}} = -\frac{1}{2}\left(\partial T_{\mathrm{g}}/\partial\omega\right)_{\omega}$, we find for the case $\omega \ll \Omega_{BL}$ that

$$\delta_I = -l/6v_{\mathrm{AL}}\Omega_{BL} \tag{12.23}$$

where l is the length of the considered magnetic flux tube, Ω_{BL} and v_{AL} are the proton gyrofrequency and the Alfvén velocity in the equatorial plane. Taking as an example the typical frequency of Pc1 pearls $f_0 = \omega_0/2\pi \sim 1$ Hz, and $L \sim 4$, we find that $l \sim \pi L R_0 \sim 8 \cdot 10^4$ km, $\Omega_{BL} \approx 42\ \mathrm{s}^{-1}$, and $\omega/\Omega_{BL} \simeq 0.15$. For the value of $v_{\mathrm{AL}} \approx 0.16\ \mathrm{km\ s}^{-1}$ we have $T_{\mathrm{g}} \sim 40$ s and $|\delta_I| \approx 0.16\ \mathrm{s}^2$. For the frequency widths of the reflection coefficient, $\Delta_{\mathrm{R}} \sim \omega_0 \sim 6\ \mathrm{s}^{-1}$ (9.135), and of the amplification coefficient $\Delta_g \sim 1.5\ \mathrm{s}^{-1}$ (9.136), the value of $\delta_r \approx 0.5\ \mathrm{s}^2$. Putting the net amplification $g \approx 2 \cdot 10^{-2}$ we find that $T_{\mathrm{p}} = 10$ s, and that $df/dt \sim 0.1\ \mathrm{Hz\ min}^{-1}$. These values do not contradict the experimental data.

12.3 Generation of hydromagnetic pulsations in the backward wave oscillator regime

A new type of hydromagnetic pulsations occurring in the Pc1 frequency range was discovered by satellite experiments – see Erlandson *et al.* (1992), Mursula *et al.* (1997), and Loto'aniu *et al.* (2005). In particular, CRRES satellite data (Loto'aniu *et al.*, 2005) showed that these Pc1 waves occur over a wide range of L values, exhibit bidirectional energy flux propagation away from the equator at low magnetic latitudes $|\text{MLat}| < 11^\circ$, and unidirectional (downward) propagation for $|\text{MLat}| > 11^\circ$ (Fig. 12.6). The situation is very similar to that for ELF/VLF Cluster observations (Fig. 11.14), which revealed the same effect for chorus emissions but with a smaller value of $|\text{MLat}| \sim 3^\circ$. The explanation of this for chorus is based on the operation of a whistler CM in the backward wave oscillator (BWO) regime. It is therefore natural to suppose that a similar regime may be realized in proton CMs too.

The key question for BWO generation is the form of the initial distribution function $\mathcal{F}_0$ of the energetic particles; it is essential to have a sharp gradient of $\mathcal{F}_0$ in velocity space. As discussed in Chapters 4 and 11, a step-like deformation of $\mathcal{F}_0$ in the field-aligned velocity component is the most likely to occur under real conditions in the magnetosphere. So we suggest that the step-like distribution function can also be formed for energetic protons, discussing some additional arguments for a step-like distribution function for energetic protons in the next section.

With an initial distribution function of energetic protons having a step-like deformation over $v_\parallel$ permits us to apply all the results on BWO generation obtained for a whistler-mode CM in Chapters 4 and 11. From the theory of BWO generation we have some key parameters for quantitative comparison with experimental data. These are:

(1) the spatial scale of the generation region;
(2) the attainability of the BWO generation threshold under realistic conditions in the magnetosphere; and
(3) the spatio-temporal dynamics of the wave packets so generated.

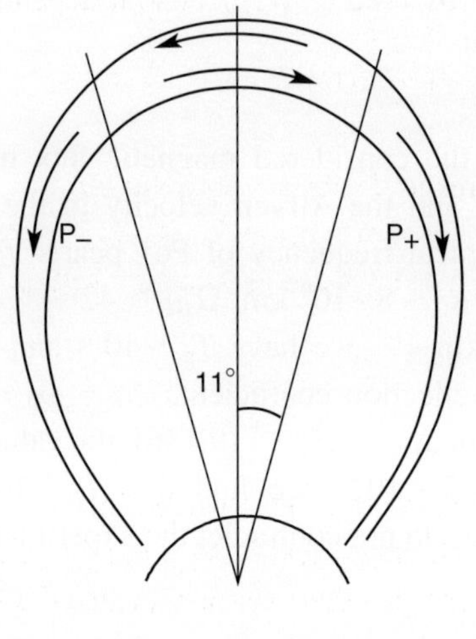

Figure 12.6 Diagram showing the observed energy flux directions of Pc1 waves along a particular magnetic flux tube.

The length scale l_{BWO} of the generation region (Fig. 11.13) in the magnetospheric BWO is determined by the geomagnetic field inhomogeneity (a parabolic variation with distance from the magnetic equator) and is given by (11.20):

$$l_{\mathrm{BWO}} \approx 2(R_0 L k_{\mathrm{A}})^{-1/3} R_0 L \tag{12.24}$$

where k_{A} is the Alfvén wavenumber, L is the L shell of the geomagnetic flux tube of interest and R_0 is the Earth's radius. When $\omega/\Omega_{BL} \ll 1$, which is considered here,

$$k_{\mathrm{A}} \simeq \frac{\omega}{c} \frac{\Omega_{\mathrm{p}L}}{\Omega_{BL}} \tag{12.25}$$

where ω, Ω_{BL}, and $\Omega_{\mathrm{p}L}$ are the wave frequency and the ion cyclotron and plasma frequencies, respectively, c is the speed of light in a vacuum, and the subscript L denotes the values in the equatorial plane.

If we fix $\omega/\Omega_{BL} = \tilde{\omega}$ and assume that the cold plasma density $n_{\mathrm{c}L}$ varies with L as

$$n_{\mathrm{c}L} = n_{\mathrm{c}L0}\, L^{-\nu}, \quad \nu \approx 3, \tag{12.26}$$

then we obtain

$$l_{\mathrm{BWO}} = l_{\mathrm{BWO}}^{(0)} (L/L_0)^{7/6} \tag{12.27}$$

where $l_{\mathrm{BWO}}^{(0)}$ corresponds to that for a given value L_0, and equals

$$l_{\mathrm{BWO}}^{(0)} \approx 2(R_0 L_0 k_{\mathrm{A}0})^{-1/3} R_0 L_0. \tag{12.28}$$

Putting $L_0 = 4$, $\omega/\Omega_{BL} = 0.2$, and $n_{\mathrm{c}L} = 30\ \mathrm{cm}^{-3}$, we obtain

$$|\mathrm{MLat}|_{L_0=4} \simeq 10^\circ, \quad k_{\mathrm{A}0} \simeq 5 \cdot 10^{-3}\ \mathrm{km}^{-1}. \tag{12.29}$$

This value is close to that observed ($\approx 11^\circ$). Also it depends very slowly on L:

$$|\mathrm{MLat}| = |\mathrm{MLat}|_0\, (L/L_0)^{1/6}. \tag{12.30}$$

The next issue is the threshold flux of energetic protons for the BWO regime. It can be estimated from threshold condition for BWO generation which is similar to BWO threshold for RB electrons (4.27) and can be written as

$$\left[\frac{\gamma_{\mathrm{stA}}\, \ell}{v_{\mathrm{gA}}\, v_{\mathrm{st}}}\right]_{\mathrm{thr}} = \frac{\pi}{2} \tag{12.31}$$

where γ_{stA} is the growth rate for a step-like distribution function of energetic protons, v_{gA} is the group velocity which coincides with the Alfvén velocity in the case of our interest $\omega/\Omega_B \ll 1$. The calculations of γ_{stA} are the same as for the case of energetic electrons (see (3.37)). According to Trakhtengerts and Demekhov (2007) we have:

$$\gamma_{\rm stA} = \left(\frac{e^2\, v_{\rm A}\, b\, S_{\rm tr}^{\rm p}}{Mc^2} \right)^{1/2} \tag{12.32}$$

where we take into account that the step velocity $v_{\rm st}$ is conected with the Alfvén velocity through the condition of cyclotron resonance:

$$v_{\rm st} = \left(\frac{\Omega_{\rm BL}}{\omega} - 1 \right) v_{\rm A}. \tag{12.33}$$

In the expression (12.32) $S_{\rm tr}^{\rm p} = n_{\rm h} v_{\rm st}$ is the flux density of energetic protons, b is the relative step height on the distribution function, and $n_{\rm h}$ is the density of energetic protons. Further we suggest that the dependence (12.26) for the cold plasma density is valid. In this case

$$v_{\rm A} = c\, \frac{\Omega_{BL}}{\Omega_{{\rm p}L}} = v_{\rm A0} \left(\frac{L_0}{L} \right)^{3/2} \tag{12.34}$$

where $v_{\rm A0}$ is the Alfvén velocity for the magnetic flux tube with $L = L_0$. From (12.31) and (12.32) we obtain:

$$S_{\rm thr} \approx \left(\frac{5\pi c}{4R_0} \right)^2 \frac{M v_{\rm A}}{e^2\, \tilde{\omega}\, b\, L^2} \tag{12.35}$$

where $\tilde{\omega} = \omega/\Omega_{BL}$ does not depend on L. Putting $L = 4$, $v_{\rm A} \approx 400$ km s^{-1}, $\tilde{\omega} \simeq 0, 2$, and $b \simeq 0.2$, we find $S_{\rm thr} \simeq 7 \cdot 10^7$ cm^{-2} s^{-1} and the particle energy $W = \frac{Mv_{\rm st}^2}{2} + \frac{Mv_{\perp T}^2}{2} \simeq 50$ keV. For higher L-values the threshold flux density decreases as $(L/L_0)^{5/2}$. For $L = 6$ its value $S_{\rm thr} \simeq 10^7$ cm^{-2} s^{-1} and the proton energy $W \geq$ 25 keV. According to the estimates made by Trakhtengerts and Demekhov (2007), these values are in agreement with satellite data.

12.4 The roles of heavy ions in the dynamics of an Alfvén cyclotron maser

Heavy ions can play very important roles in the CI development. This is because the addition of only a small proportion of heavy ions to the background plasma density can drastically change the Alfvén refractive index near the gyrofrequencies of these ions. The expression for the refractive index $N_{\rm i}$ for the cold plasma with two sorts (proton (p) and helium or oxygen (i)) of ions is written following Roux *et al.*

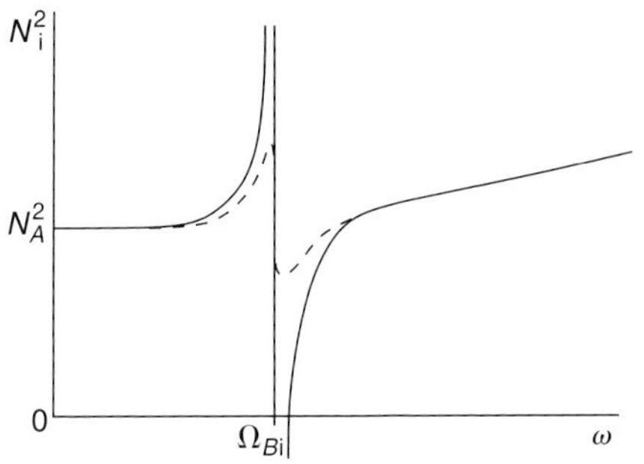

Figure 12.7 Dependence of the refractive index N_i on frequency ω for a ULF wave, with $k \| B$ near the ion gyrofrequency in the case of a cold plasma (solid line) and with thermal motions being taken into account (dashed line).

(1984), as

$$N_i = \left[1 + \frac{\Omega_p^2}{\Omega_{Bp}(\Omega_{Bp} - \omega)} + \frac{\Omega_i^2}{\Omega_{Bi}(\Omega_{Bi} - \omega)} \right]^{1/2} \tag{12.36}$$

where $\Omega_{p,i}^2$ is the square of proton or ion plasma frequency, and $\Omega_{Bp,i}$ is the proton or ion gyrofrequency. The dependence of $N_i^2(\omega)$ is shown in Fig. 12.7 by a thick line.

The expression (12.36) is valid for field-aligned propagation $\vec{k} \| \vec{B}$, which is the most interesting case for Alfvén masers. According to (12.36), both a non-propagating zone with $N_i^2 < 0$ and a resonance zone with $N_i^2 \to \pm\infty$ appear; these lead to the very important processes of reflection and absorption of Alfvén waves, respectively, accompanied by the redistribution of density and heating of the background plasma (Suvorov and Trakhtengerts, 1987).

In the real magnetosphere thermal effects in the background plasma (Ginzburg, 1970, Sazhin, 1993) change the expression (12.36) for N_i^2, the non-propagating zone disappears and gyroresonance absorption by heavy ions appears (Lequeau and Roux, 1992). Such a change of N_i^2 is shown in Fig. 12.7 by the dashed line, and is given by the formula:

$$N_i^2 = 1 + \frac{\Omega_p^2}{\Omega_{Bp}(\Omega_{Bp} - \omega)} - \frac{\Omega_i^2}{\Omega_{Bi}\omega} + \frac{\Omega_i^2 \, i\sqrt{\pi}}{\omega k v_{Ti}} w(\xi_i), \tag{12.37}$$

$$w(\xi_i) = \exp\left\{-\xi_i^2\right\}\left(1 + \frac{2i}{\sqrt{\pi}}\int_0^{\xi_i} \exp(x^2)\,dx\right), \quad \xi_i = \frac{\omega - \Omega_{Bi}}{k v_{Ti}}. \tag{12.38}$$

Using (12.37) and (12.38) we can obtain the criterion for when a small addition of heavy ions does not influence the propagation and polarization of waves, and fulfills the role of an absorbing filter at the frequency $\omega \simeq \Omega_{Bi}$. For that it is enough to estimate the maximum contribution of the heavy ions in the real part of N.

This contribution may be neglected if

$$\delta n < \delta_{\rm cr} = \frac{2}{3\sqrt{3\pi}} N_{\rm A}\beta_{\rm i} \tag{12.39}$$

where $N_{\rm A} = \Omega_{\rm p}/\Omega_{B\rm p}$, $\beta_{\rm i} = v_{T\rm i}/c$, $v_{T\rm i}$ is the thermal velocity of heavy ions, and $\delta n = (n_{\rm i}/n_{\rm p})_{\rm R} \ll 1$ is the ratio of the ion density $n_{\rm i}$ to the proton density $n_{\rm p}$ in the background plasma at the resonance point (R) $\omega = \Omega_{B\rm i}(z)$. The presence of heavy ions with their density satisfying the inequality (12.39) does not influence the propagation of Alfvén waves significantly. However, it can suppress the development of the CI at frequencies $\omega \gtrsim (\Omega_{B\rm i})_L$, where $(\Omega_{B\rm i})_L$ is the gyrofrequency in the equatorial plane of the geomagnetic flux tube specified by its L shell. This suppression is determined by the imaginary part of the refractive index Im $N_{\rm i}^2$, which is equal to:

$$\mathrm{Im}\, N_{\rm i}^2 = -\sqrt{\pi}\,(\Omega_{\rm p}^2/\Omega_{B\rm p}^2)(N_{\rm A}\beta_{\rm i})^{-1}\exp\{-\xi_{\rm i}^2\}. \tag{12.40}$$

Now we can estimate the optical depth

$$\tau_{\rm i} = 2\int |æ_{\rm i}|\, dz \tag{12.41}$$

which characterizes the exponential (by $\exp\{-\tau_{\rm i}\}$ times) decrease of wave intensity during one crossing by a wave packet of the gyroresonance region $\omega = \Omega_{\rm i}(z_{\rm R})$. The spatial damping rate $æ_{\rm i}$ in (12.41) is related to Im $N_{\rm i}^2$ by the expression:

$$æ_{\rm i} = -\frac{\omega}{2cN_{\rm A}}\,\mathrm{Im}\, N_{\rm i}^2. \tag{12.42}$$

Substituting (12.42) into (12.41) we find (for regions not too close to the equatorial plane) that

$$\tau_{\rm i} = 3\pi k_{\rm A} a_{\rm i}\delta n \tag{12.43}$$

where $k_{\rm A} = \omega N_{\rm A}/c$, $a_{\rm i} = |\partial \ln B/\partial z|_{z\rm R}^{-1}$ for each type of ions, and δn is given by (12.39).

It is interesting to estimate $\delta_{\rm cr}$ (12.39) and $\tau_{\rm i}$ (12.43) for the conditions of the Earth's proton radiation belts. Putting $N_{\rm A} = 5\cdot 10^2$, and $\tau_{\rm i} \simeq 10$ eV ($\beta_{\rm i} \sim 10^{-4}$ for He), we find that $\delta_{\rm cr} \sim 2\cdot 10^{-2}$. During the process of heating the heavy ions $\delta_{\rm cr}$ grows and reaches the value $\delta_{\rm cr} \sim 0.1$ for $T_{\rm i} \sim 200$ eV. However, for $T_{\rm i} \sim 10$ eV, $\omega \sim 3\,{\rm s}^{-1}$ ($k_{\rm A} \sim 5\cdot 10^{-3}$ km), $a_{\rm p} \sim 10^4$ km, the optical depth $\tau_{\rm i} \sim 1$ already for a value $\delta n \sim 2\cdot 10^{-3}$.

The cyclotron amplification of Alfvén waves is determined by three parameters:

- the concentration (density) of the energetic protons $n_{\rm h}$;
- the anisotropy coefficient $\alpha = |\overline{W}_\perp| / |\overline{W}_\parallel|$; and
- the parameter $\beta_* = \Omega_{\rm p\it L}\,\beta_0/\Omega_{BL}$, $\beta_0 = v_0/c$, where v_0 is the characteristic velocity of energetic protons.

When $\alpha \gtrsim 1$ and $\beta_* \gg 1$ the maximum one-hop amplification Γ_m occurs at the frequency $\omega \sim \Omega_{BL}/\beta_*$ (see (9.101)), and is given by:

$$\Gamma_m \simeq \frac{\pi}{3c} a(\delta n)_{pL} N_A^2 \beta_0 \tag{12.44}$$

where a is determined by the relation (2.8), and $(\delta n)_{pL} = n_{hL}/n_{cL}$, the ratio of hot to cold proton densities in the equatorial plane. If we take $\beta_0 \sim 10^{-2}$ ($W_p \sim 10^2$ keV) and $\omega \sim \Omega_{BHe}$ we find that $\Gamma_m \sim \tau_i$ under the condition that $n_{hL} \sim n_i$, n_{hL} and n_i being the densities of energetic protons and cold ions, respectively. Thus, if $n_i \ll n_{hL}$, the influence of heavy ions can be neglected.

A more typical case for the Earth's proton radiation belts corresponds to the opposite inequality, when

$$n_{hL} < n_i(z_R) \tag{12.45}$$

where $n_i(z_R)$ is the density of heavy ions at the gyroresonant point $\omega = \Omega_{Bi}(z_R)$. In the case (12.45), $\tau_i > \Gamma_m$ at the frequencies $\omega \geq \Omega_i(z_R)$, and multiple amplification of cyclotron waves is possible only at smaller frequencies $\omega < \Omega_{Bi}(z_R)$. A schematic picture of such a generation mechanism is shown in Fig. 12.8. For low frequencies $\omega < \Omega_{Bi}(z_R)$, the CI develops in the ordinary way. Here the propagation in a magnetospheric waveguide and reflection from the ionosphere are important.

As is the case for the electron radiation belts, the noise-like hydromagnetic emission is generated at this stage with a frequency drift toward higher frequencies. This drift is due to the quasi-linear relaxation of the distribution function of the energetic protons, when a relatively sharp gradient (a step-like deformation) appears on the boundary between resonant and non-resonant protons just as with the electron CI (see Chapter 9). But this similarity vanishes when the maximum wave frequency reaches the value $\omega_m \simeq \Omega_{Bi}(z_R)$. A very sharp edge is formed in the frequency spectrum close to ω_m due to the strong absorption of frequencies $\omega \geq \omega_m$; that is favourable for the further development of a step-like deformation in the distribution function.

This step can be a real source of BWO generation of hydromagnetic Pc1 waves, such as were observed in the CRRES data. We can also associate the quasi-linear

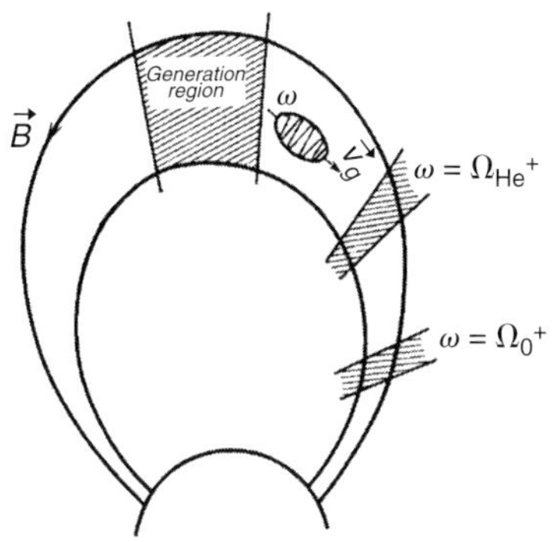

Figure 12.8 Schematic picture of Pc1 wave generation; the different gyroresonant layers with different heavy ions are shown shaded.

stage of the CI, which should be accompanied by noise-like emissions, with IPDP hydromagnetic pulsations.

12.5 Summary

Energetic protons of the ring current precipitate out of the radiation belts when they drift into a region of high density plasma, e.g. at the plasmapause. The location and shape of this precipitation region shown diagrammatically in Fig. 12.3 are determined by (12.16) and (12.17), under the assumption of strong pitch-angle diffusion.

During the ring current formation phase of a geomagnetic storm, irregular pulsations of diminishing periods (IPDP) are generated in the dusk sector as protons drift into regions of plasma detached from the plasmasphere. Rising frequency emissions $\lesssim 1$ Hz are generated as relaxation oscillations of the Alfvén mode cyclotron maser. The duration of an IPDP event, ~ 30 minutes, is satisfactorily explained by the theory.

In the recovery phase of a magnetic storm, highly structured continuous pulsations (Pc1, termed 'pearls') are produced as one Alfvén mode wave packet bounces back and forth between magnetically conjugate ionospheres at $L \sim 4$. The fact that the reflection coefficients of the ionospheric mirrors are frequency selective and change with the flux of precipitating protons explains the special spectral characteristic of these ~ 1 Hz pulsations (Fig. 12.4b and Equation 12.22). They are generated in the auto-oscillation regime of the Alfvén maser's operation.

Pc1 waves observed by the CRRES satellite (Loto'aniu *et al.*, 2005) propagating away from the magnetic equator at magnetic latitudes exceeding 11° are explained as being amplified in the periodic generation regime of a backward wave oscillator (BWO). This regime can be generated by a step $\sim 10\%$ in the proton distribution function at the gyroresonant proton velocity along the geomagnetic field line of interest ($L \sim 4$). The flux of protons with energies $\gtrsim 100$ keV has to exceed a threshold value of $8 \cdot 10^7$ cm^{-2} s^{-1} to generate a wave of period ~ 6 s and having an amplitude ~ 100 pT.

The presence of a small percentage of heavy ions, such as helium and oxygen ions, dramatically changes the refractive index of the plasma for Alfvén wave propagation at frequencies close to those ion gyrofrequencies. Gyroresonant absorption occurs; the optical depth is given by (12.43). This is important when the number density of heavy ions whose velocity equals the proton gyroresonant velocity exceeds the density of hot (energetic) protons in the magnetosphere (12.45). Then amplification is only possible at frequencies below the heavy ion gyrofrequency. Rising frequency emissions are generated, but only up to the heavy ion gyrofrequency. Such an effect leads to the further development of a step-like deformation of the distribution function, which further enhances the wave growth at just less than the heavy ion gyrofrequency.

Chapter 13

Cyclotron masers elsewhere in the solar system and in laboratory plasma devices

In previous chapters we have analysed both whistler and Alfvén cyclotron masers under the conditions appropriate for the Earth's magnetosphere. We have seen that the main role of these CMs is the regulation of the energetic charged particle population in the radiation belts. It is natural to anticipate that similar processes take place for other planets in the solar system, which have their own magnetic fields. Wave–particle measurements taken by spacecraft visiting Jupiter and Saturn have confirmed the possibility of whistler CM operation there. Most information has been obtained about Jupiter's magnetosphere; we shall consider in Section 13.1 some new effects of whistler CM operation which are due to the specific features of the Jovian radiation belts.

Whistler-mode cyclotron masers operate in the solar corona. The gigantic scales of solar magnetic loops and the very dense cold plasma background, which determines the collisional damping of whistler waves, lead to the very effective exchange of energy between cold and hot plasma components and to the explosive character of the cyclotron instability.

Another important problem concerns the modelling of CM operation in laboratory magnetic traps, which allow us to check many theoretical predictions and conclusions. Section 13.3 gives a short review of this topic.

13.1 Whistler-mode cyclotron maser and the Jovian electron radiation belts

The Jovian magnetosphere has many similar features to the Earth's magnetosphere, including the interaction with the solar wind and the formation of the radiation belts. The strong Jovian magnetic field (4 G against 0.3 G (or 30,000 nT) for the

Earth, on the planetary 'surface') introduces important quantitative differences. In particular the same wave–particle acceleration mechanisms, filling the radiation belts with energetic particles, are much stronger in Jupiter's magnetosphere, resulting in larger fluxes of relativistic electrons there than in the Earth's radiation belts.

The important difference of the properties of Jupiter's magnetosphere (Dessler, 1983) is due to the presence of the Jovian satellites, especially Io, which injects a huge number of neutral particles into its volume. These particles become ionized and form a dense, cold plasma torus surrounding Jupiter. The electrodynamic interaction of Io with the Jovian magnetic field serves as an additional acceleration mechanism for the radiation belt electrons. While these processes alter the detailed dynamics, they do not change the basic scenario of radiation belt formation, discussed in Chapter 10 for the Earth's radiation belts.

The wave–particle observations made by the Voyager spacecraft in the Jovian magnetosphere confirm the universality of processes operating in the radiation belts. The spatial distribution and the energetic spectra of the energetic charged particles in the Jovian radiation belts demonstrate features common with the Earth's radiation belts, and a new injection process of energetic electrons which is accompanied by an enhanced level of whistler-mode turbulence. An example of such universal processes is radiation belt formation under the joint action of diffusion across L shells and pitch-angle diffusion. Applied to the Earth's radiation belts this problem was considered in detail in Chapter 10. Similar considerations for the Jovian radiation belts were presented by Barbosa and Coroniti (1976a, b). They provided the relativistic generalization, but the problem was solved in a non-self-consistent approach to pitch-angle diffusion. Such a self-consistent approach was developed here in Chapter 10, and it would be interesting to apply it to the Jovian radiation belts. We leave the quantitative consideration of that problem to future investigators. Another universal process concerns the dynamics of the radiation belts. For the Earth's radiation belts we have shown that a very important dynamical regime is the relaxation oscillations of the cyclotron wave intensity and energetic electron fluxes (Chapters 9 and 10). We can imagine that similar processes will operate in the Jovian radiation belts too.

Based on these universal features we discuss an interesting effect, which was discovered in the process of making measurements of energetic electrons and waves in Jupiter's magnetosphere. The flux and spectrum of relativistic electrons with energies $\sim$3–30 MeV exhibit 10-hour variations close to Jupiter's rotation period, and can be observed both inside and outside Jupiter's magnetosphere (Chenette *et al.*, 1974, Simpson *et al.*, 1975). More detailed investigation by Simpson *et al.* (1992) showed that these modulations are a global temporal phenomenon, which can be termed 'clock variations'. Voyager and Ulysses observations have shown that very low frequency ($\sim$1–20 kHz) electromagnetic emissions observed inside and outside Jupiter's magnetosphere also exhibit 'clock-like' temporal behaviour (Kurth *et al.*, 1986, Kaiser *et al.*, 1993). That is, the emissions exhibit global 'pulses' at the planetary period, rather than being due to a corotating radio source.

Now, following the methods developed in Chapters 9–11 we try to explain this 'clock' effect on the basis of relaxation oscillations occurring as cyclotron instability develops – see Bespalov (1985, 1996), Bespalov and Savina (2005), and Bespalov *et al.* (2005). At first we should show that the frequency Ω_R of relaxation oscillations depends weakly on L-value for the typical source I of energetic electrons in the Jovian radiation belts, and is close to the planet's rotation rate $\Omega_J = 2\pi/T_J$, $T_J \approx 10$ hour. Such a commonly accepted source in the radiation belts is charged particle transport across L shells. According to Bespalov *et al.* (2005) and Equation (10.4), this source can be written as

$$I = L^2 \frac{\partial}{\partial L}\left(\frac{D_{LL}}{L^2}\frac{\partial N_0}{\partial L}\right) \tag{13.1}$$

where N_0 is the stationary distribution of the radiation belt electrons and D_{LL} is the cross-L-shell diffusion coefficient. We recall that the frequency Ω_R and the damping rate ν_R of relaxation oscillations are connected to the source I by the expressions (9.30):

$$\Omega_R = (hI)^{1/2}, \quad 2\nu_R = (hI/\nu), \tag{13.2}$$

where h is given by the relation (9.19a), and the damping of whistler waves ν is determined by the relation (9.12a). It was shown by Bespalov (1985, 1996) that for the Jovian magnetosphere Ω_R depends weakly on L, and is equal to

$$\Omega_R = \left(\frac{cD\,|\ln R|}{R_J}\right)^{1/2} \tag{13.3}$$

where the coefficient D comes into the expression for the diffusion coefficient $D_{LL} = DL^4$, c is the velocity of light, R_J is the radius of Jupiter, and R is the reflection coefficient of whistler waves from the ionosphere. According to the estimates made by Bespalov (1985, 1996), Ω_R (13.3) is close to $\Omega_J \approx 1.76 \times 10^{-4}$ rad s^{-1}. From (13.2) we can find ν_R. Putting $|\ln R| \sim 3$, we find that

$$\nu_R \approx DL^2 \approx 10^{-10} L^2 \text{ s}^{-1}. \tag{13.4}$$

Comparing $\Omega_R \sim \Omega_J$ with ν_R we conclude that we are dealing with relaxation oscillations having a high quality factor.

The coincidence of Ω_R and Ω_J does not create of itself the 'clock' effect; it is necessary to synchronize the oscillations occurring on different magnetic flux tubes. This synchronization is supplied by the magnetic drift of energetic electrons and by the rotation of the planet. We shall illustrate these effects using the two-level approach for the description of cyclotron maser operation (see Chapter 9). In this approach the number of the energetic electrons N and the wave energy flux density $\mathcal{E}$ in the magnetic flux tube with unit cross-section at its foot are described by the following system of equations (see (9.19)):

$$\frac{\partial N}{\partial t} - \Omega_{\rm d}\frac{\partial N}{\partial \varphi} = -\delta(\varphi)\,\mathcal{E}N - \frac{N}{T_{\rm s}} + I(t,\varphi), \tag{13.5a}$$

$$\frac{\partial \mathcal{E}}{\partial t} = h(\varphi)\,\mathcal{E}N - \nu(t,\varphi)\,\mathcal{E} \tag{13.5b}$$

where $\Omega_{\rm d}$ is the averaged angular velocity of the magnetic drift, in azimuth φ, and the term $-N/T_{\rm s}$ in (13.5a) characterizes particle losses due to synchrotron emission.

We consider the system (13.5) in the inertial frame, in which the Jovian ionosphere is rotating with angular velocity Ω_J. The most important term in (13.5), which is responsible for the synchronization of different magnetic flux tubes, is the damping rate $\nu(t,\varphi) = |\ln R|/T_{\rm g}$ of the whistler-mode waves, where the reflection coefficient from the ionosphere changes with time at Ω_J, and the group period $T_{\rm g}$ depends on the length of the magnetic flux tube, which is a function of φ in the asymmetric Jovian magnetosphere. Thus, we can present $\nu(t,\varphi)$ in the form:

$$\nu(t,\varphi) = f_1(\varphi - \Omega_J t)\, f_2(\varphi), \tag{13.6}$$

where $f_1(\varphi - \Omega_J t)$ and $f_2(\varphi)$ are periodic functions with period 2π. Using their Fourier representation we can write $\nu(t,\varphi)$ as

$$\nu(t,\varphi) = \nu_{\sim}(t) + a\, f_2(\varphi) + \Phi(t,\varphi), \tag{13.7}$$

where $\nu_{\sim}(t)$ is the sum of the periodic functions with angular frequency $n\Omega_J$ ($n = 1, 2, \ldots$), a is a numerical coefficient, and the function $\Phi(t,\varphi)$ satisfies the condition

$$\int_0^{2\pi} \Phi(t,\varphi)\, d\varphi = 0. \tag{13.8}$$

We consider the simplest case, when the modulation of ν and other parameters is weak, and N and $\mathcal{E}$ can be represented as

$$N = N_0 + N_{\sim}, \quad \mathcal{E} = \mathcal{E}_0 + \mathcal{E}_{\sim}, \quad N_{\sim}/N_0 \ll 1, \quad \mathcal{E}_{\sim}/\mathcal{E}_0 \ll 1. \tag{13.9}$$

In this case, after averaging (13.5) over φ, we obtain the following equation for the modulation $\mathcal{E}_{\sim}$ in the whistler energy flux:

$$\frac{\partial^2 \mathcal{E}_{\sim}}{\partial t^2} + 2\nu_{\rm R}\frac{\partial \mathcal{E}_{\sim}}{\partial t} + \Omega_{\rm R}^2 \mathcal{E}_{\sim} = -\mathcal{E}_0\Omega_{\rm R}^2\left[\frac{\nu_{\sim}(t)}{\nu_0} + \frac{1}{2\nu_{\rm R}}\frac{\partial}{\partial t}\left(\frac{\nu_{\sim}(t)}{\nu_0}\right)\right]. \tag{13.10}$$

It is evident from this equation that variations of the whistler damping rate with the frequency Ω_J act effectively as a source for the 'driven' global oscillations in the radiation belt parameters. It is also seen from (13.10) that, at the resonant condition, when the frequency of the driven force Ω_J is close to $\Omega_{\rm R}$, the modulation is amplified by a factor $Q^2 = (\Omega_R/\nu_R)^2 \gg 1$. More strict numerical simulations of (13.5), provided by Bespalov *et al.* (2005), have confirmed the above simple estimations.

So, we come to the conclusion that the relaxation oscillations of the cyclotron instability in the Jovian electron radiation belts can be a good explanation for the 'clock' effect.

13.2 Whistler cyclotron maser and eruptive phenomena in the solar corona

Up to now we have dealt with planetary magnetospheres filled by a not too dense cold plasma together with energetic charged particles constituting the radiation belts. The cold plasma component played an important role in the formation of the eigenmodes of cyclotron masers, but did not participate in the energy exchange between particles and waves.

A different situation occurs in stellar atmospheres, in particular, in the solar corona. Radiation belts can also exist in the solar corona; they are associated with active magnetic loops, which develop before a solar flare, i.e. during the preflare stage. There are several very important differences between solar whistler and Alfvén mode cyclotron masers and planetary maser systems. First, we mention the huge dimensions of coronal loops, where the whistler cyclotron instability is realized; the linear and nonlinear characteristic spatial scales of the instability are much less than the scale of a magnetic loop. This means that only local parameters determine the development of the instability. Secondly, the cold plasma component is very dense in the active magnetic loop, and the damping of waves due to electron–ion collisions determines the instability threshold.

The source of energetic charged particles in the solar radiation belts, which we associate with active magnetic loops, is due to the magnetic reconfiguration of the preflare stage, which includes the appearance of new magnetic loops, the generation and dissipation of strong electric currents, magnetic reconnection and magnetic compression. All these processes lead to a compression of the background plasma, to the production of energetic charged particles and to the development of a solar flare. The main unsolved problem is to estimate the separate contributions of these various processes. Without pretending to provide a detailed picture of a solar flare, we consider a particular type of eruptive phenomenon in the solar corona, where a whistler-mode cyclotron maser plays a decisive role (Trakhtengerts, 1996). We suggest that, in the process of the magnetic reconfiguration during the preflare stage, magnetic compression of the active loop takes place, and that this is accompanied by additional radiative cooling, by a cold plasma density increase and by the accumulation of energetic electrons in a radiation belt (see Fig. 13.1).

The reality of such a scenario is confirmed by the analytical and numerical calculations of Trakhtengerts and Shalashov (1999). They considered the compression of a coronal magnetic loop, filled by cold plasma, taking into account the radiative

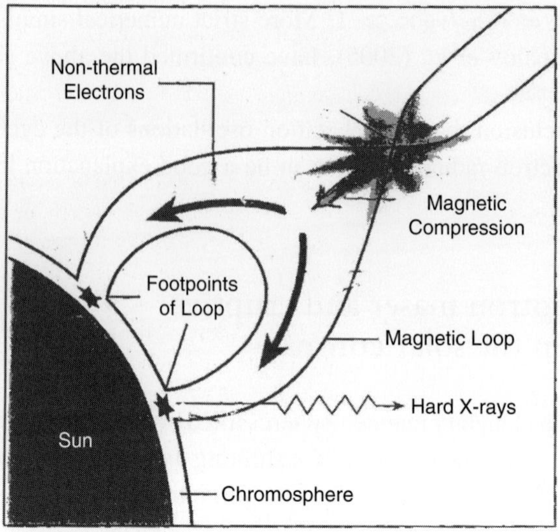

Figure 13.1 Diagram illustrating the magnetic compression which occurs before a solar flare.

cooling, recombination, electron–ion collisions and the acceleration of electrons by the induction electric field due to the runaway effects. The behaviour of the plasma depends on the temporal rate of the magnetic compression and on the initial parameters.

Figure 13.2 illustrates the results of a numerical simulation of this problem, when the magnetic field growth rate was $(t_B)^{-1} \sim 1/(30\text{ s})$, the recombination time $t_\text{p} = 60$ s, the initial density and electron temperature were $n_0 = 3 \cdot 10^9\ \text{cm}^{-3}$ and $T_0 = 2 \cdot 10^5$ K, respectively, the spatial dimension of the loop $L = 10^8$ m, and the initial magnetic mirror ratio of the magnetic loop is $\sigma = 10^3$. The lower curve in Fig. 13.2 demonstrates the background plasma cooling, the middle curve shows the growth of the energy $n_\text{h} W_\text{h}$ of energetic electrons ($W_\text{h} > 50$ keV), and the upper curve characterizes the ratio of the whistler cyclotron instability growth rate γ to the whistler collisional damping rate ν. The instability threshold is determined by the equality

$$\gamma = \nu. \tag{13.11}$$

Taking into account the expressions for γ (3.21) and ν, we shall have, instead of (13.11),

$$0.2(\alpha - 1)\omega(n_\text{h}/n_\text{c}) = (\omega/\omega_B)\,\nu_{ei} \tag{13.12}$$

where α is the anisotropy coefficient, n_h and n_c are the densities of energetic electrons and background plasma, respectively, ω_B is the electron gyrofrequency, and the electron – ion collision frequency ν_{ei} is given by Ginzburg (1970) as:

$$\nu_{ei} \approx 50\, n_\text{c} T^{-3/2} \tag{13.13}$$

where we must substitute n_c in cm^{-3} and T in degrees, K.

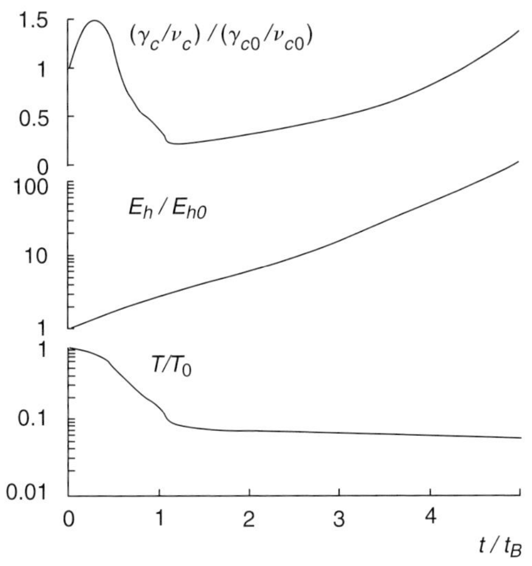

Figure 13.2 The threshold parameter γ/ν, the total energy of the hot electrons $n_h W_h$, and the background plasma temperature T as functions of time, measured in units of the magnetic compression characteristic time t_B. All quantities are normalized by their initial values (subscript 0). To determine the increment of the cyclotron instability, only electrons with energies $W > 80$ keV (as sufficiently anisotropic particles) were included, while all the electrons were used to determine the total energy. The calculations were performed for the parameters (defined in the text) $t_B = 30$ s, $t_p = 60$ s, $n_0 = 3 \cdot 10^9$ cm^{-3}, $T_0 = 2 \cdot 10^5$ K, $\sigma = 10^3$, and $L_\parallel = 10^8$ m.

The relationships (13.12) and (13.13) give the maximum density of hot electrons, n_{hm}, which can be accumulated in the magnetic trap before the instability is switched on:

$$n_{hm} = \frac{50\, n_c^2}{0.2\,(\alpha - 1)\,\omega_B T^{3/2}}\,. \tag{13.14}$$

Substituting, as an example, $n_c \sim 3 \cdot 10^{10}$ cm^{-3}, $T \sim 6 \cdot 10^4$ K, $\alpha \sim 1.5$, $B \sim 20$ G, and $W_0 \sim 20$ keV into (13.14), we find that

$$\left(\frac{n_h}{n_c}\right) \sim 10^{-3}, \quad \beta_0 = \frac{8\pi n_h W_0}{B^2} \sim 0.07, \quad P_0 \sim 7 \cdot 10^9 \text{ erg/cm}^2\text{s}, \tag{13.15}$$

where P_0 is the energy flux density of energetic electrons.

Now we come to the key element in the flare model proposed by Trakhtengerts (1996). The point is that after the instability threshold (13.11) is reached through a relatively slow accumulation of fast electrons in the magnetic trap, the subsequent instability becomes explosive. This is due to plasma heating by waves as the instability develops and to the decrease in ν, which corresponds to the nonlinear increase of the resultant instability growth rate $(\gamma - \nu)$. In a simple magnetic loop configuration, the instability threshold is reached first in the central part of the magnetic flux tube where n_h is a maximum and n_c is a minimum. Since the instability rise time is very short ($\gamma(t = 0) \geq 10^4$–10^5 s^{-1} for $n_c \geq 10^{10}$ cm^{-3} under real conditions), the instability area is localized rapidly (over a time $\sim\gamma^{-1}$) in the immediate vicinity of the centre where the wave energy is accumulated and where the plasma temperature increases. Then heat is propagated away, by heat conduction, in the form of a nonlinear thermal wave from a plane source which is instantaneously switched on, mainly in the magnetic field direction.

The source is active as long as fast electrons with an anisotropic velocity distribution feed the thermal wavefront, i.e. energy is accumulated from the entire loop volume in a very small local area between two temperature jumps which broaden, away from the magnetic trap centre. Depending on the initial conditions, the kinetic energy density in the background plasma confined between the thermal wavefronts can exceed the external magnetic field pressure; this will cause the expansion and floating up of plasma in the decreasing magnetic field direction. At this stage, hot plasma is ejected into interplanetary space, as a coronal mass ejection (CME) event.

The quantitative description of this scenario of a solar flare is based on a system of quasi-linear equations for the distribution function of fast electrons and for the spectral energy density of whistler-mode waves, and a system of magnetohydrodynamic equations for the temperature, density and magnetic field of the background plasma where the wave energy is used as the source. Solving these coupled systems of equations in rigorous form is very complicated. We restrict ourselves to an analysis of the development of the cyclotron instability and to the generation of the nonlinear thermal wave, leaving aside the problems of distortion of the plasma density profile and magnetic field structure, and of shock wave generation.

Under these assumptions, the initial conditions include a kinetic equation for the velocity distribution function $\mathcal{F}$ of fast electrons, (8.22)

$$\frac{\partial \mathcal{F}}{\partial t} + v_{\parallel} \frac{\partial \mathcal{F}}{\partial z} = \hat{D}(v_{\parallel}, v_{\perp}, \mathcal{E}_{\omega})\, \mathcal{F}, \tag{13.16}$$

a transport equation (8.26) for the spectral energy flux density $\mathcal{E}_{\omega}$ of whistler-mode waves, propagating in both directions, $\pm z$, along the magnetic field line

$$\frac{1}{v_{\rm g}} \frac{\partial \mathcal{E}_{\omega}^{\pm}}{\partial t} \pm \frac{\partial \mathcal{E}_{\omega}^{\pm}}{\partial z} = \frac{\gamma - \nu}{v_{\rm g}} \mathcal{E}_{\omega}^{\pm}, \tag{13.17}$$

and a hydrodynamic equation for the background plasma temperature T,

$$\frac{3}{2} n_{\rm c} \frac{\partial T}{\partial t} = q - Q n_{\rm c}^2 + \frac{\partial}{\partial z} \chi \frac{\partial T}{\partial z}. \tag{13.18}$$

In (13.16)–(13.18), $v_{\parallel}$ and $v_{\perp}$ are the longitudinal and transverse velocities of a fast electron in the magnetic field, $\hat{D}$ is a quasi-linear diffusion operator in the magnetic field (see Chapter 8), $v_{\rm g}$ is the group velocity of whistler-mode waves, and Q is the loss function which determines plasma energy losses in the solar corona (Priest, 1982); the heating source q and the electron thermal conductivity χ are given by

$$q = \int (\nu \mathcal{E}_{\omega} / v_{\rm g})\, d\omega, \quad \chi = n_{\rm c} v_{Te}^2 / \nu_{ei}, \quad v_{Te}^2 = 2T/m. \tag{13.19}$$

The loss function Q is important in the preflare stage when the parameters of the background plasma and fast electrons evolve slowly enough (over a time of more than 1 min); the background plasma temperature is kept at a low level despite the energy injections. At this stage, the instability growth rate γ increases slowly (e.g. due to magnetic compression), reaching the instability threshold (13.17). For a simple magnetic field configuration (Fig. 13.1), the instability threshold is achieved, first of all, in the central region of the magnetic trap; after that the instability develops in an explosive fashion. Since the temporal scale of instability development is very small, it is sufficient to take into account only two main processes in the initial stage: the slow growth of γ in space in the preflare evolution stage and the inhomogeneous heating of the background plasma by the wave generated thereby. The nonlinear saturation of the whistler wave amplitude and the spatial effects (group wave propagation and heat conduction) in Equations (13.17) and (13.18) can be neglected. In this approximation, the development of the instability is described by two equations

$$\frac{\partial \mathcal{E}}{\partial t} = \gamma_0(z, t)\,\mathcal{E} - \left(\frac{\omega \nu_{ei}}{\omega_H}\right)\mathcal{E}, \tag{13.20}$$

$$\frac{\partial T}{\partial t} = \left(\frac{2\omega \nu_{ei}}{3\omega_H n_c v_g}\right)\mathcal{E}, \tag{13.21}$$

where ν_{ei} is a function of T according to (13.13). For simplicity's sake, we consider only the case where $(T - T_0)/T < 1$ and $\partial\gamma/\partial t =$ constant. In this case, Equations (13.20) and (13.21) transform into one equation:

$$\frac{\partial \ln \mathcal{E}}{\partial t} = \gamma_0(z, t) - \nu_0 + \frac{\nu_0^2}{v_g n_c T_0}\int_0^t \mathcal{E}\, dt, \tag{13.22}$$

where T_0 and $\nu_0(T_0)$ are the plasma temperature and wave damping when the instability threshold is reached for $\gamma_0(z = 0,\ t = 0) = \nu_0$. It should be noted that Equation (13.21) describes correctly the wave generation only when the required instability threshold is reached at a particular point z. In this case, $\partial\gamma_0/\partial t \sim$ constant, and the solution of (13.22) has the form

$$\int_0^e \frac{dx}{\sqrt{ax + b(\exp\{x\} - 1)}} = \nu_0\, t\,\sqrt{2}, \tag{13.23}$$

where $e = \ln(\mathcal{E}/\mathcal{E}_0)$, $a = (\gamma_0\tau)^{-1}$, $\tau^{-1} = \partial \ln \gamma_0/\partial t$, $b = \mathcal{E}_0/(n_c T_0 v_g)$, and $\mathcal{E}_0$ is the initial wave energy flux density (in a stable plasma); x and t are the variable of integration and time, respectively.

According to (13.23), the flare proceeds in two conventional stages. If $ax > b(\exp\{x\} - 1)$, we have, from (13.23)

$$\frac{\mathcal{E}}{\mathcal{E}_0} \sim \exp\left\{\frac{\nu_0 t^2}{2\tau}\right\}. \tag{13.24}$$

On the other hand, for $b(\exp\{x\}-1) > ax$, the wave energy increases in an explosive manner:

$$\mathcal{E}^{-1/2} = \mathcal{E}_1^{-1/2} - \nu_0 t (2 n_c T_0 v_g)^{-1/2}, \quad \mathcal{E}_1 = n_c T_0 v_g/(\nu_0 \tau). \tag{13.25}$$

According to (13.25), the wave energy density becomes infinite over a finite time, the time scale for the explosive phase,

$$t_{\rm exp} = (2\tau/\nu_0)^{1/2}. \tag{13.26}$$

It is obvious that the waves are concentrated in the vicinity of the point where the instability threshold is reached first and the explosive time (13.26) determines the leading edge of the flare. Assuming, in particular, that $\nu_0 \sim 3 \cdot 10^4\ {\rm s}^{-1}$ (which corresponds to $n_c \sim 3 \cdot 10^{10}\ {\rm cm}^{-3}$, $T_0 \sim 6 \cdot 10^4$ K, and $\alpha \sim 1.5$) and $\tau \sim 10^3$ s, we find that $t_{\rm exp} \sim 0.3$ s.

After the instability is initiated the flare develops as a thermal explosion from a plane source, which is localized in the equatorial region of the flaring loop. We are interested in the asymptotic behaviour of the wave when the two thermal wavefronts moving away from the centre are far enough apart from each other so that we can use a self-similar solution. A schematic diagram of this thermal wave is shown in Fig. 13.3. As was mentioned above the thermal wave is fed until fast electrons with a non-equilibrium (anisotropic) velocity distribution enter the interior of the thermal wave through the wavefronts. Thus, energy builds up in the thermal wave from the entire volume of the flaring loop. The source is still active after one complete bounce period of fast electrons through the magnetic trap, since anisotropy is restored due to the loss cone. The degree of secondary anisotropy is small ($\alpha \sim \sigma^{-1}$) when the mirror ratio σ is large, but this process can be significant in the total energy balance.

Thus, our objective is to find a law expressing the evolution of the nonlinear thermal wave, allowing for the energy being fed in by fast electrons coming through the wave fronts. We are interested in the strong heating case in which the gas temperature T behind the front is much greater than the gas temperature T_0 ahead the front and we can assume that the latter is equal to zero. Heat is fed in due to the dissipation

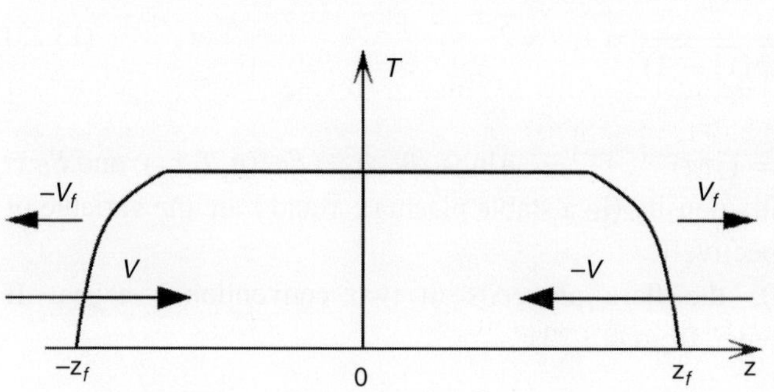

Figure 13.3 Schematic picture of a nonlinear thermal wave in the form of two thermal wavefronts propagating at velocity V_f away from region at $z = 0$ where the cyclotron instability switches on; v is the velocity of newly energetic electrons having an anisotropic velocity distribution which feed the thermal wavefront.

of the waves excited by fast electrons in a very narrow region in the vicinity of the thermal wavefront ($\Delta z \sim v_g/\nu_0 \leq 1$ km). That heat is distributed rapidly inside the thermal wave owing to the high thermal conductivity. We shall seek a self-similar solution of Equation (13.18), and the source and the heat fed through the wave fronts will be taken into account by using the energy conservation law for fast electrons plus the background plasma system.

In this approach the nonlinear thermal wave is described by the equation

$$\frac{\partial \Theta}{\partial t} = \frac{\partial}{\partial z} \left(\chi_0 \, \Theta^{5/2} \right) \frac{\partial \Theta}{\partial z} \tag{13.27}$$

where we use the expression (13.19) for the electron thermal conductivity χ, the dimensionless variable $\Theta = T/T_0$, $\chi_0 = 2v_{T0}^2/3\nu_{ei0}$, and the subscript '0' indicates the initial values of the quantities at $t = 0$. Equation (13.27) should be solved together with the energy conservation law

$$\frac{3}{2} n_c T_0 \int_{-z_f}^{z_f} \Theta(z) \, dz = 2 \, |P_0 - P_I| \, t \tag{13.28}$$

where $\pm z_f$ are the coordinates of the thermal fronts (Fig. 13.3), P_0 is the energy flux density of energetic electrons having an anisotropic distribution function, and P_I is the same quantity but after the distribution function becomes isotropic due to the development of the cyclotron instability. The energy fed into the cold plasma $|P_0 - P_I|$ can be expressed through the anisotropy coefficient $\alpha = v_{0\perp}^2/v_{0\parallel}^2$. In the limit $\omega \ll \omega_B$, Trakhtengerts (1996) showed that it is given by

$$|P_I - P_0| = \frac{\alpha - 1}{\alpha + 1} P_0 \equiv \delta P_0. \tag{13.29}$$

The solution of (13.27) can be written in the self-similar variable $\xi = z - \int v_f \, dt \equiv z - z_f$ in the following form (as $T_0 \to 0$):

$$\Theta = \left(\frac{5 v_f}{2 \chi_0} |z_f - z| \right)^{2/5} \tag{13.30}$$

where v_f is related to z_f by $dz_f/dt = v_f$. After substituting (13.30) into (13.28) and integrating over z we find the relationship:

$$z_f^{7/2} \frac{dz_f}{dt} = \frac{2\chi_0}{5} (\delta P_0)^{5/2} \, t^{5/2}. \tag{13.31}$$

The solution of (13.31) can be represented in the convenient form:

$$\eta \equiv \frac{z_f}{v_{0\parallel} t} = \left(\frac{\delta_1 n_h}{n_c} \right)^{5/9} \left(\frac{n_c v_{0\parallel}^3}{\omega_p^4 t} \right)^{2/9} \tag{13.32}$$

where $\omega_p = (4\pi e^2 n_c/m)^{1/2}$ is the background plasma frequency and $\delta_1 = 0.5(\alpha - 1)$. The electron temperature T and the ratio of gas-kinetic plasma pressure to magnetic pressure, $\beta = (8\pi n_c T/B^2)$, can be expressed through the variable η. In the centre, $z = 0$,

$$T = \frac{\delta_1}{\eta}\frac{n_h}{n_c}W_0, \quad W_0 = \frac{mv_0^2}{2}, \tag{13.33}$$

$$\beta = \frac{\delta_1}{\eta}\beta_0, \quad \beta_0 = \frac{8\pi n_h W_0}{B^2}. \tag{13.34}$$

From relationships (13.32)–(13.34) we can ascertain the dynamic and energetic features of the flare. The flare is manifested, first of all, by hard X-ray radiation coming from the feet of the flaring loop in the chromosphere. The source of that radiation is fast electrons which, as the cyclotron instability develops, undergo strong pitch-angle scattering and precipitate through the loss cone into the dense solar atmosphere. The background plasma heating at the centre of the magnetic trap is delayed with respect to fast electron precipitation. The general principles which follow directly from the initial system of Equations (13.16)–(13.19) suggest that the precipitated flux of energetic electrons is proportional to the wave energy density and the plasma temperature increases as a wave energy integral with respect to time. There is a corresponding delay between the emission of soft X-ray radiation from the heated plasma at the top of the field line and the hard X-ray radiation produced via the interaction of precipitating fast electrons with the solar atmosphere.

As the flare progresses, β increases and can, in principle, exceed unity owing to the high energy density in the small thermal wave volume. However, as long as $\beta < 1$ the flare is one-dimensional and is extended along the magnetic flux tube. We now estimate T and β using relationships (13.32)–(13.34). Choosing the same initial parameters as we used before (see, e.g. (13.15)), and with t being the time, the dimensionless variable η can be written in the form

$$\eta = 0.01\,(50/t)^{2/9}. \tag{13.35}$$

In accordance with (13.34), β will reach unity in a short time $t_{cr} \sim 0.15$ s. In this case, $T \sim 3.6 \cdot 10^6$ K and the dimension of the thermal wave $2z_f \sim 10^3$ km. The temperature $T \sim 7 \cdot 10^6$ for $t \sim 2$ s.

Let us estimate the degree of anisotropy in the distribution function of fast electrons which is due to the existence of the loss cone. Given the mirror ratio σ, this anisotropy is equal to

$$\alpha_c = \frac{\sigma + 0.5}{\sigma - 1}, \quad \delta = \frac{\alpha - 1}{\alpha + 1} \approx \frac{3}{2\sigma}. \tag{13.36}$$

When $\sigma \gg 1$ this source is not as intense but is active for a longer period of time. The activity of the source depends on the lifetime of fast electrons, which is given by (8.63)

$$\tau_{1t} \approx (l/v_0)\sigma, \tag{13.37}$$

where l is the dimension of the flaring loop. For $\sigma \sim 10$ and $l/v_0 \sim 2$ s, $\tau_{1t} \sim 20$ s.

In the second stage of the flare, a shock wave is excited and the flare area is extended across the magnetic field lines if β is greater than unity. Fast electrons are involved in the thermal explosion from the neighbouring magnetic flux tubes, and hot plasma is ejected into interplanetary space. Examination of the details of this stage is beyond the scope of our consideration here, and we give only a rough estimate of the energy budget of the flare. In accordance with (13.32), the longitudinal dimension of the thermal wave in the magnetic field will reach $2z_f \sim 10^5$ km and the temperature will reach $T \sim 8 \cdot 10^6$ K over a time $t \sim 10$ s. The transverse dimension $l_\perp$ can be estimated by assuming that $l_\perp \sim v_A \cdot t$ for $\beta > 1$, where v_A is the Alfvén velocity. For $v_A \sim 2 \cdot 10^3$ km/s we have $l_\perp \sim 2 \cdot 10^4$ km. Then the total flare energy is equal to

$$W_\Sigma \sim n_c T_m 2\pi l_\perp^2 z_{fr} \sim 4 \cdot 10^{30} \text{ erg}. \tag{13.38}$$

This dynamical picture and the quantitative estimates broadly agree with the experimental data on the initial stage of a typical solar flare – see Dennis (1991), and Enome and Hirayamo (1993).

The main acceleration of charged particles and the storage of energy in a flare loop occurs during the preflare stage, over an interval of some minutes. At this stage, a magnetic trap with a cold dense background plasma is filled by energetic electrons having energies from 10 keV up to 10 MeV (13.15). The actual acceleration mechanism is adiabatic magnetic compression and runaway effects, which create a tail of energetic electrons with a transverse velocity anisotropy. Non-adiabatic processes due to the magnetic reconnection play an important role at this stage, injecting superthermal electrons with energies from 1 to 10 keV into the magnetic trap.

The active stage of the solar flare is due to the explosive evolution (Trakhtengerts, 1996) of the cyclotron instability (13.25), with a considerable portion of the energy of the energetic electrons being transferred to the background plasma, while the electrons themselves precipitate through the loss cone into the lower chromosphere, generating hard X-rays. Since such thermal explosions evolve in a small volume near the minimum magnetic field region near the equatorial plane and involve the energy of particles from the entire trap volume, the kinetic energy of the heated region ($\sim 4 \times 10^{30}$ erg) can exceed the magnetic pressure confining the plasma. In this case, coronal material, together with some fraction of the energetic charged particles, will be injected into interplanetary space.

13.3 Cyclotron masers in laboratory magnetic traps

Investigations of axial plasma magnetic traps in the laboratory have a long history. We consider devices which possess an axial magnetic field with magnetic mirrors, filled with plasma, including a cold component with the addition of energetic electrons. As a rule the density n_c of the cold component is much greater than the energetic (hot) electron density n_h

$$n_h/n_c \ll 1. \tag{13.39}$$

The early experiments were conducted during the 1960s – see Ard *et al.* (1966), Perkins and Barr (1966), Alikaev *et al.* (1968), Ikegami *et al.* (1969), Jacquinot *et al.* (1969), and Booske *et al.* (1985). Although the particular conditions and plasma sources were different in each experiment, they all revealed some common characteristics. For example, bursts of electromagnetic emissions at frequencies $\omega < \omega_{BL}$, where ω_{BL} is the electron gyrofrequency in the central region of the magnetic trap, were observed; these were accompanied by the precipitation of trapped energetic electrons through the loss cone. The electromagnetic emissions were whistler type waves whose wave vector $\vec{k}$ was close to the direction of the magnetic field $\vec{B}$.

Understanding these experimental results came later (Gaponov-Grekhov *et al.*, 1981, Demekhov and Trakhtengerts, 1986) on the basis of physical mechanisms developed for the theory of magnetospheric whistler cyclotron masers, considered in Chapters 3, 8 and 9, and in the previous sections of this chapter. Below we consider some results of these laboratory experiments in more detail.

We select the group of experiments (Alikaev *et al.*, 1968, Ikegami *et al.*, 1969, Jacquinot *et al.*, 1969), where the whistler burst and electron precipitation were observed during the decaying phase of the plasma in the device. This stage appeared after switching off the source which created both plasma components, when the cold plasma density n_c decreased faster than did n_h. The burst of electromagnetic emission appeared when

$$\omega_p \simeq \omega_{BL}$$

where $\omega_p = \left[\dfrac{4\pi e^2(n_c + n_h)}{m}\right]^{1/2}$ is the plasma frequency and ω_{BL} is the gyrofrequency at the central region of the magnetic trap. This condition seems to be a specific feature of laboratory magnetic traps which are very short in comparison with space magnetic traps, and multiple cyclotron amplification of a whistler-mode pulse is needed with the reflection coefficient R at the magnetic mirrors being close to unity. The sharp increase of R when $\omega_p \leq \omega_{BL}$ can be due to the specific behaviour of the whistler refractive index, as function of ω_p, as shown in Fig. 13.4. When a whistler-mode wave is generated in the region where $\omega_p > \omega_{BL}$ it transforms near $\omega_p \sim \omega_B$ to resonant plasma waves (branch ‘b’). In this case the reflection coefficient is very

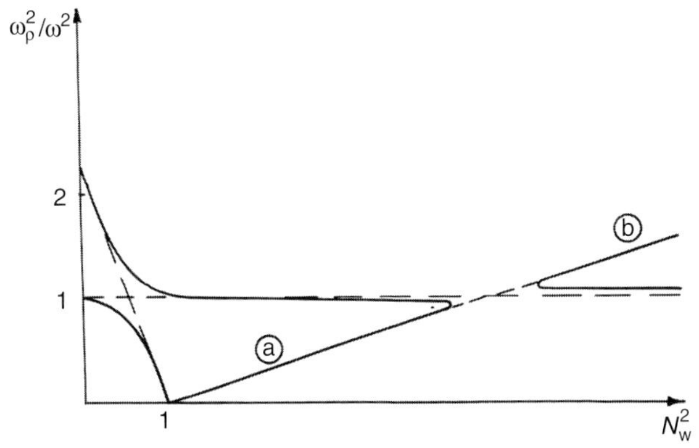

Figure 13.4 Dependence of the refractive index N_W on the plasma frequency ω_p for whistler waves with wave vector k close to the B-direction.

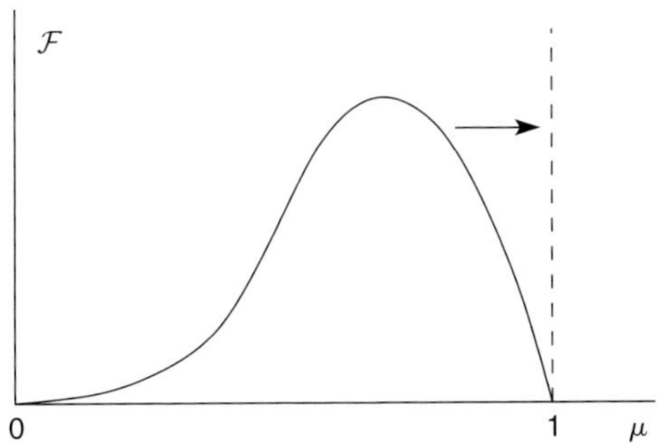

Figure 13.5 Diagram illustrating a butterfly-like pitch angle dependence of the electron distribution function; the arrow shows the direction of pitch-angle diffusion.

small in the geometrical optics approximation. When $\omega_p < \omega_{BL}$ the transformation to plasma waves does not occur, and the reflection coefficient can be close to unity.

Qualitatively the development of the cyclotron instability in this case can be described in the following way. As $\omega_p \simeq \omega_{BL}$ during the process of a cold plasma relaxation, the whistler reflection coefficient R grows and the instability threshold

$$\gamma \simeq \frac{|\ln R|}{T_g} \tag{13.40}$$

is reached; in (13.40) γ is the cyclotron instability growth rate, and T_g is the period of oscillations of a whistler-mode wave packet between the ends of the device along the magnetic field. The growth of the wave intensity leads to the precipitation of energetic electrons into the loss cone (see Fig. 13.5) and a further rapid decrease of the plasma frequency due to the decrease of n_h. As a result a very short (with the characteristic time $\tau \sim \gamma^{-1}$) intense burst of whistler-mode waves is generated. Detailed quantitative estimates for the experiment (Alikaev *et al.*, 1968) showed that this model (Gaponov-Grekhov *et al.*, 1986, Demekhov and Trakhtengerts, 1986) was valid.

The interesting results obtained in the experiments of Ard *et al.* (1966), where a non-varying plasma was produced in the magnetic trap by the external pumping wave during the process of electron cyclotron resonance discharge. The point of cyclotron resonance for the pump wave could be changed relative to the central region of the magnetic trap. As a result a two-component plasma was created, with $n_h/n_c \ll 1$, and the energetic electrons had a butterfly-like distribution function over velocity with a shifted pitch-angle maximum relative the pitch-angle $\theta = \pi/2$. A succession of electromagnetic pulses was observed at the frequency $\omega < \omega_{BL}$, which was determined by the loss cone anisotropy, and these pulses were accompanied by bursts of precipitated energetic electrons. A characteristic feature of this succession of pulses was the decrease of the pulse repetition period when the point of cyclotron resonance for the pump wave moved towards the central cross-section of the device. Here the pulse generation becomes a continuous generation. The physical explanation of these results is based on the fact that the change of butterfly-like distribution function as the cyclotron instability develops leads to the growth of the number of energetic electrons with pitch-angle $\theta = \pi/2$ (Fig. 13.5). As a result the CI growth rate is increasing at its initial stage; this causes auto-oscillation of the wave generation process when a stationary source produces energetic electrons. Apparently this effect disappears when the pump wave frequency is close to ω_{BL}, and the butterfly distribution is not formed. Quantitative considerations published by Demekhov and Trakhtengerts (1986) confirm this qualitative picture.

For the experiment of Perkins and Barr (1966), the cyclotron instability developed under adiabatic plasma compression. A strong optical emission was observed from the volume, which they connected with heating of the background plasma during magnetic compression. However, the estimates made by Demekhov and Trakhtengerts (1986) showed that the interaction of electrons with neutrals left the electron temperature the same as it was at the beginning of the compression. Rapid heating by cyclotron waves as the CI develops, considered in the previous Section and applied to the solar corona, can produce the high temperature of the background plasma and hence explain the optical emission.

Special experiments have recently been undertaken for laboratory modelling of the dynamics of space whistler-mode cyclotron masers (Vodopyanov *et al.*, 2005). The plasma was created by a pump wave during the process of electron cyclotron resonance discharge in the central region of a magnetic trap and comprises two electron populations (background and energetic), whose number densities and temperatures were, respectively, $n_c \sim 10^{13}$ to 10^{14} cm^{-3} and $T_c \sim 300$ eV and $n_h \sim 10^{10}$ cm^{-3} and $T_h \approx 10$ keV. These ratios of T_h/T_c and n_h/n_c are rather close to the corresponding values in the Earth's magnetospheric radiation belts. Quasi-periodic pulsed precipitation of energetic electrons from the trap, accompanied by pulses of electromagnetic emission at frequencies below the electron gyrofrequency at the centre of the device, were detected. This electromagnetic emission corresponds to whistler-mode waves whose wave vector $\vec{k}$ lies close to the direction of the magnetic field. An example of observed correlated pulses of whistler-mode waves and precipitated electrons is given in Fig. 13.6. Vodopyanov *et al.* (2005) have shown that this burst-like regime of the cyclotron instability could be associated with relaxation oscillations, as considered in Chapters 9 and 11. The nonlinear deformation of the distribution function of energetic electrons, such as the relaxation of a butterfly-like distribution or the involvement of newly resonant electrons (Chapters 9 and 11) can

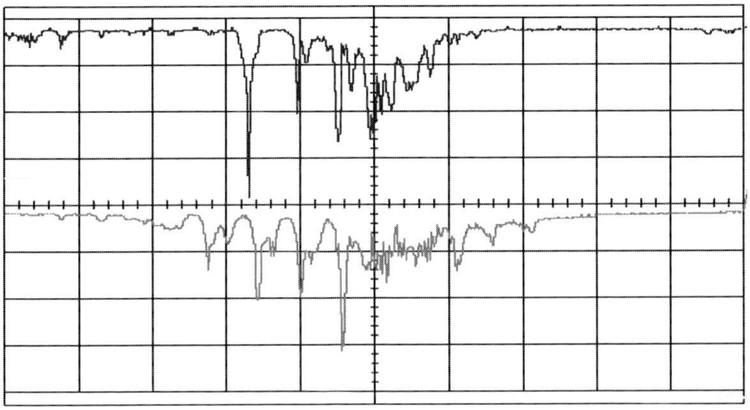

Figure 13.6 An example of the burst-like regime of the cyclotron instability in a laboratory experiment; whistler-mode wave generation (upper curve) and electron precipitation (lower curve) are seen simultaneously (from Vodopyanov *et al.*, 2005).

be responsible for auto-oscillations of the wave intensity and for the precipitating electron flux.

13.4 Controlled thermonuclear fusion aspects

Much plasma physics research today has the objective of designing a practical device in which the thermonuclear fusion of hydrogen nuclei (deuterium/tritium) may be controlled (Kadomtsev, 1958; Rudakov and Sagdeev, 1958). Fusion occurs in an uncontrolled manner in a hydrogen bomb. It also occurs naturally in the centre of stars, such as the Sun.

In order to produce more energy by fusion than is needed to heat the plasma and to provide energy for the bremsstrahlung radiation emitted (i.e. lost from the system) when electrons collide with ions, the product of the plasma density and the time for which the plasma is confined must exceed a certain value. This is termed the Lawson criterion (see Chen, 1974, Pease, 1993). The plasma may be contained by magnetic fields of suitable geometry, in either an open system using magnetic mirrors (Post *et al.*, 1973) or a closed system, a torus, such as a pulsed tokamak or a steady state stellarator (Chen, 1974).

In a magnetic mirror device, there are many instabilities which can occur, over a wide range of frequencies, causing 'anomalous transport', i.e. the loss of charged particles from the device through the loss cone at the mirrors. For investigations of a number of these instabilities and their possible suppression, the theoretical techniques discussed in this book could be particularly valuable. Another application of the methods introduced here is a consideration of plasma heating, for example by lower or upper hybrid waves, or at the fundamental or higher harmonics of the electron cyclotron frequency, and including the effects of Doppler frequency shifts.

Besides the Joule heating associated with the current flowing in a plasma which produces the desired magnetic field geometry in toroidal plasma devices, wave heating is also important there. This may be electron or ion cyclotron resonance heating. For fusion, the energy must, of course, eventually be transferred to the ions (Chen, 1974). It should be noted that it is ions in the high energy tail of the distribution which undergo fusion (Mikhailovskii, 1998).

Cyclotron maser excitation of the fast Alfvén wave by energetic ion populations at their cyclotron frequency and low-order harmonics has been observed (Cottrell *et al.*, 1993, Dendy *et al.*, 1995) for alpha particles (helium ions) produced by deuterium-tritium fusion reactions in the Tokamak Fusion Test Reactor (TFTR) or in the Joint European Torus (JET). Ion cyclotron emission lines are generated. In JET, these can be explained by the anisotropic ring structure of the ion distribution function (Cottrell *et al.*, 1993). That is found to be linearly unstable to relaxation to waves propagating across the magnetic field on the fast Alfvén – ion Bernstein branch of the dispersion relation at the first 10 cyclotron harmonics of the energetic ions. The linear growth rates at the ion cyclotron harmonics are $\sim$0.02 times the harmonic of the ion cyclotron frequency. Dendy (1994) has reviewed this area of work, and also considered similar ion cyclotron spectra observed in the magnetosphere (McClements and Dendy, 1993).

13.5 Summary

Whilst we have concentrated in this book on cyclotron masers at work in the Earth's magnetosphere, we close with a brief overview to introduce their roles in the Jovian magnetosphere, the solar corona, and in laboratory devices.

As for Earth's radiation belts, Jupiter's are formed by the joint action of inward diffusion across L shells and pitch-angle diffusion, a self-consistent treatment of which remains to be performed for Jupiter. The electrodynamic interaction of Io, which releases large amounts of neutral atoms into Jupiter's magnetosphere, with the Jovian magnetic field serves as an extra acceleration mechanism for radiation belt electrons to high energies.

The rotational ($\sim$10 hour) periodicity of the fluxes of charged particles inside and outside Jupiter's magnetosphere, and also the intensity of VLF emissions, termed 'clock variations', can be explained as being relaxation oscillations of the cyclotron instability of the Jovian radiation belt electrons (Bespalov *et al.*, 2005).

A novel model for the development of a solar flare (Fig. 13.1) is presented. The main acceleration of charged particles and the storage of energy in a flare loop occurs during the preflare stage, over an interval of some minutes. At this stage, a magnetic trap with a cold dense background plasma is filled by energetic electrons having energies from 10 keV up to 10 MeV (13.15). The actual acceleration mechanism is adiabatic magnetic compression and runaway effects, which create a tail of energetic electrons with a transverse velocity anisotropy. Non-adiabatic processes due to the magnetic reconnection play an important role at this stage, injecting superthermal electrons with energies from 1 to 10 keV into the magnetic trap.

The active stage of the solar flare is due to the explosive evolution (Trakhtengerts, 1996) of the cyclotron instability (13.25), with a considerable portion of the energy of the energetic electrons being transferred to the background plasma, while the electrons themselves precipitate through the loss cone into the lower chromosphere, generating hard X-rays. Since such thermal explosions evolve in a small volume near the minimum magnetic field region near the equatorial plane and involve the energy of particles from the entire trap volume, the kinetic energy of the heated region ($\sim 4 \times 10^{30}$ erg) can exceed the magnetic pressure confining the plasma. In this case, coronal material, together with some fraction of the energetic charged particles, will be injected into interplanetary space.

In many laboratory plasma devices containing magnetic traps, bursts of whistler-mode radiation have been found to be accompanied by the precipitation (loss) of energetic electrons from the trap via the loss cone. Such results can be understood using the theoretical concepts presented earlier in this book. Vodopyanov *et al.* (2005) have undertaken laboratory experiments with parameters simulating conditions in the Earth's magnetosphere. Pulses of whistler-mode waves and precipitating electrons (Fig. 13.6) have been discussed in terms of a burst-like regime of the cyclotron instability associated with relaxation oscillations, as investigated in Chapters 9 and 11.

Finally, mention is made of some topics discussed in this book concerning instabilities and heating in devices which are designed for controlled thermonuclear fusion.

Epilogue

Summarizing the content of this book, we conclude that whistler and Alfvén mode cyclotron masers in space parallel the family of ground-based maser systems in the laboratory, ranging from optical quantum generators to gyrotrons and free-electron lasers. They all require an active medium and a spatial geometry with mirrors at both ends. Many features of whistler and Alfvén maser operations reveal similarities to the operation of optical quantum generators; these include relaxation oscillations, self-oscillation generation regimes, and passive and active mode synchronization. Such regimes occur in cyclotron masers when the energetic electrons have a broad velocity distribution (i.e. the electron velocity range, or dispersion, is similar to the mean electron velocity).

CMs operating in the Earth's magnetosphere explain quantitatively features of many phenomena observed in the ELF/VLF frequency range, such as whistler-mode hiss and quasi-periodic emissions bouncing between conjugate hemispheres; they also determine the dynamics of electrons in the van Allen radiation belts and of pulsating auroral patches. Similar regimes operate for the proton CM and Alfvén waves, but with changes to the frequencies and characteristic time scales corresponding to the proton to electron mass ratio of 1836. Thus cyclotron masers control both the spatial and temporal distributions of the energetic charged particle populations, the Earth's radiation belts, trapped in the dipolar geomagnetic field. These have practical consequences, one example being to specify the radiation dose suffered by instruments aboard satellites in a circular low Earth orbit, in an elliptical orbit or in geostationary orbit.

Very interesting effects concerning the generation of ELF/VLF and ULF discrete emissions have recently been discovered in satellite measurements. It has been found that ELF/VLF chorus is generated in a small region near the equatorial plane of a certain geomagnetic flux tube; the length of this region is only $\sim 2\%$ of the entire length of the magnetic flux tube. Similar effects are observed in the Pc-1 frequency range ($\sim$ 1 Hz) for the proton CM. When we remember that there are no mirrors reflecting these waves and that the distribution function of energetic protons initially has a wide velocity range, this generation phenomenon is extremely surprising. A possible scenario for such generation is considered in Chapters 9 and 11.

It is shown there that the generation of noise-like emissions can be accompanied by the formation of sharp gradients (step-like deformations) to the distribution function over the field-aligned velocity component of energetic charged particles. Such gradients will be almost impossible to observe directly from a satellite, because of their localized nature in space, time and pitch-angle. However, they permit the development of an absolute instability for length-limited systems even when the waves are not reflected from the ionosphere. For whistler and Alfvén mode CMs this absolute instability corresponds to the generation of waves in the backward wave oscillator regime. This regime is well-known in electronics, where sharp gradients in the distribution function (such as a delta-function) can be made artificially. Here we can appreciate the close analogy between whistler and Alfvén mode CMs in space and in laboratory electronic devices.

This scenario, exhibiting the transition from the generation of noise-like emissions to the generation of discrete emissions, serves as a universal mechanism for the generation of discrete emissions in space plasmas. Many topical problems of whistler CMs, especially in the case of discrete emissions, are closely connected with triggered emissions which demonstrate the crucial nonlinear aspects of cyclotron wave–particle interactions. The advantage of triggered signals for detailed study is that they have a well-known source – a quasi-monochromatic whistler-mode wave generated by a ground-based radio transmitter. Modelling the triggered emissions satisfactorily, self-consistently, gives us very important information on details of the cyclotron wave–particle interaction, which is required to construct the nonlinear theory of CM operation. In particular, this recent modelling provides the fullest picture to date of the mechanism whereby a discrete VLF emission has a particular dynamic spectral form, whether it be a riser, faller or hook.

We have not considered another important type of electron cyclotron maser, which occurs in rarefied space plasmas when the electron plasma frequency is much less than the electron gyrofrequency. In this case the electromagnetic wave eigenmodes have frequencies very close to the electron cyclotron frequency, and their wave vectors are almost perpendicular to the magnetic field direction. This radiation can emerge from a planet's magnetosphere into space. Such conditions are realized in the auroral regions of the Earth's magnetosphere, where the cyclotron maser manifests itself as the generator of auroral kilometric radiation (AKR). It is probable that this type of cyclotron maser determines the characteristic radio emissions from Jupiter and the other giant planets.

The theory of the operation of electron cyclotron masers in rarefied space plasmas, especially the field of linear theory and the electromagnetic properties of the generation region, is developing nowadays (see Bingham and Cairns, 2000; LaBelle and Treumann, 2006). However, the nonlinear theory has not yet been developed;

we hope that the achievements of the nonlinear theory of whistler-mode CMs presented here will be useful in this other field of investigation. Another theme on which further theoretical research is needed is the development of the cyclotron instability leading to the acceleration of relativistic electrons. Such effects definitely occur in the Earth's magnetosphere (see Meredith *et al.*, 2002, 2003; Summers *et al.*, 2002), leading to the so-called 'killer' electrons which harm microcircuits aboard satellites. Some models of electron acceleration by whistler-mode waves have been discussed in the case of discrete (Chapter 6) and noise-like (Chapter 8) emissions, but they demand generalization to the relativistic case and further development of the self-consistent nonlinear theory of CM dynamics.

For Alfven masers a very real problem is maser operation in a multi-ion background plasma such as constitutes the magnetospheric plasma. The presence of heavy ions, whose gyrofrequencies lie in the frequency range of the unstable waves, can drastically change the CM operation, and lead to both the heating and the acceleration of the separate ion species. The preliminary estimates made in Chapter 12 demonstrate the importance of this issue.

Whistler-mode CMs can perform a special function in very dense and active stellar atmospheres, in particular, in the solar corona. For the solar corona, the threshold for the cyclotron instability to occur is determined by Coulomb collisions between electrons and ions, and decreases when the plasma is heated by cyclotron waves. This leads to the explosive development of the cyclotron instability, and to the local – and significant – heating of the background plasma at the centre of a coronal magnetic loop. We have considered only the initial stages of this process, when a thermal wave is formed. It would be very interesting to investigate the further development of this 'fireball', taking into account the plasma density redistribution and the distortion of the magnetic field.

Whilst it is evident that much has already been accomplished in the realms of the complicated plasma theory necessary to explain the operation of whistler and Alfv́en mode cyclotron masers in space, it is also clear that much still remains to be done. That is the challenge presented to the readers of this book.

Systems of units, conversion factors and useful numerical values

The equations in this book are presented in the c.g.s. Gaussian system of units. Nowadays, however, the Système International (SI), or rationalized MKSA system of units, is more generally used.

In order to convert quantities given in this book into SI units (shown below), they should be divided by the conversion factors given below (see Jackson, 1962, p. 620).

Quantity, symbol	SI units		Gaussian units
Length, l	metre, m	10^2	cm
Mass, m	kilogram, kg	10^3	gm
Time, t	second, s	1	s
Force	newton, N	10^5	dynes
Energy, W	joule, J, Nm	10^7	ergs
Charge, e	coulomb, C	3×10^9	
Current	ampere, A, C/s	3×10^9	
Potential, V	volt, V	1/300	
Electric field, E	volt/metre, V/m	$10^{-4}/3$	
Resistance	ohm	$10^{-11}/9$	
Electric displacement, D	C/m	$12\pi \times 10^5$	
Capacitance	farad, F	9×10^{11}	cm
Magnetic flux density, B	tesla, T	10^4	gauss
Magnetic flux	weber, Wb	10^8	maxwells
Electric permittivity of free space	F/m	4π	
Magnetic permeability of free space	henry/m	$9 \times 10/4\pi$	

Here, the factor 3 should, more correctly, be 2.9979; the velocity of light in free space, $c = 2.9979 \times 10^8$ m/s.

Some numerical values of useful constants are shown below (Woan, 2000).

Charge on electron, e	1.6×10^{-19} C
One electron volt, 1 eV	1.6×10^{-19} J
Electron mass, m_e	9.1×10^{-31} kg
Proton mass, m_p	1.67×10^{-27} kg
Boltzmann's constant, k	1.38×10^{-23} J/K
ϵ_0	$1/\mu_0 c^2 = 8.85 \times 10^{-12}$ F/m
μ_0	$4\pi \times 10^{-7}$ henry/m
Electron plasma frequency	$9\sqrt{n_e}$ Hz, with plasma density in m^{-3}
Electron gyrofrequency	28×10^9 B Hz, with B in T
Alfvén velocity, v_A	2.2×10^{16} $B\, n_e^{-1/2}$, with in n_e m^{-3} and B in T
π	3.141
e	2.718

Glossary of terms

a, a_L, a_0	scale length of magnetic field variation away from the equator, (2.8)
a	$v\|_0 - v_\|$, coefficient in Taylor series expansion, (3.2)
a, b	constants coming into (9.139)
a	numerical coefficient, (13.7)
a	charged particle gyroradius, (2.26)
a_i	term, for each type of ion, defined by (12.43)
$\vec{a}_l$	normalized polarization vector of the l-th eigenmode, (2.60)
$\alpha_{\alpha m}, \alpha_{\beta l}$	components of normalized polarization vector of the m-th or l-th eigenmode, (2.79)
a_x	component of $\vec{a}$ along x-axis (2.61)
a_y	component of $\vec{a}$ along y-axis (2.62)
a_z	component of $\vec{a}$ along z-axis (2.63)
amu	amplitude of elliptical function, (A.5.87)
$a_{M,E}$	parameters, determining the specific dependence of D^M_{LL} and D^E_{LL} on æ, (10.6)
a, b	parameters defined by (13.23)
A	real part of amplitude of wave electric field, (2.111)
A_k	amplitude of wave electric field of k-th wave packet, (8.4)
A_y	y component of amplitude of wave electric field, (2.114)
A_1, A_2, A_3	terms in (9.77)
$\mathcal{A}$	complex amplitude of slowly varying wave electric field, (2.127)
$\mathcal{A}_\Omega$	complex amplitude of wave electric field in BWO, (4.3)
$\mathcal{A}_{ent}$	complex amplitude of wave electric field at entry to generator region, (4.14)
$\mathcal{A}_{ext}$	complex amplitude of wave electric field at exit from generator region, (4.7)

b	$v_{\perp 0} - v_\perp$, coefficient in Taylor series expansion, (3.3)
b	amplitude of magnetoacoustic wave, (9.39)
b	field-aligned magnetic component of MHD wave, (10.50)
b	relative height of step in distribution function, (12.32)
b_ℓ	amplitude of wave magnetic field of a particular wave packet, (8.1)
$b_{\rm w}$	whistler-mode wave magnetic field amplitude, (Section 10.5)
b_*	term defined by (6.113)
$b_p(t)$	wave magnetic field pulse for Alfvén sweep maser, (12.20)
b_k^2	spectral intensity of wave magnetic field of k-th wave packet, (8.3)
B	ambient (geo)magnetic field strength, (2.3)
$B_{\rm in}$	initial amplitude of triggering signal, (Section 11.2)
B_I	magnetic field strength in ionosphere, (12.4)
B_k	complex wave magnetic field of k-th wave packet, (8.2)
B_L	minimum (lowest) value of magnetic field on a particular magnetic flux tube, in the equatorial plane for a dipole (geo)magnetic field, at a particular L-shell, (2.16)
B_M	maximum value of magnetic field on a particular magnetic flux tube, on the planet's surface, (2.7)
B_0	external magnetic field, (see Section 6.6)
$B_{\rm p}$	wave magnetic field of 'pearl' pulse, (12.20)
$B_{\rm R}$	magnetic field strength at magnetic reflection point, (2.17)
$B_{\rm s}$	sideband wave magnetic field, (see 5.38)
$B_\sim$	wave magnetic field $= NA$, (2.57)
$B_{\sim\,\rm ant}$	wave magnetic field driven by beam antenna current, (6.112)
$\tilde{B}$	B/B_L, (8.10)
$\tilde{B}_{zR}$	$B(z_R)/B_L$, (8.45)

c	velocity of light in free space
$c_{1,2}$	constants of motion, (6.71)
c_ν	normalization constant, (3.72)
cn	Jacobian elliptic function, (5.20)
subscript c	at boundary (edge) of loss cone, (9.14a)
subscript cr	critical value, (Section 13.2)
C	normalization constant, (9.105)
C	normalized constant, (10.19)
$C_{\rm erg}$	ergodic velocity distribution function, (6.85)
C_j	constant, (4.40)
C_k	amplitude of cos component to $\mathcal{F}_k$, (8.13)
C_1	motion integral, (5.5)
$C_{1,2}$	constants determined by BWO boundary conditions, (4.9)
$C_{1,2}$	constants, (8.51)
C^2	term defined by (9.162)

d	$(1-\mu\tilde{B})^{1/2}/\mu\tilde{B}^{-1/2}$, (8.20)
d	waveguide diameter, (Section 10.5)
d_k	$\omega_{Bk}(1-\mu\tilde{B})^{1/2}/\mu\tilde{B}^{-1/2}$, (8.10)
dn	Jacobian elliptic function, (5.20)
dB	decibels
$d^3\vec{v}$	element of phase space, (3.60)
$\vec{D}$	electric displacement, (2.50)
D	term defined by (8.23)
D_{ij}	diffusion coefficient defined by (10.1)
D	diffusion coefficient defined by (8.29)
D	diffusion coefficient, (9.102)
D_0	term defined by (9.1)
D_1	diffusion coefficient over energies (8.76)
D_{LL}	cross L-shell diffusion coefficient, (10.4)
$D_{\mu\mu}$	pitch-angle diffusion coefficient, (8.22)
$D_{ææ}$	pitch-angle diffusion coefficient in variables $æ = \mu^{1/2}$, (10.4)
D	coefficient related to diffusion, (13.3)
$\bar{D}$	bounce-averaged (over z) diffusion coefficient, (8.34)
$\hat{D}$	quasi-linear diffusion operator, (13.16)
$\mathcal{D}_l$	dispersion relation, (2.88)
$\mathcal{D}_{\alpha\beta}$	matrix, (2.59)
e	electric charge on particle, (2.14)
e	magnitude of charge on (negatively charged) electron
e	logarithmic energy flux value (13.23)
e_α	charge on particle of type α, (2.13)
e_i	charge on positive ion i, (2.89)
subscript e	electron, (2.90)
subscript ent	entrance of wave packet, where electron enters, its forward edge, (6.30)
subscript erg	ergodic, (6.85)
subscript ext	exit of wave packet, where accelerated electron escapes, its trailing edge, (6.33)
$\vec{E}$	electric field, (2.21)
$E_\sim$	wave electric field, including phase, (2.111)
E_l	wave electric field, for eigenmode l, or spectral component, (2.60)
E_{w}	whistler wave electric field, (2.93)
$E(\varphi, æ)$	elliptic integral of second kind, (A.5.85)
$E(æ)$	complete elliptic integral, (5.59)
$E(æ')$	$E\left(\sqrt{1-æ^2}\right)$, (5.59)
E_0	dawn to dusk electric field across magnetosphere, (2.45)
$E_0(t)$	short duration pulse electric field, (9.121)

E_1, E_2, E_n	amplified (or damped) pulse electric field after 1, 2, n-hop propagation, (9.123)
$E_{\perp L}$	electric field perpendicular to ambient magnetic field, in equatorial plane, (2.24)
f	wave frequency, of triggered emission, at any instant, (11.8)
f_0	frequency of pump wave, (11.8)
$f_{1,2}$	periodic functions, (13.6)
$F(\varphi, æ)$	elliptic integral of first kind, (A.5.84)
$\mathcal{F}$	distribution function, (Section 2.3), (2.132)
$\mathcal{F}_\mathrm{e}$	distribution function external to (outside) loss cone, (8.38a)
$\mathcal{F}_\mathrm{i}$	distribution function inside loss cone, (8.38a)
$\mathcal{F}_\mathrm{in,out}$	distribution function of electrons entering, leaving duct, (9.99)
$\mathcal{F}_k$	contribution of k-th wave packet to $\mathcal{F}_\sim$, (8.13)
$\mathcal{F}_0$	initial distribution function, (2.38)
$\mathcal{F}_\alpha$	distribution function for particles of type, (2.49)
$\mathcal{F}_\mathrm{sm,sm1}$	smooth part of distribution function, (3.34)
$\mathcal{F}'_v$	differential operator, (5.28)
$\mathcal{F}_L$	distribution function on certain L-shell, for strong diffusion, (12.1)
$\mathcal{F}_*$	distribution function defined by (9.12a)
$\mathcal{F}_\sim$	rapidly changing (at frequency ω) part of distribution function, (8.12)
$\mathcal{F}_\mathrm{A}$	change in distribution function due to Alfvén waves, (9.62)
g	off diagonal component of permittivity tensor, (2.54)
g	term defined by (4.44)
g_ik	metric tensor for dipolar geomagnetic field line, (9.66)
G	total logarithmic amplification, (7.32)
G	quantity given in (9.62)
$G(\omega)$	transmission coefficient, (9.112)
$G_{1,2}$	real, imaginary parts of transmission coefficient, (9.112)
$G_{10,20}$	defined by (9.125)
$G_{\vec{k},s}$	spectral power radiated by a single charged particle at the s-th harmonic of its gyrofrequency (8.67)
h	$B_\sim/B$, ratio of wave magnetic field to ambient magnetic field, (5.1)
h	parameter defined by (9.19a)
h	thickness of ionospheric F-layer, (9.146)

$\mathrm{He}(x)$	Heaviside operator, equal to 0 for $x < 0$, and unity for $x > 0$, (3.34)
$\mathcal{H}$	Hamiltonian, (2.32)
i	$\sqrt{-1}$
$i_\perp$	change of first adiabatic invariant due to a wave–particle interaction, (3.55)
subscript i	ion, (2.90)
I	integral given by (3.67)
I	energetic charged particle source, (13.1)
Im	imaginary part of, (2.115)
I_0, I_1	modified Bessel functions (8.51)
subscript I	value of parameter in ionosphere, (6.66)
j	electric current
$j_{1,2}$	resonant particle current, (5.62)
j_{c}	electric current due to cold plasma motions, (2.46)
j_{h}	electric current due to hot (energetic) charged particle motions, (2.46)
$j_\parallel$	field-aligned current, (12.9)
$j_{E_{\sim s}}$	current in direction of electric field of secondary wave, (6.97)
j_0	dimensionless characteristic of source, (Section 9.3.1)
J, J_Ω	complex amplitude of wave electric field in BWO, (4.1)
$J_n(t)$	source, in eigenfunction notation, (9.8)
$J_n(y)$	Bessel functions, (5.74)
J	resonance current, (4.21)
J	Jacobian, (5.26)
J	charged particle source, (8.34)
J_0	source in stationary state, (9.107)
J_0	zeroth-order Bessel function, (9.15)
J_{Im}, J_i	resonant current 90° out of phase with wave electric field, (Fig. 11.4)
J_{Re}, J_r	resonant current in phase with wave electric field, (Fig. 11.4)
$J_\perp$	first adiabatic invariant (magnetic moment), (2.25)
$J_{\perp\alpha}$	first adiabatic invariant of charged particle of type α, (2.12)
$J_\parallel$	second adiabatic invariant of charged particle, (2.20)
$J_{j=1,2,3}$	three adiabatic invariants, (10.1)
J_*, $J_{\perp 0}$	maximum value of $J_\perp$, (7.2)
J_*	current source defined by (9.12a)
J_Σ	total source power, (8.62)
J_Σ	number of energetic particles entering flux tube in unit time, (9.19)
$J_{\Sigma\sim}$	total effective source power due to MHD wave, (9.39)

$\mathcal{J}$	$JB^{-3/2}$, (7.19)
subscript J	Jupiter
$\vec{k}$	wave normal, wave vector, $-\partial\Theta/\partial\vec{r}$
k_α, k_β	components of wave vector, (2.59)
$k_{\rm m}$	wave vector corresponding to maximum frequency $\omega_{\rm m}$, (3.35)
$k_{\rm w}$, k_0	cold plasma contribution to whistler wave vector, (2.128)
k_0	wavenumber for gyroresonance, (see 9.63)
$k_{n,m}$	eigenvalues, (9.44)
subscript k	k-th wave packet, (8.2)
K	constant, (2.45)
K	function given in (9.62)
K	quantity determined by (5.47)
$K(æ)$	complete elliptic integral of first kind, (5.23)
$K(æ')$	$K\left(\sqrt{1-æ^2}\right)$, (5.93)
$\mathcal{K}_1(x)$	modified Bessel function, (8.80)
l	definition specifying an eigenmode, (2.60)
l	characteristic spatial (length) scale of adiabatic trap, magnetic trap
l	length of wave packet, (5.61)
l	working length, (11.18)
l	length of BWO system, (4.7)
$l_{\rm BWO}$	length of BWO system, (Fig. 11.13), (12.24)
$l_{\rm c}$	mean free path between collisions of energetic charged particles
$l_{\rm tr}$	trapping length, (6.9)
$l_{\rm eff}$	k/α_B, (8.36)
$\bar{l}_{\rm eff}$	average value of $l_{\rm eff}$, (10.10)
$l_{\rm eff}$	effective length, (11.18)
$l_\perp$	transverse dimension of geomagnetic flux tube, (Section 9.2.1)
$l_\perp$	transverse dimension of solar magnetic flux tube, (13.38)
$l_\parallel$	dimension of proton precipitation region along drift path, (Fig. 12.3), (12.15)
$l_\perp$	dimension of proton precipitation region across drift path, (Fig. 12.3), (12.15)
L	L-value of particular magnetic flux tube, McIlwain parameter, (2.6)
L	spatial dimension of magnetic loop in solar corona, (Section 13.2)
L_0	particular starting value of L, (2.26)
$L_{\max}$	maximum L-value, (12.14)
L_F	characteristic scale of electron density decrease above F-layer maximum density (9.146)
subscript L	value of parameter in magnetic equatorial plane, (6.39)
m	parameter specifying the eigenmode, (2.80)
m	mass of charged particle, or mass of electron, (2.14)

m_α	mass of charged particle of type α, (2.13)
subscript m, or max	maximum value
M	mass of proton, (see 9.89)
M	magnetic moment of a planet, (2.5), or of charged particle, (10.1a)
M_k	term involving magnetic moment of charged particle and magnetic field parameters (8.9)
subscript M	mirror point
subscript M	MHD contribution, (9.50)

n	integer, (see 13.7)
$\vec{n}$	unit wave normal vector, (2.79)
n_α, n_β	components of $\vec{n}$, (2.79)
n	number density (per unit volume) of charged particles, (2.30)
$n_{1,2}$	number density of energetic electron beams, (6.76)
n_b	total number density of energetic electrons in beam, n_1+n_2, (6.78)
n_c	cold plasma number density, (2.105)
n_e	electron density
n_h	hot plasma number density, (2.105)
n_0	energetic electron density, (9.106)
n_i	ion density, (12.39)
n_p	proton density, (12.39)
n_st	number density of electrons at step in distribution function, (5.68)
$n_{\mathrm{c}L}$	cold plasma density in equatorial plane, (9.104)
$n_{\mathrm{h}L}$	density of energetic protons in equatorial plane, (9.149)
n_x	number of points used in computation, (Fig. 9.16)
n_E	E-region electron density, (9.10)
n_F	F-region peak electron density, (see 9.146)
n_Σ	$\int n\,dz$ for F-layer, (9.147)
subscript n	eigenmode, (Section 9.2.1)
$N_{1,2}$	refractive index, (2.66)
N_A	refractive index for Alfvén waves, (2.95)
N_i	Alfvén refractive index for a plasma containing two types of ions, (12.36)
N_k	refractive index for k-th wave packet, (8.4)
N_l	refractive index for l-th eigenmode, (2.64)
$N_{m(l)}$	refractive index for $m(l)$-th eigenmode, (2.81)
N_w	refractive index for whistler-mode waves, (2.92)
N_h	number of energetic charged particles in a magnetic flux tube of unit cross-sectional area in the ionosphere, (8.59)
N_0	number of energetic charged particles in such a flux tube outside the duct (see 10.52)
$\tilde{N}$	initial value of N, (9.21)
$N_\sim$	MHD contribution to N, (9.55)
$N_n(t)$	amplitude in expansion of $\mathcal{F}(\mu, t)$, (9.4)

N_0	zeroth-order Neumann function, (9.15)
N_0	stationary value of $N_{\rm h}$, (9.26)
superscript (o)	zeroth-term in approximation, (5.39)
p	$(\omega_{\rm p}^2\omega/c^2\omega_B)^{1/2}$, useful parameter relating to whistler-mode refractive index, (6.32)
$p(u)$	parameter of wave packet internal structure, (9.163)
p	proton kinetic pressure, (12.5)
$p_\parallel$	proton kinetic pressure along geomagnetic flux tube, (12.7)
$p_\perp$	proton kinetic pressure across geomagnetic flux tube, (12.7)
p_0	proton pressure at plasmapause bulge in equatorial plane, (12.11)
p_0	$(a\omega_{BL}/v_*) \gg 1$, parameter defined by (7.34)
p_1	first eigenvalue, ~ 0.15, (9.16)
p_n	eigenvalues, (9.5)
$p(t)$	function defined by (5.83)
subscript p	proton
P	principal part of the integral, (3.14)
P	gas kinetic pressure, (see 9.79)
$P_\perp$	energetic electron pressure perpendicular to geomagnetic field, (9.42)
$P_\sim$	oscillating component of pressure, (9.79)
P_0, P_I	energy flow density of energetic electrons, (13.15)
q	$(\alpha_{\rm int}/v_1v_2)^{1/2}l$, threshold for BWO, (4.17)
q	Jacobi's parameter (nome), (5.30)
q	dispersive characteristics for whistler, (6.63)
q	harmonic number of Alfvén wave modulational instability, (9.85)
q	coefficient defined by (9.149)
$q^{\rm w}$, $q^{\rm s}$	fractional increase of trapped electron flux at 'cliff', (10.44)
q	azimuthal parameter involved in energetic proton pressure, (12.10)
q	heating source, (13.18)
q_E	production rate of E-region ionization, (9.10)
Q	parameter involving wave frequency/gyrofrequency, (6.42)
Q	loss function, (13.18)
Q	quality factor of a resonator
Q_J	quality factor for relaxation oscillations, (9.40)
Q_k	term defined by (8.19)
$Q_{1,2}$	slowly changing functions taken at stationary point $z_{\rm st}$, (3.65)
Q_{n_1,n_2}	coefficients for untrapped and trapped particles, respectively, (5.30)

$Q(\xi, \eta)$	integral involved in (7.23)
r	spatial coordinate, (Section 2.3)
r	radial coordinate, e.g. of any point in magnetosphere measured from centre of planet, (2.3)
r	power law index in (10.44)
r	dimensionless parameter, defined by (11.7)
r_{B0}	energetic proton gyroradius, (12.10)
$\vec{r}_0$	unit vector of $\vec{r}$, (2.3)
R	magnetic reflection point (see 2.17)
R	wave reflection coefficient, (8.85)
$R_{1,2}$	wave reflection coefficient at two conjugate hemispheres, (8.39)
R_c	LR_0, (2.45)
R_c	radius of curvature of geomagnetic field line, (Section 9.2.1)
$\vec{R}_c$	geocentric radial distance to charged particle's guiding centre, (2.10)
R_J	radius of Jupiter, (Section 13.1)
R_0	radius of Earth, $\sim$ 6370 km, (2.5)
$\dot{R}_{\parallel c}$	velocity of guiding centre along magnetic field line, (2.13)
$\dot{R}_{\perp c}$	velocity of guiding centre across magnetic field line, (2.13)
Re	real part of, (2.114)
subscript R	at the resonance point, (12.39)
s	z/a, parameter specifying variation of B with distance z away from magnetic equator, (6.41), (Section 11.3)
s	order of cyclotron harmonic, (8.67)
sn	Jacobian elliptic function, (5.27)
$\vec{s_0}$	unit vector along magnetic field line
s_*	maximum value of $s = z/a$ for electron acceleration to occur, (6.42)
s_ω	value of $s = z/a$ where wave of angular frequency ω is generated, (7.73)
subscript s	sideband, (subsection 5.4)
subscript s	secondary wave, (6.96)
subscript s	eigenvalue, (Section 9.2.1)
subscript sl	slot region, between inner and outer radiation belts, (Fig. 10.1)
subscript sm	smooth distribution function, (Section 11.3)
subscript st	step in distribution function, (Section 12.3)
superscript s, or S	from pag316 strong diffusion, (8.63)
superscript $*$	complex conjugate
S_k	amplitude of sin component of $\mathcal{F}_k$, (8.13)
S_0	cross sectional area of duct, (9.99)
S_A, S_{pr}	precipitated energetic charged particle flux, (Fig. 9.25, Section 12.1)

S_{tr}	trapped energetic charged particle flux, (Fig. 10.3, Section 12.1)
S_E	energetic charged particle flux density into ionosphere, (9.10)
$S_\sim$	amplitude of precipitated flux modulation, (10.50)
$S_L(W, L)$	integral flux density of radiation belt electrons on a particular L shell, (10.36)
$S(x, t)$	area of duct filled with energetic electrons at time t, (10.52)
t	time
t_i	duration of flash, (10.53)
$\tilde{t}$	modified time, (see 7.13)
t_{exp}	time scale for explosive phase, (13.26)
t_{g}	time of generation of secondary signal, (11.4)
t_{i}	duration of emission spike, (10.53)
t_{p}	characteristic time scale of envelope soliton, (9.141)
t_B	time scale for magnetic compression, (Fig. 13.3)
t_D	time for energetic electrons to drift through duct, (10.52)
t_*	time at which two beams appear, (see 6.78)
subscript t	trailing edge of wave packet, (see 6.93)
subscript thr	threshold, (4.16)
subscript tr	trapped, (5.15)
subscript tr	transition, (1.2.24)
subscript $*$	refering to point where electrons escape from wave, (6.85)
T	temperature of electrons, (13.13)
$T_\parallel$	temperature of energetic component of plasma along magnetic field line, given in energy units, (3.12)
$T_\perp$	temperature of energetic component of plasma across magnetic field line, given in energy units, (3.12)
T	transmission coefficient for BWO device, (4.14)
T	time given by $l(1/v_0 + 1/v_g)$, (5.80)
T	repetition period, period of pulsations, $T_p + t_i$, (Section 10.5)
T_{BWO}	repetition period for chorus, (Section 11.4, (11.19))
T_{ch}	duration of chorus event, (Section 11.4)
T_{g}	two-hop whistler-mode propagation time
T_ℓ	lifetime of trapped particle, (8.58)
T_0	duration of pulse, (9.164)
T_0	temperature at instability threshold, (13.22)
T_{p}	period of pulse repetition, (9.129)
T_{p}	duration of one 'pearl' pulse, (12.21)
T_{r}	ionospheric relaxation time, (9.91)
T_{s}	characteristic loss time for synchrotron processes, (13.5a)
T_{sp}	spatial period, (6.16)
T_t	trigger pulse length, (Section 11.3)

T_B	bounce period of charged particles in a magnetic trap, (2.20)
$\bar{T}_B$	value of T_B averaged over μ, (8.63)
$T_{\mathcal{D}}$	time scale of pitch-angle diffusion, (8.48)
$T_{\rm IPDP}$	duration of IPDP event, (Section 12.2)
T_J	period of relaxation oscillations, (9.108)
T_M	self-modulation period, (Section 9.5)
$T_{\rm MD}$	period of magnetic drift (in longitude, around Earth), (section 10.7)
$T_{\rm QL}$	time scale of quasi-linear process, (11.13)
u	variable similar to ξ, (6.97)
u	$\tau - \tau_0$, (9.160)
u	v^2/B_L, (see 10.4)
$\tilde{u}$	term proportional to u (10.8)
$u_{\rm g}$	$v_{\rm g}/v_0$, dimensionless group velocity, (7.61)
u_*	v_*/v_0, (7.61)
$u_\parallel$	$v_{\rm R} - v_\parallel$, canonical variable, (5.10)
$u_{1,2}$	roots of (7.68)
u_Σ	$u_* + u_{\rm g}$, (7.63)
subscript un	untrapped, (5.16)
$U_{\rm eff}$	effective potential, (see 6.15)
U_0	ionization potential for air, (see 9.148)
v	velocity of charged particle, (2.15)
v_e	electron velocity, (see 2.93)
$v_{\rm g}$	group velocity, $\partial\omega/\partial k$, (2.101)
v_i	mean velocity of energetic ions (protons), (9.89)
$v_{\rm in}$	initial velocity of electrons, (6.73)
$v_{\rm min}$	minimum velocity for cyclotron resonance in equatorial plane, (Section 9.3)
v_0	cyclotron resonance velocity, corresponding to the lowest energy electrons in cyclotron resonance with wave at maximum frequency, (3.35)
v_0	velocity of electron beam wave, (4.25)
v_0	characteristic velocity, (9.11)
v_0	electron velocity along magnetic field, at step in distribution function, (3.36)
v_{0*}	velocity of electrons at front, (6.3.12)
$v_{\rm ph}$	phase velocity, c/N, (3.73)
$v_\parallel$	velocity of charged particle parallel to magnetic field, (2.13)
$v_\perp$	velocity of charged particle across magnetic field, (2.11)
$v_{\parallel 0}$	energetic electron velocity along magnetic field line, for gyroresonance, (3.2)

$v_{\perp 0}$	energetic electron velocity across magnetic field, at gyroresonance condition, (3.46)
$v_{\parallel L}$	maximum value of $v_{\parallel}$, at equator, (3.75)
$v_{\perp L}$	velocity of charged particle perpendicular to magnetic field, in equatorial plane, (2.26)
$v_{\perp T}$	mean thermal velocity, (5.34)
$v_{\perp 1,2}$	$v_{\perp}$ interval for trapping by whistler-mode wave, (6.37)
$v_{1,2}$	group velocity of two waves in BWO, (4.1)
$v_{1,2}$	field-aligned velocity of electron beams, (6.75)
v_{st}	step velocity in proton distribution function, (see 12.31)
$\bar{v}$	mean energetic proton velocity, (12.8)
$\bar{v}_{\perp}$	characteristic value of $v_{\perp}$, (5.72)
v_{*}	maximum value of $v_{\parallel L}$, corresponding to step in distribution function, (3.75)
$v_{\parallel *}$	parallel velocity at injection point, (7.1)
v_{*i}	step velocity at injection point, (see 7.3)
v_{A}	Alfvén velocity, (9.41)
v_{D}	drift velocity of charged particles across magnetic field, (2.21)
v_{R}	velocity corresponding to cyclotron resonance condition, (3.53)
v_{RL}	minimum value of v_{R}, at magnetic equator, (3.68)
$v_{T\mathrm{i}}$	ion thermal velocity, (12.38)
v_{Σ}	$v_{\mathrm{g}} + \lvert v_{\parallel}\rvert$, (6.108)

w	sum of kinetic and potential energies of charged particle, (2.14)
w	change of particle energy due to wave–particle interaction, (3.54)
w	volume of magnetic flux tube, (12.6)
$w(\xi_i)$	expression defined by (12.38)
superscript w, or W	weak diffusion
W	kinetic energy of charged particle, (2.14)
W_h	energy of hot electrons, (Fig. 13.3)
W_{m}	maximum value of kinetic energy, (9.119)
W_{min}	minimum value of kinetic energy for gyroresonance, (8.43)
W_0	characteristic energy, (3.71)
W_{p}	kinetic energy of protons, (Section 12.1)
W_L	kinetic energy of charged particle in magnetic equatorial plane, (2.20)
W_{RL}	minimum (at magnetic equator) energy for gyroresonance, (3.68)
$W_{\parallel L}$	charged particle kinetic energy along magnetic field, in equatorial plane, (2.29)
$W_{\perp L}$	charged particle kinetic energy across magnetic field, in equatorial plane, (2.29)

W_* parallel energy of step electrons, at injection point, (7.15)
W_Σ total kinetic energy, (13.38)

x, y, z Cartesian coordinates of a point
x^1, x^2, x^3 curvilinear coordinates specifying dipolar magnetic field line, (9.66)
x variable defined by (3.21)
x v/v_0, dimensionless variable, (8.77)
x dimensionless variable defined by (9.93)
x $\sin\theta_{\mathrm{L}}$, (Fig. 9.16)
$\vec{x}_0$ unit vector of Cartesian coordinate across magnetic field, (2.11)

y $(1 - 1/\alpha)$, where α is temperature anisotropy, (3.19)
y argument of Bessel function, (5.74)
$\vec{y}_0$ unit vector of Cartesian coordinate across magnetic field, (2.11)
y magnetoionic variable, ω_B/ω, (6.63)
y dimensionless variable defined by (9.93)
y $= 1$, specifying transition between low and high energy electrons, (10.31)
Y expression in dispersion relation for modulational instability due to Alfvén waves, (9.82)
$Y(\theta_L)$ term in second adiabatic invariant, (10.1a)

z Cartesian coordinate along magnetic flux tube, measured from magnetic equator, (2.8)
$\vec{z}_0$ unit vector along magnetic field, (2.93)
z_f coordinate of thermal wave front, (Fig. 13.4)
z_{st} stationary point, (3.65)
z_{R} root of exact cyclotron condition, $\omega - \omega_B - kv_\parallel = 0$, (8.36a)
z_* value of z where electrons escape from wave packet, (6.85)
z_* injection point of beam, (7.3)
z_* electron mirror points, (8.30)
$Z_1(æ)$ first eigenfunction, (10.15)
$Z_n(\mu)$ n-th eigenfunction, (9.5)

α temperature anisotropy, $T_\perp/T_\parallel$, anisotropy parameter, (Chapter 1), (10.58)
α electron acceleration (deceleration) parameter, (6.73)
α $(8\mu_{\mathrm{c}} W_0/D_1)^{1/2}$, dimensionless parameter determining efficiency of electron acceleration, (8.81)
α dimensionless source amplitude, $1/vt_{\mathrm{thr}}$, (Fig. 9.15)
α_{eff} interaction coefficient, defined by (7.26)

α_{int}	interaction coefficient, (4.4)
α_n	coefficient in expansion, (5.48)
α_0	$\alpha_{\text{eff}}/\delta^{1/2}$, (7.56)
$\alpha_{x,y,z}$	parameters defining polarization ellipse, (2.68)
α_B	factor defining inhomogeneity of magnetic field, (6.12)
α_E	E-region recombination coefficient, (9.10)
α_N	new inhomogeneity factor for whistler wave packet of changing frequency, (6.51)
$\alpha_{\Delta\omega}$	equivalent inhomogeneity factor due to changing frequency, (6.53)
α_1	ratio $W_\perp/W_\parallel$ for charged particle, (2.28)
$\alpha_{1,2}$	feedback interaction coefficients in BWO, (4.1)
α_{10}	coefficient given by (5.72)
subscript α	specifying type of charged particle, (2.11)
α, β	parameters determining the structure of envelope soliton, (9.141)
β	v/c, ratio of electron velocity to velocity of light in free space, (2.108)
β	$(8\pi nT/B^2)$, ratio of kinetic (thermal) pressure to magnetic pressure, (9.61)
β_*	$(8\pi n_c T_\parallel/B^2)$, ratio of kinetic plasma pressure, including cold plasma density and temperature of energetic electrons, to magnetic pressure, (3.19)
β	$1/2\lvert\alpha_0\rvert\ell_{tr}^2$, inhomogeneity parameter, (9.137)
$\beta_\perp$, $\beta_\parallel$	$(2T_\perp/mc)^{1/2}$, $(2T_\parallel/mc)^{1/2}$, (3.38)
$\beta_\parallel$	$v_\parallel/c$, (6.5)
$\beta_{\perp 0}$	$v_{\perp 0}/c$, (3.26)
γ	wave growth rate, (2.106)
γ_{lin}, γ_L	linear growth rate, (Section 11.2), (5.29)
γ_{sm}	growth rate for smooth distribution function, (11.14)
γ_{st}	growth rate with step in distribution function, (3.41)
γ_{HD}	growth rate for hydrodynamic stage of instability, (3.33)
$\tilde{\gamma}$	normalized linear growth rate, (3.19)
$\tilde{\gamma}$	initial growth rate, (9.23)
γ_δ	growth rate for distribution function as delta-function, (3.33)
γ_{A}	growth rate for Alfvén wave modulational instability, (9.84)
γ_{BWO}	growth rate of BWO device, (4.28), (Table 11.1)
γ_{stA}	growth rate with step in distribution function of energetic protons, (12.32)
γ_N	nonlinear growth rate, (5.27)
γ_{SB}	sideband growth rate, (5.53)
Γ	amplification factor, (Fig. 2.8)
Γ^+	one-hop whistler-mode wave amplification, (3.63)
Γ_0	whistler amplification for specific conditions, (9.110)

$\Gamma_{\rm m}$	maximum amplification in the case of second order CR, (see 7.41)
$\Gamma_{\rm w}$	total whistler-mode amplification for two-hop propagation, (8.41)
$\Gamma_{\rm sm}$	amplification for smooth distribution function, (Section 11.3)
$\Gamma_{\rm st}$	amplification with step in distribution function, (Section 11.3)
Γ_A	cyclotron amplification for Alfvén mode waves, (9.86)
Γ_*	amplification term defined in (9.104)
$\Gamma(\omega)$	defined in (9.136)
$\Gamma(\nu+1)$	gamma-function, (3.72)
δ	delta-function, (see 2.43)
δ	$n_{\rm h}/n_{\rm c}$, ratio of hot to cold plasma densities, (see 2.124)
δ	small parameter, (5.1)
δ	frequency sweep rate in lightning signal, (6.59)
δ	dimensionless parameter, ~ 1, (7.78)
δ	general charged particle loss rate, (8.35)
δ	parameter defined by (9.19a)
δ	parameter characterizing pitch-angle distribution function, (9.112)
δ	term defined by (9.126)
$\delta_{\rm p}$	$T_{\rm p} - T_{\rm g}$, (see (9.130))
$\delta_{\rm Re}, \delta_{\rm Im}$	defined by (9.134)
$\delta_\parallel$	parameter characterizing butterfly distribution function, (9.113)
δ_0	$2/T_B(\mu_{\rm c})$, (8.35)
δ_v	dispersion of parallel velocities of energetic ions in beam, (6.79)
$\delta_{\alpha\beta}$	unit diagonal tensor, (2.59)
δ_{lm}	unit diagonal tensor, (2.85)
δk	$(k - k_{\rm s})$, where $k_{\rm s}$ is sideband wave vector, (5.36)
$\delta\omega$	$(\omega - \omega_{\rm s})$, where $\omega_{\rm s}$ is sideband frequency, (5.36)
$\delta\tilde{\omega}$	shifted $\delta\omega$, (5.40)
δn	$(n_{\rm i}/n_{\rm p})$, (12.39)
$(\delta n)_{\rm pL}$	ratio of hot to cold proton densities in equatorial plane, (12.39)
$\delta(\Delta)$	delta-function, (3.14)
Δ	$\frac{1}{2}(\varepsilon^2 - g^2 - \varepsilon\eta)$, (2.77)
Δ	$\omega - \omega_B - kv_\parallel$, (3.3)
Δ	determinant, (4.45)
Δ	parameter given by (9.40)
Δ	parameter characterizing butterfly distribution function, (9.113)
$\Delta n_{\rm h}$	step height, (11.14)
$\Delta_{\rm g}$	frequency width of amplification, (9.136)
Δ_0	frequency mismatch, (6.95)
$\Delta v_{\rm st}$	step width, (11.15)
Δ_w	$\omega_w - \omega_B - kv_\parallel$, (3.28)
Δ_w	$\omega_w - \omega_B + kv_0 + ka(\varphi)$, (4.21)
Δ_{w0}	$\omega_w - \omega_B - kv_0$, (4.22)
Δ_R	frequency width of reflection coefficient, (9.135)

$\Delta_\perp$	perpendicular Laplacian operator, (9.47)
$\Delta(\xi, \eta)$	$\omega - \omega_B + kv_*$, (7.18)
Δn_E	change of E-region density, (9.90)
$\Delta f_{\rm trig}$	$f - f_0$, frequency difference of triggered emission from triggering signal, (see (11.8))
Δt	spike duration, (9.36)
$\Delta v_{\rm R}$	range of velocities over which electrons are trapped in potential well of wave, (11.1)
ΔB	$B(z") - B_*$, (7.14)
$\Delta\varphi$	range of magnetic longitudes, i. e. of magnetic local times, (12.15)
$\Delta\varphi_{\rm F}$	longitudinal width of 'cliff' front, (Fig. 10.3)
$\Delta\varphi_{\rm rel}$	relaxation width of 'cliff' front, (Fig. 10.3)
$\Delta\omega$	frequency width of wave packet, (8.1)
$\Delta\omega_i$	frequency separation between wave packets, (8.1)
$\Delta\omega_{R0}$	defined by (9.157)
$\Delta\Phi$	difference of electric potential between two points, (2.20)
$\Delta\Omega$	$(\Omega - s\Omega_{\rm d})$ d (9.71)
$\Delta_{1,2,3}$	coefficients defined by (9.12a)
ε	small quantity, (5.77)
ε	B_k/B, small quantity, (see 8.11)
ε	$\mathcal{E}_\omega/v_{\rm g}$, (9.104)
ε_1	$B_\sim/B$, small quantity, (see 6.12)
ε_2	$l_{\rm tr}/l$, small quantity, (see 6.12)
ε	two diagonal components of permittivity tensor, (2.54)
$\varepsilon_{\alpha\beta}$	components of cold plasma permittivity tensor, (2.53)
$\varepsilon^0_{\alpha\beta}$	components of permittivity tensor, with ambient magnetic field along z-axis, (2.54)
ε_{ij}	components of permittivity tensor, for k along z-axis and B in xz-plane, (2.65)
$\mathcal{E}_w$	whistler-mode wave energy density, (5.54)
$\mathcal{E}$	$\int \mathcal{E}_\omega \, d\omega$, wave energy flux density, (8.64)
$\mathcal{E}_k$	wave spectral energy density, (8.5)
$\mathcal{E}_\omega$	whistler-mode wave energy flux through cross-section of duct, (8.26)
$\mathcal{E}_\Sigma$	wave energy flux for narrow spectrum, corresponding to D_1, (8.84)
$\mathcal{E}_\Sigma$	total wave energy flux density, (9.3)
$\tilde{\mathcal{E}}$	initial value of $\mathcal{E}$, (9.21)
$\mathcal{E}_\sim$	amplitude of modulation of $\mathcal{E}$, (9.51)
$\mathcal{E}(t)$	whistler-mode wave envelope, (9.127)
ζ	variable, $t_0 = t - z/v_0$, (5.64)
ζ	new variable, following particle trajectories, (6.107)
ζ	variable $\ln(\mathcal{E}/\mathcal{E}_0)$, (9.33)

η	third diagonal component of permittivity tensor, (2.54)
η	$l(æ + \Omega/v_0)/2\pi$, (4.42)
η	new variable following waves, (6.107)
η	mean number of electron–ion pairs per unit volume produced by one energetic proton, (9.148)
η	$z_f/v_{0\parallel}t$, solution to (13.31)
$\eta_{\rm st}$	stationary phase point, (7.42)
η_E	mean number of electron–ion pairs per unit volume produced by one energetic electron, (9.10)
$\eta_{xx}, \eta_{yx}, \eta_{yy}$	components given by (2.67)
$\eta(\omega)$	frequency dependence of growth rate for step distribution function, (3.41)
$\eta_0(\xi)$	value of η, (7.31), corresponding to initial value $t(\xi, \eta) = 0$
θ	pitch-angle, angle between v and B for a charged particle moving at velocity v in magnetic field B, (2.15)
$\theta_{\rm c}$	pitch-angle at edge of loss cone, (2.18)
θ_L	pitch-angle in equatorial plane, (2.18)
$\theta_R = \pi/2$	pitch-angle at mirror point, 90°, (2.17)
Θ	full phase of wave, (2.113)
Θ_0	initial phase of whistler wave, (2.111)
$\Theta_{\rm w}$	$(\omega_{\rm w}t - k_{\rm w}z)$, cold plasma contribution to phase of whistler-mode wave, (2.127)
Θ'	time derivative of phase (9.154)
Θ	T/T_0, dimensionless variable, (13.27)
æ	wavenumber for wave complex amplitude of BWO, (4.3)
æ	energy integral, (5.14)
æ	$\mu^{1/2}$, (9.12)
$æ_0$, $æ_1$	characteristics of butterfly distribution, (9.113)
$æ_{1,2}$	eigenvalues for BWO system, (4.8)
$æ_i$	spatial damping rate, (12.41)
$æ_j$	eigenvalues for BWO system, for j = 1–3, (4.40)
$æ_0$	$\mu_{\rm c}^{1/2}$, (9.112)
$æ_0$	wavenumber for Alfvén disturbances, (see 9.89)
$æ_s$	longitudinal eigenvalues, (9.43)
$æ_{\rm w}$	$\gamma_{\rm w}/v_{\rm g}$, wave amplification coefficient for whistler waves, (8.26)
$æ_{\rm A}$	$\gamma_{\rm A}/v_{\rm g}$, wave amplification coefficient for Alfvén waves, (9.86)
$æ = 1$	condition for trapping of electrons in potential well of wave, (5.14)
$æ'$	$1/æ$, (6.24)
$æ(z, t)$	contribution of beam electrons to wave vector k, (7.9)

λ	magnetic latitude, e. g. of any point in planetary magnetosphere, (2.4)
$\vec{\lambda}_0$	unit vector of λ, (2.3)
λ	wavelength of electromagnetic wave, (Section 10.5)
λ	$\omega_{ps}^2 v_0^2/2c^2$; if this parameter >1, BWO generation starts, (5.82)
λ	complex amplification ratio (9.27)
$\lambda_{1,2,3}$	three solutions of (9.27)
Λ	$F - F'$, difference between two elliptic functions, (see 5.47)
Λ	coefficient defined by (10.17)
$\hat{\Lambda}$	operator defined by (8.69b)
$\hat{\hat{\Lambda}}$	operator defined by (8.69a)
$\Lambda_\perp$	operator defined in (8.74)
μ	$2B_L J_\perp/W = B_L \sin^2\theta(z)/B(z)$, parameter of charged particle, (2.16)
μ_c	value of μ at edge of loss cone, (see 8.33)
μ_m	maximum value of μ of charged particle, (see 8.38)
ν	damping rate, (3.14)
ν	interaction coefficient, (5.70)
ν	parameter specifying inhomogeneity of magnetic field along field line, (10.2)
ν_{ei}	electron–ion collision frequency, (13.12)
ν_0	value of interaction coefficient, (5.82)
ν_J	damping rate of relaxation oscillations, (9.27)
ν_R	damping rate of relaxation oscillations, (13.2)
$\nu_\sim$	oscillatory part of $\nu(t, \varphi)$, (13.7)
ξ	radial displacement of magnetic field line due to Alfvén wave, (9.68)
ξ	variable, (3.54)
ξ	new variable, (6.3)
ξ	new phase, (5.12)
ξ	$(1 - \omega_0/\omega)$, (9.101)
ξ	variable of integration, (5.17)
ξ_{st}	stationary point, (7.41)
ξ_*	value of ξ where electrons escape from wave packet, (6.85)
Π	integral defined in (5.45)
Π	product of n terms, (7.36)

ρ_c	density of electric charge due to cold plasma, (2.48)
ρ_h	density of electric charge due to energetic particles, (2.48)
ρ_B	gyroradius of energetic charged particle in magnetic field
$\rho_{B\alpha}$	gyroradius of energetic charged particle of type α, (2.11)
σ	magnetic mirror ratio, ratio of maximum to minimum magnetic field in magnetic trap, (2.18)
σ	B_M/B, (8.27)
σ_A	cross-sectional area of magnetic flux tube, (6.39)
τ	characteristic time scale, (2.19)
τ	duration of wave packet trapping electrons, (6.46)
τ	$v_0 t/a$, dimensionless time, (7.61)
τ	parameter defined by (9.19a)
τ_i	optical depth, (12.41)
τ_j	characteristic time of charged particle source, (Section 9.3.1)
τ_{1t}	lifetime of fast electrons, (13.37)
τ_0	trapping period, (5.14)
τ	$\int_{-\infty}^{t} \|E\|^2 dt'$, 'new' time, (9.140)
τ_D	shifted time delay $(3/4)T_g$, (9.150)
τ_0	characteristic decay time for electrostatic pulses, (10.7)
$\tau_0^*(v_{\perp T})$	trapping period for mean velocity, $v_{\perp T}$, (5.34)
τ_{opt}	optimum value of wave packet duration for maximum electron acceleration, (6.47)
τ_{st}	$v_{st0} t/a$, (11.21)
τ_ω	dimensionless time where wave of frequency ω is generated, (7.73)
τ_L	charged particle lifetime, (see 10.1)
τ_R	$(\alpha_E n_E)^{-1}$, recombination lifetime, (9.148a)
φ	gyrophase angle, (2.11)
φ	azimuthal angle, i. e. magnetic longitude, (10.38)
φ_0	initial gyrophase, (3.1)
φ_0	magnetic longitude defined in (10.42)
$\varphi_0(L)$	magnetic longitude of plasmapause bulge in equatorial plane, (12.11)
$\varphi_1(L)$	magnetic longitude where CI is switched off, (12.13)
$\varphi(\omega)$	function of wave frequency and gyrofrequency, (6.68)
$\phi(\tilde{\omega}\tau_0)$	frequency dependence of sideband growth rate, (Figure 5.4)
Φ	electric potential, (2.14)
Φ	$\Delta v_g/v_\Sigma$, (7.55)
Φ	magnetic flux, (section (9.2.2)

Φ	distribution function over μ integrated over duct cross section, (9.98)
Φ	function defined by (13.8)
$\Phi(L, u)$	distribution function of radiation belt electrons, (10.15)
$\Phi(J_\perp, z)$	distribution over first adiabatic invariant, (7.1)
$\Phi(\xi, \eta'')$	phase mismatch, (7.27)
χ	angle between wave normal k and ambient magnetic field B, (2.65)
χ	R parameter characterising change of reflection coefficient, (9.27)
χ	angle defined in Figure 12.3, (12.15)
χ	electron thermal conductivity, (13.19)
ψ	$\Theta - \varphi$, phase difference between wave electric field and electron perpendicular velocity, (2.113)
ψ	longitude, (Section 9.2.2)
ψ	arbitrary constant, (9.129)
ψ_0	initial relative phase, (5.67)
ψ_s	phase of sideband wave, (5.43)
ψ_{LT}	local time angle, measured from midnight, (Fig. 2.4)
ψ_{LT0}	local time angle of source of energetic charged particles, injection point, (2.43)
$\vec{\psi}$	unit vector corresponding to local time angle, (2.23)
$\psi(\Omega)$	normalized amplification, (see 9.89)
ω	angular frequency of electromagnetic wave, $\partial\Theta/\partial t$, (Section 2.4)
$\omega(\vec{k}, \vec{r})$	dispersion relation, (2.103)
ω_{bk}	gyrofrequency of charged particle in wave magnetic field of k-th wave packet, (8.20)
ω_0	ω_{BL}/β_*, (9.101)
ω_0	Alfvén maser resonant frequency, (12.22)
ω_0	frequency defined (9.137)
ω_g	frequency corresponding to maximum amplification, (9.136)
ω_h	gyrofrequency of charged particle in wave magnetic field, (5.2)
ω_{hs}	gyrofrequency of charged particle in magnetic field of sideband wave, (5.36)
ω_m	maximum frequency, (3.35)
ω_p	plasma frequency, $(4\pi ne^2/m)^{1/2}$, (Section 2.4)
ω_{ph}	hot plasma frequency, (3.27)
ω_{pL}	plasma frequency at equator (Fig. 11.8)
ω_{ps}	plasma frequency for step electrons, (5.76)
ω_s	frequency of sideband, (5.36)
ω_w	frequency of whistler wave in cold plasma, (2.128)

$\tilde{\omega}$	ω/ω_B, (3.33)
ω_*	maximum frequency for cyclotron instability, (3.18)
ω_B	electron gyrofrequency, (2.96)
$\omega_{B\alpha}$	eB/mc, $\partial\varphi/\partial t$, gyrofrequency of charged particle of type α, (2.12)
$\omega_{B\sim}$	gyrofrequency of charged particle in wave magnetic field, (5.1)
ω_{B_*}	electron gyrofrequency at point z_*, (7.3)
ω_R	frequency corresponding to maximum reflection coefficient, (9.135)
Ω	$(\omega - \omega_B - kv_\parallel)$, frequency deviation from gyroresonant condition, (3.28)
Ω	addition to wave frequency characterizing BWO generation, (4.3)
Ω	phase space domain, (6.87)
Ω	addition to wave frequency due to beam of energetic particles, (7.9)
Ω	self-modulation frequency, $\sim q(v_A/l)$, (9.58)
Ω	$(\omega - \omega_0)$, (9.126)
Ω_{or}	cross-over frequency, (Fig.12.7)
Ω_{d}	angular drift velocity of energetic charged particle, (9.70)
$\bar{\Omega}_{\mathrm{d}}$	mean angular drift velocity of energetic charged particle, (9.62)
Ω_0	mean frequency generated during spike development, (9.88)
Ω_{p}	ion plasma frequency, (12.25)
Ω_{tr}	trapping frequency of electrons in wave magnetic field, (5.14)
Ω_A	Alfvén wave resonant frequency, (9.81)
Ω_{Bp}	proton gyrofrequency, (9.85)
Ω_{Bi}	ion gyrofrequency, (12.36)
Ω_J	frequency of relaxation oscillation, (9.30)
Ω_{JN}	Ω_{J} for highly nonlinear situation, (9.35)
Ω_J	Jupiter's rotation frequency, (Section 13.1)
Ω_M	frequency of MHD wave, (9.45)
Ω_R	frequency of relaxation oscillations, (13.2)
Ω_*	Larmor drift frequency, $\sim v^2/\omega_B\, l_\perp^2$, (see 9.70)

Abbreviations and acronyms

AKR	auroral kilometric radiation
AM	Alfvén maser
BWO	backward wave oscillator
c.g.s.	centimetre, gram, second system of units
CI	cyclotron instability
CM	cyclotron maser
CME	coronal mass ejection
CR	cyclotron resonance
DM	dipole magnetic field
ELF	extremely low frequency, 3 Hz to 3 kHz
ERB	(Earth's) electron radiation belts
FCM	flow cyclotron maser
G	gauss, unit of magnetic flux density, in c.g.s. units
GEOS-1	first European geostationary satellite for scientific research
GEOS-2	second European geostationary satellite for scientific research
HD	hydrodynamic
IAR	ionospheric Alfvén resonator
IPDP	irregular pulsations of diminishing period
JASTP	*Journal of Atmospheric and Solar-Terrestrial Physics*
JEF	Jacobian elliptic function
JET	Joint European Torus
L	*L*-shell, *L*-value, McIlwain's (1961) parameter
LT	local time
maser	microwave amplification by the stimulated emission of radiation
MHD	magnetohydrodynamic
MKSA	metre, kilogram, second, ampere system of units

MLat	magnetic latitude
MSP	modified stationary phase
NAA	a US military VLF radio transmitter
NOAA	National Oceanic and Atmospheric Administration (of USA)
NPG	a US military VLF radio transmitter
Pc1	type of continuous pulsations of geomagnetic field
QL	quasi-linear
QP	quasi-periodic
RB	radiation belt
SCM	space cyclotron maser
SD	strong diffusion
SI	Système International
TFTR	tokamak fusion test reactor
ULF	ultra low frequency, < 3 Hz
VLF	very low frequency, 3 to 30 kHz
WD	weak diffusion

Bibliography

Akasofu, S.-I., and Chapman, S. 1972. *Solar-Terrestrial Physics*. Oxford: Clarendon Press.

Albert, J.M. 2000. Gyroresonant interactions of radiation belt particles with a monochromatic electromagnetic wave. *J. Geophys. Res.* **105**, 21191–21209.

Alpert, Ia.L., Guseva, A.G., and Fligel, D.S. 1967. *Propagation of Low Frequency Electromagnetic Waves in the Earth Ionosphere Waveguide* (in Russian). Moscow: Nauka.

Alikaev, V.V., Glagolev, V.M., and Morosov, S.A. 1968. Anisotropic instability in a hot electron plasma, contained in an adiabatic trap. *Plasma Phys.* **10**, 753–774.

Altshul, L.M., and Karpman, V.I. 1965. Theory of nonlinear oscillations in a collisionless plasma. *Sov. Phys. JETP* **49**, 515.

Andronov, A.A., and Trakhtengerts, V.Y. 1963. Instability of one-dimensional packets and absorption of electromagnetic waves in a plasma. *Sov. Phys. JETP* **45**, 1009–1015.

Andronov, A.A., and Trakhtengerts, V.Y. 1964. Kinetic instability of outer Earth's radiation belts. *Geomagn. Aeron.* **4**, 233.

Angerami, J.J. 1970. Whistler duct properties deduced from VLF observations mode with the OGO-3 satellite near the magnetic equator. *J. Geophys. Res.* **75**, 6115–6135.

Ard, W.B., Dandl, R.A., and Stetson, R.F. 1966. Observations of instabilities in a hot-electron plasmas. *Phys. Fluids* **9**, 1498–1503.

Arnol'd, V.I. 1978. *Ordinary Differential Equations*. Cumberland, Rhode Island: MIT Press.

Ashour-Abdalla, M. 1970. Nonlinear particle trajectories in whistler mode wave packet. *Planet. Space Sci.* **18**, 1799–1812.

Ashour-Abdalla, M. 1972. Amplification of whistler waves in the magnetosphere. *Planet. Space Sci.* **20**, 639–662.

Barbosa, D.D., and Coroniti, F.V. 1976a. Relativistic electrons and whistlers in Jupiter's magnetosphere. *J. Geophys. Res.* **81**, 4531–4536.

Barbosa, D.D., and Coroniti, F.V. 1976b. Lossy radial diffusion of relativistic Jovian electrons. *J. Geophys. Res.* **81**, 4553–4560.

Baumjohann, W., and Treumann, R. 1996. *Basic Space Plasma Physics*. London: Imperial College Press.

Bekefi, G., Hirshfield, J.L., and Brown, S.C. 1961. Cyclotron emission from plasmas with non-Maxwellian distributions. *Phys. Rev.* **122**, 1037–1042.

Bell, T.F. 1965. Nonlinear Alfvén waves in a Vlasov plasma. *Phys. Fluids* **8**, 1829–1839.

Bell, T.F., Inan, U.S., Helliwell, R.A.S and Scudder, J.D. 2000. Simultaneous triggered VLF emissions and energetic electron distributions observed on POLAR with PWI and HYDRA. *Geophys. Res. Lett.* **27**, 165–168.

Belyaev, P.P., Polyakov, S.V., Rapoport, V.O., and Trakhtengerts, V.Y. 1990. The ionospheric Alfvén resonator. *J. Atmos. Terr. Phys.* **52**, 781–788.

Belyaev, S.T. 1958. *Physics of Plasmas and the Problem of Controlled Thermonuclear Fusion*. Moscow: AN SSSR **3**, 50–65.

Benz, A.O., Conway, J., and Gudel, M. 1998. First VLBI images of a main-sequence star. *Astron. Astrophys.* **331**, 596–600.

Bespalov, P.A., and Trakhtengerts, V.Y. 1976a. Dynamics of the cyclotron instability in a Magnetic trap. *Fizika Plazmi* **2**, 396–406.

Bespalov, P.A., Trakhtengerts, V.Y. 1976b. On nonlinear oscillatory processes in the earth magnetosphere. *Izv. Vuzov. Radiofizika* **19**, 801–811.

Bespalov, P.A., and Trakhtengerts, V.Y. 1978a. Theory of generation of the chorus-type ELF/VLF radiation. *Geomagn. Aeron.* **18**, 627.

Bespalov, P.A., and Trakhtengerts, V.Y. 1978b. Auto modulation of cyclotron instability of Alfvén waves. *Fizika Plazmi* **4**, 177–183.

Bespalov, P.A., and Trakhtengerts, V.Y. 1979. About regimes of the pitch-angle diffusion in the geomagnetic trap. *Fizika Plazmi* **5**, 383–390.

Bespalov, P.A. 1981. Self-modulation of emission in a plasma cyclotron maser. *Sov. Phys. JETP Letters* **33**, 192–195.

Bespalov, P.A. 1982. Self-excitation of periodic cyclotron instability regimes in a plasma magnetic trap. *Phys. Scripta* **T2/2**, 576–579.

Bespalov, P.A., and Koval, L.N. 1982. Saturation of periodic regimes of cyclotron instability in plasma mirror traps. *Fizika Plazmi* **8**, 1136–1144.

Bespalov, P.A. 1984. Passive mode locking in masers with non-equidistant spectra. *Sov. Phys. JETP* **87**, 1894–1905.

Bespalov, P.A. 1985. A global resonance in Jovian radiation belts. *Sov. Astron. Lett.* **11**, 30.

Bespalov, P.A., Trakhtengerts, V.Y. 1986a. *Alfvén Masers* (in Russian). Gorky: Inst. of Appl. Phys.

Bespalov, P.A., and Trakhtengerts, V.Y. 1986b. Cyclotron instability of the Earth's radiation belts. In *Reviews of Plasma Physics*, New York: Plenum Press **10**, 155–292.

Bespalov, P.A. 1996. Self-consistent model of clock event in outer Jovian electron radiation belts. *Planet. Space Sci.* **44**, 565–568.

Bespalov, P.A., Savina, O.N., and Mizonova, V.G. 2003. Magnetospheric VLF response to the atmospheric infrasonic waves. *ADV Space Res.* **31**, 1235–1240.

Bespalov, P.A., and Savina, O.N. 2005. Global Synchronization of oscillations in the level of whistlers near Jupiter as a consequence of the spatial detection of the

Q factor of a Magnetospheric resonator. *JETP Lett.* **81**, 151–155.

Bespalov, P.A., Savina, O.N., and Cowley, S.W.H. 2005. Synchronized oscillations in energetic electron fluxes and whistler wave intensity in Jupiter's middle magnetosphere. *J. Geophys. Res.* **110**, A09209, doi: 10.1029/2005JA011147.

Beutier, T., and Boscher, D. 1995. A three-dimensional analysis of the electron radiation belt by the SALAMMBO code. *J. Geophys. Res.* **100**, 14853–14861.

Beutier, T., Boscher, D., and France, M. 1995. SALAMMBO: A three-dimensional simulation of the proton radiation belt. *J. Geophys. Res.* **100**, 17181–17188.

Bingham, R., and Cairns, R.A. 2000. Generation of auroral kilometric radiation by electron horseshoe distributions. *Phys. Plasmas* **7**, 3089–3092.

Bingham, R., Cairns, R.A., and Kellett, B.J. 2001. Coherent cyclotron maser radiation from UV Ceti. *Astron. and Astrophys.* **370**, 1000–1003.

Bingham, R., Kellett, B.J., Cairns, R.A. *et al.*, 2004. Cyclotron maser radiation in space and laboratory plasmas. *Contrils. Plasma Phys.* **44**, 382–387.

Bogomolov, Y.L., Demekhov, A.G., Trakhtengerts, V.Y., *et al.*, 1991. On the dynamics of fast particles quasi-linear relaxation in a plasma with particle and wave sources and sinks. *Sov. J. Plasma Phys.* **17**, 402–407.

Booske, J.H., Getty, W.D., Gilgenbach, R.M., and Jong, R.A. 1985. Experiments on whistler mode electron-cyclotron resonance plasma startup and heating in an axisymmetric magnetic mirror. *Phys. Fluids* **28**, 3116–3126.

Borwein, J.M. and Borwein, P.B. 1987. *Pi and the AGM: a study in analytic number theory and computational complexity.* New York: Wiley.

Bourdarie, S., Boscher, D., Beutier, T., Sauvaud, J.-A., and Blank, M. 1996. Magnetic storm modeling in the Earth's electron belt by the SALAMMBO code. *J. Geophys. Res.* **101**, 27171–27176.

Brice, N.M. 1963. An explanation of triggered very-low-frequency emissions. *J. Geophys. Res.* **68**, 4626–4628.

Brice, N.M. 1964. Fundamentals of very low frequency emission generation mechanisms. *J. Geophys. Res.* **69**, 4515–4522.

Brinca, A.L. 1972. Whistler side-band growth due to nonlinear wave-particle interaction. *J. Geophys. Res.* **77**, 3508–3523.

Brinca, A.L. 1973, Whistler modulational instability. *J. Geophys. Res.* **78**, 181–190.

Brossier, C. 1964. Non-linear model of progressive electromagnetic waves in the presence of a magnetic field. *Nucl. Fusion* **4**, 137–144.

Budden, K.G. 1964. *Lectures on Magnetoionic Theory*. London: Blackie.

Budko, N.I., Karpman, V.I., and Shklyar, D.R. 1971. On plasma stability in the longitudinal monochromatic wave field. *Sov. Phys. JETP* **61**, 778.

Budko, N.I., Karpman, V.I., and Pokhotelov, O.A. 1972. The nonlinear theory of whistler monochromatic waves in the magnetosphere. *Cosmic Electrodynamics* **2**, 165–183.

Burtis, W.J., and Helliwell, R.A. 1976. Magnetosphere chorus: Occurrence patterns and normalised frequency. *Planet. Space Sci.* **24**, 1007–1024.

Byrd, P.F., and Friedman, M.D. 1954. *Handbook of Elliptic Integrals for Engineers and Physicists.* Berlin: Springer.

Carlson, C.R., Helliwell, R.A., and Inan, U.S. 1990. Space-time evolution of whistler mode wave growth in the magnetosphere. *J. Geophys. Res.* **95**, 15073–15089.

Chen, F.F. 1974. *Introduction to Plasma Physics*. New York: Plenum Press.

Chenette, D.L., Conlon, T.F., and Simpson, J.A. 1974. Bursts of relativistic electrons from Jupiter observed in interplanetary space with the time variation of the planetary rotation period. *J. Geophys. Res.* **79**, 3551–3558.

Church, S.R., and Thorne, R.M. 1983. On the origin of plasmaspheric hiss: ray path integrated amplification. *J. Geophys. Res.* **88**, 7941–7957.

Cornilleau-Wehrlin, N., Chanteur, G., Perraut, S., *et al.*, 2003. First results obtained by the Cluster STAFF experiment. *Ann. Geophys.* **21**, 437–456.

Cornwall, J.M., Coroniti, F.V., Thorne, R.M. 1970. Turbulent loss of ring current protons. *J. Geophys. Res.* **75**, 4699–4709.

Coroniti, F.V., and Kennel, C.F. 1970. Electron precipitation pulsations. *J. Geophys. Res.* **75**, 1279–1289.

Cottrell, G.A., Bhatnagar, V.P., Da Costa, O., Dendy, R.O., Jacquinot, J., McClements, K.G., McCune, D.C., Nave, M.F.F., Smeulders, P., and Start, D.F.H. 1993. Ion cyclotron emission measurements during JET deuterium-tritium experiments. *Nucl. Fusion* **33**, 1365–1387.

Cuperman, S., and Landau, R.W. 1974. On the enhancement of the whistler instability in the magnetosphere by cold plasma injection. *J. Geophys. Res.* **79**, 128–134.

Cuperman, S., and Salu, Y. 1974. Optimum cold plasma density for maximum whistler instability: Numerical versus analytical. *J. Geophys. Res.* **79**, 135–137.

Das, A.C. 1968. A mechanism for VLF emissions. *J. Geophys. Res.* **73**, 7457–7471.

Davidson, G.T. 1979. Self-modulated VLF wave–electron interactions in the magnetosphere: A cause of auroral pulsations. *J. Geophys. Res.* **84**, 6517–6523.

Davidson, G.T. 1986. Pitch-angle diffusion in morningside aurorae. 2. The formation of repetitive auroral pulsations. *J. Geophys. Res.* **91**, 4429–4436.

Davidson, G.T., and Chiu, Y.T. 1986. A closed nonlinear model of wave-particle interactions in the outer trapping and morningside auroral regions. *J. Geophys. Res.* **91**, 13705–13710.

Davidson, G.T., and Chiu, Y.T. 1991. An unusual nonlinear system in the magnetosphere: A possible driver for auroral pulsations. *J. Geophys. Res.* **96**, 19353–19362.

Demekhov, A.G., and Trakhtengerts, V. Yu. 1986. Several questions of radiation dynamics in magnetic traps. *Radiophys. Quantum Electron.* **29**, 848–857.

Demekhov, A.G. 1991. On the role of the loss cone in the formation of the spike-like regime of the whistler cyclotron instability. *Geomagn. Aeron.* **31**, 1099–1101.

Demekhov, A.G., and Trakhtengerts, V.Y. 1994. A mechanism of formation of pulsating aurorae. *J. Geophys. Res.* **99**, 5831–5841.

Demekhov, A.G., Trakhtengerts, V.Y., Hobara, Y., and Hayakawa, M. 2000. Cyclotron amplification of whistler waves by non-stationary electron beams in an inhomogeneous magnetic field. *Phys. Plasmas* **7**, 5153–5158.

Demekhov, A.G., and Trakhtengerts, V.Y. 2001. Theory of generation of discrete ELF/VLF emissions in the Earth's magnetosphere. *Radiophys. Quantum Electron.* **44**, 103–116.

Demekhov, A.G., Nunn, D., and Trakhtengerts, V.Y. 2003. Backward wave oscillator regime of the whistler cyclotron instability in an inhomogeneous magnetic field. *Phys. Plasmas* **10**, 4472–4477.

Demekhov, A.G., and Trakhtengerts, V. Yu. 2005. Dynamics of the magnetospheric cyclotron ELF/VLF maser in the backward-wave-oscillator regime. I. Basic equations and results in the case of a uniform magnetic field. *Radiophys. Quantum Electron.* **48**, 639–649.

Demekhov, A.G., Trakhtengerts, V. Yu., Rycroft, M.J., and Nunn, D. 2006. Electron acceleration in the magnetosphere by whistler-mode waves of varying frequency. *Geomag. Aeron.* **46**, 711–716.

Dendy, R.O. 1994. Interpretation of ion cyclotron emission from fusion and space plasmas. *Plasma Phys. Control. Fusion* **36**, B163–B172.

Dendy, R.O., McClements, K.G., Lashmore-Davies, C.N., Cottrell, G.A., Majeski, R., and Cauffman, S. 1995. Ion cyclotron emission due to collective instability of fusion products and beam ions in JET and TFTR. *Nucl. Fusion* **35**, 1733–1742.

Dennis, B.R. 1991. The controversial relationship between hard X-ray and soft X-ray flares: Causal or non-causal? In *Flare Physics in Solar Activity Maximum 22, Proceedings of the International SOLAR-A Science Meeting Held at Tokyo, Japan, 23–26 October 1990*. Uchida, Y., Canfield, R.C., Watanabe, T., Hiei, E. (Eds) Lecture Notes in Physics, Berlin: Springer **387**, 89–95.

Dessler, A.J. 1983. *Physical of the Jovian Magnetosphere.* New York: Cambridge University Press.

Dungey, J.W. 1963. Resonant effects of plasma waves on charged particles in a magnetic field. *J. Fluid Mech.* **15**, 74–82.

Dysthe, K.B. 1971. Some studies of triggered whistler emissions. *J. Geophys. Res.* **76**, 6915–6931.

Engel, R.D. 1965. Nonlinear stability of the extraordinary wave in a plasma. *Phys. Fluids* **8**, 939–950.

Enome, S., and Hirayama, H., eds. 1993. *Proceedings of Kofu Symposium*, NRO Report No. 360.

Erlandson, R.E., Anderson, B.J., and Zanetti, L.J. 1992. Viking magnetic and electric field observations of periodic Pc 1 waves: Pearl pulsations. *J. Geophys. Res.* **97**, 14,823–14,832.

Erokhin, N.S., and Mazitov, R.K. 1968. The nonlinear theory of electromagnetic wave attenuation in plasmas. *Zh. Prikl. Mekh. Tekh. Fiz.* **5**, 11–17.

Galeev, A.A. 1967. Quasi-linear theory of the loss-cone instability. *J. Plasma Phys.* **1**, 105–112.

Galeev, A.A., and Sagdeev, R.Z. 1973. Nonlinear plasma theory. In *Voprosi teorii plasmi*, issue, **7**, ed. M.A. Leontovich. Moscow: Atomizdat, p. 3–145.

Gaponov, A.V. 1959. Interaction of non-straight electron flows with electromagnetic waves in transmission lines. *Izv. Vyssh. Uchebn. Zaved. Radiofiz.* **2**, 450–462.

Gaponov, A.V. 1960. Instability of a system of excited oscillators to electromagnetic disturbances. *Sov. Phys. JETP* **39**, 326–331.

Gaponov, A.V., and Yulpatov, V.K. 1962. Interaction of electron beams with closed electromagnetic field in resonators. *Radiotekh. and Elektronika.* **7**, 631–642.

Gaponov, A.V., Petelin, M.I., and Yulpatov, V.K. 1967. Stimulated radiation of excited classical oscillators and its application to the high-frequency electronics. *Radiophys. Quantum Electron.* **10**, 794–813.

Gaponov-Grekhov A.V., and Petelin, M.I. 1981. Cyclotron resonance Masers (in Russian). In *Science and Humanity.* Moscow: Znanie, 283–298.

Gaponov-Grekhov, A.V., Glagolev, V.M., and Trakhtengerts, V.Y. 1986. Cyclotron resonance maser with background plasma. *Sov. Phys. JETP* **53**, 1981.

Gary, S.P. 1993. *Theory of Space Plasma Microinstabilities*. Cambridge: Cambridge University Press.

Gendrin, R. 1968. Pitch-angle diffusion of low energy protons due to gyroresonant interaction with hydromagnetic waves. *J. Atmos. Terr. Phys.* **30**, 1313–1330.

Gendrin, R., and Troitskaya, V.A. 1965. Preliminary results of a micropulsation experiment at conjugate points, *Radio Sci.* 69D, 1107–1116.

Ginzburg, V.L. 1970. *Propagation of Electromagnetic Waves in Plasmas*. Oxford: Pergamon Press.

Ginzburg, N.S., and Kuznetsov, S.P. 1981. Periodic and stochastic automodulational regimes in electron generators with volume interaction. *Relativistic High-Frequency Electronics* (in Russian). Gorky: IAP Press, 101–104.

Gokhberg, M.B., Karpman, V.I., and Pokhotelov, O.A. 1972. On the nonlinear theory of evolution of pearls (Pc1) *Dokl. Akad. Nauk SSSR* **204**, 848–850.

Goldman, M.V. 1970. Theory of stability of large periodic plasma waves. *Phys. Fluids* **13**, 1281–1289.

Gradstein, I.S., and Ryzhik, I.M. 1965. *Table of Integrals, Series and Products*. New York: Academic Press.

Guglielmi, V.G., and Troitskaya, V.A. 1973. *Geomagnetic Pulsations and Diagnostics of the Magnetosphere*. Moscow: Nauka.

Gurevich, A.V., and Dimant, Ya.S. 1987. Kinetic theory of convective transfer of fast particles in Tokamak. *Voprosi Teorii Plasmi* **16**, 3.

Gurnett, D.A. and Bhattacharjee, A. 2005. *Introduction to Plasma Physics: With Space and Laboratory Applications*. Cambridge: Cambridge University Press.

Hansen, H.J., Mravlag, E., and Scourfield, M.W.J. 1988. Coupled 3- and 1.3-Hz components in auroral pulsations. *J. Geophys. Res.* **93**, 10029–10034.

Hattori, K., Hayakawa, M., Logoutte, D., *et al.*, 1991. Further evidence of triggering chorus emission from wavelets in the hiss band. *Planet. Space Sci.* **39**, 1465–1472.

Hayakawa, M., and Sazhin, S.S. 1992. Mid-latitude and plasmaspheric hiss: A review. *Planet. Space Sci.* **40**, 1325–1338.

Helliwell, R.A., Katsufrakis, J.P., Trimpi, M., and Brice, N. 1964. Artificially stimulated very-low-frequency from the ionosphere. *J. Geophys. Res.* **69**, 2391–2394.

Helliwell, R.A. 1965. *Whistlers and Related Ionospheric Phenomena*. Palo Alto, Calif: Stanford University Press.

Helliwell, R.A. 1967. A theory of discrete VLF emissions from the magnetosphere. *J. Geophys. Res.* **72**, 4773–4790.

Helliwell, R.A. 1969. Low-frequency waves in the magnetosphere. *Rev. Geophys.* **7**, 281–303.

Helliwell, R.A. 1970. Intensity of Discrete VLF Emissions. In *Particles and Fields in the Magnetosphere*. B.M. McCormack, and Alvio Renzini (Eds). Dordrecht, Reidel. 292–391.

Helliwell, R.A., and Crystal, T.L. 1973. A feedback model of cyclotron interaction between whistler-mode waves and energetic electrons in the magnetosphere. *J. Geophys. Res.* **78**, 7357–7371.

Helliwell, R.A., and Katsufrakis, J.P. 1974. VLF wave injection into the magnetosphere from Siple Station, Antarctica. *J. Geophys. Res.* **79**, 2511–2519.

Helliwell, R.A. 1983. Controlled stimulation of VLF emissions from Siple Station, Antarctica. *Radio Sci.* **18**, 801–814.

Helliwell, R.A. 1988. VLF wave injection experiments from Siple Station, Antarctica. *ADV. Space Res.* **8**, 279–289.

Helliwell, R.A. 1993. 40 years of whistlers. *Mod. Radio Sci.* 189–212.

Helliwell, R.A. 2000. Triggering of whistler mode emissions from Siple Station, Antarctica. *Geophys. Res. Lett.* **27**, 1455–1458.

Hobara, Y., Trakhtengerts, V.Y., Demekhov, A.G., and Hayakawa, M. 2000. Formation of electron beams under the interaction of a whistler wave packet with the radiation belt electrons. *J. Atmos. Sol.-Terr. Phys.* **62**, 541–552.

Horne, R.B., Thorne, R.M., Shprits, Y.Y., *et al.*, 2005. Wave acceleration of electrons in the Van Allen radiation belts. *Nature* **437**, 227–230.

Huang, C.Y., Goertz, C.K, and Anderson, R.R. 1983. A theoretical study of plasmaspheric hiss generation. *J. Geophys. Res.* **88**, 7927–7940.

Huang, L., Hawkins, J.G., and Lee, L.C. 1990. On the generation of the pulsating aurora by the loss cone driven whistler instability in the equatorial region. *J. Geophys. Res.* **95**, 3893–3906.

Hughes, A.R.W., Ferencz, C., and Gwal, A.K. (Eds.) 2003. *Very Low Frequency (VLF) Phenomena*. New Delhi: Narosa Publishing House.

Ikegami, H., Ikezi, H., Hosokawa, M., Takayama, K., Tanaka, S. 1968. Microwave burst at triggered instability in a hot electron plasma. *Phys. Fluids.* **11**, 1061–1064.

Inan, U.S., Bell, T.F. 1977. The plasmapause as a VLF wave guide. *J. Geophys. Res.* **82**, 2819–2827.

Istomin, Ya.N., and Karpman, V.I. 1973. Nonlinear evolution of a quasi-monochromatic packet of helical waves in a plasma. *Sov. Phys. JETP* **36**, 69–74.

Istomin, Ya.N., Karpman, V.I., and Shklyar, D.R. 1975. Drag effects on resonant interaction between particles and Langmuir wave in an inhomogeneous plasma. *Sov. Phys. JETP* **69**, 909–920.

Ivanov, A.A., and Rudakov, L.I. 1966. Dynamics of quasi-linear relaxation of a collisionless plasma. *Sov. Phys. JETP* **24**, 1027.

Ivanov, A.A. 1977. *Physics of Strongly Non-equilibrium Plasmas* (in Russian). Moscow: Atomizdat.

Jackson, J.D. 1962. *Classical Electrodynamics*. New York: John Wiley and Sons.

Jacobs, J.A. 1970. *Geomagnetic Micropulsations*. Berlin: Springer-Verlag.

Jacquinot, J., Leloup, C., Poffe, J.P., de Pretis, M., Waelbroeck, F., Evrard, P., Ripault, J. 1969. Etude de microinstabilites dans un plasma d'electrons chauds confine. *Proc. Int. Conf. Plasma Phys. and Controlled Nucl. Fusion Res.* IAEA, Vienna, **2**, 347–358.

Jiricek, F., and Triska, P. 1982. Ducted and non-ducted propagation of Omega signals within the plasmasphere. *Adv. Space Res.* **2**, 231–234.

Johnson, H.R. 1955. Backward-wave oscillators. *Proc. IRE* **43**, 684–697.

JASTP, Special Issue. 2000. *Ionospheric Alfvén Resonator*, **62**, 231–322.

Kadomtsev, B. B. 1958. *Physics of Plasmas and the Problem of Controlled Thermonuclear Fusion*. Moscow: AN SSSR, **4**, 370–379.

Kaiser, M.L., Desch, M.D., and Farrell, W.M. 1993. Clock-like behavior of Jovian continuum radiation. *Planet. Space Sci.* **41**, 1073–1077.

Kamke, E. 1959. *Differentialgleichnungen: Lösungsmethoden and Lösungen*. Leipzig: Academic-Verlag.

Kangas, J., Guglielmi, A., and Pokhotelov, O. 1998. Morphology and physics of short-period magnetic pulsations: A review. *Space Science Reviews* **83**, 435–512.

Karpman, V.I., and Shklyar, D.R. 1972. Nonlinear damping of potential monochromatic waves in an inhomogeneous plasma. *Sov. Phys. JETP* **35**, 500–505.

Karpman, V.I., Istomin, Ya.N., and Shklyar, D.R. 1974a. Nonlinear theory of a quasi-monochromatic whistler mode packet in an inhomogeneous plasma. *Plasma Phys.* **16**, 685–703.

Karpman, V.I., Istomin, Ya.N., and Shklyar, D.R. 1974b. Nonlinear frequency shift and self-modulation of quasi-monochromatic whistlers in the inhomogeneous plasma (magnetosphere). *Planet. Space Sci.* **22**, 859–871.

Karpman, V.I., and Shklyar, D.R. 1975a. Nonlinear Landau damping in an inhomogeneous plasma. *Sov. Phys. JETP* **40**, 53–56.

Karpman, V.I., Istomin, Ya.N., and Shklyar, D.R. 1975b. Effects of nonlinear interaction of monochromatic waves with resonant particles in an inhomogeneous plasma. *Physica Scripta* **11**, 278–284.

Karpman, V.I., and Kaufman, R.N. 1982. Whistler wave propagation in density ducts. *J. Plasma Physics* **25**, 225–238.

Kennel, C.F., and Engelman, F.E. 1966. Velocity space diffusion from weak plasma turbulence in a magnetic field. *Phys. Fluids* **9**, 2377–2388.

Kennel, C.F., and Petschek, H.E. 1966. Limit on stably trapped particle fluxes. *J. Geophys. Res.* **71**, 1–28.

Kennel, C.F. 1969. Consequences of a magnetospheric plasma. *Rev. Geophys.* **7**, 339–419.

Khanin, Ya.I. 1995. *Principles of Laser Dynamics*. Amsterdam: Elsevier.

Kimura, I. 1966. Effects of ions on whistler-mode ray tracing. *Radio Science* **1** (new series), 269–283.

Kimura, I. 1967. On observations and theories of the VLF-emissions. *Planet. Space Sci.* **15**, 1427–1462.

Kimura, I. 1968. Triggering of VLF magnetospheric noise by a low-power (100 W) transmitter. *J. Geophys. Res.* **73**, 445–447.

Knox, F.B. 1969. Growth of a packet of finite amplitude VLF waves, with special reference to the magnetosphere. *Planet. Space Sci.* **17**, 13–30.

Kondratyev, I.G., Kudrin, A., and Zaboronkova, T.M. 1999. *Electrodynamics of Density Ducts in Magnetized Plasmas*. Amsterdam: Gordon and Breach.

Kovner, M.S., Kuznetsova, V.A., and Likhter, I.I. 1977. On the mid-latitude modulation of fluxes of energetic electrons and VLF emissions. *Geomagn. Aeron.* **17**, 867.

Kruer, W.L., Dawson, J.M., and Sudan, R.N. 1969. Trapped-particle instability. *Phys. Rev. Letters* **23**, 838–84.

Kurth, W.S., Gurnett, D.A., and Scarf, F.L. 1986. Periodic amplitude variations in Jovian continuum radiation. *J. Geophys. Res.* **91**, 13,523–13,530.

Kuznetsov, S.P., and Trubetskov, D.I. 1977. Nonlinear transients during interaction between the electron beam moving in crossed fields and the backward electromagnetic wave. *Radiophys. Quantum Electron.* **20**, 204–213.

LaBelle, J.W., and Treumann, R.A. 2006. *Geospace Electromagnetic Waves and Radiation*. Berlin: Springer.

Laird, M.J., and Knox, F.B. 1965. Exact solution for charged particle trajectories in an electromagnetic field. *Phys. Fluids* **8**, 755–756.

Laird, M.J., and Nunn, D. 1975. Full-wave VLF modes in a cylindrically symmetric enhancement of plasma density. *Planet. Space Sci.* **23**, 1649–1657.

Lemaire, J.F., and Gringauz, K.I. 1998. *The Earth's Plasmasphere*. Cambridge: Cambridge University Press.

Le Queau, D. and Roux, A. 1992. Heating of oxygen ions by resonant absorption of Alfvén waves in a multicomponent plasma. *J. Geophys. Res.* **97**, 14929–14946.

Liemohn, M.W., Kozyra, J.U., Thomsen, M.F., Roeder, J.L., Lu, G., Borovsky, J.E. and Cayton, T.E. 2001. Dominant role of the asymmetric ring current in producing the stormtime Dst, *J. Geophys. Res.*, **106**, 10883–10904.

Loto'aniu, T.M., Fraser, B.J., Waters, C.L. 2005. Propagation of electromagnetic ion cyclotron wave energy in the magnetosphere. *J. Geophys. Res.* **110** (A07), A07214, doi:10.1029/2004JA010816.

Lutomirski, R.F., and Sudan, R.H. 1966. Exact nonlinear electromagnetic whistler modes. *Phys. Rev.* **147**, 156–165.

Lyons, L.R., Thorne, R.M., and Kennel, C.F. 1972. Pitch-angle diffusion of radiation belt electrons within the plasmasphere. *J. Geophys. Res.* **77**, 3455–3474.

Lyons, L.R., and Thorne, R.M. 1973. Equilibrium structure of radiation belt electrons. *J. Geophys. Res.* **78**, 2142–2149.

Lyons, L.R., and Williams, D.J. 1984. *Quantitative Aspects of Magnetospheric Physics*. Dordrecht: Reidel.

Malmberg, J.H., Wharton, C.B., Gould, R.W., and O'Neil, T.M. 1968. Plasma wave echo experiment. *Phys. Rev. Lett.* **20**, 95–97.

Maltseva, O.A., Molchanov, O.A. 1987. *Propagation of Low Frequency Waves in the Earth's Magnetosphere* (in Russian). Moscow: Nauka.

Manninen, J., Turunen, T., *et al.*, 1996. *Atlas of VLF Emissions Observed at Porojarvi*. SGO Technical report, Sodankylä, Finland.

Mazitov, R.K. 1965. Damping of plasma waves. *Dokl. Akad. Nauk. SSSR* **204**, 22–25.

Matsumoto, H., and Kimura, I. 1971. Linear and nonlinear cyclotron instability and VLF emissions in the magnetosphere. *Planet. Space Sci.* **19**, 567–608.

Matsumoto, H. 1979. Nonlinear whistler-mode interaction and triggered emissions in the magnetosphere: a review. In *Wave instabilities in space plasmas*. P.J. Palmadesso and K. Papadopoulos (Eds). Dordrecht: Reidel.

Matsumoto, H., and Omura, Y. 1981. Cluster and channel effect phase bunchings by whistler waves in the non-uniform geomagnetic field. *J. Geophys. Res.* **86**, 779–791.

McClements, K.J., and Dendy, R.O. 1993. Ion cyclotron harmonic wave generation by ring protons in space plasmas. *J. Geophys. Res.* **98**, 11689–11700.

McIlwain, C.E. 1961. Coordinates for mapping the distribution of magnetically trapped particles. *J. Geophys. Res.* **66**, 3681–3691.

Melrose, D.B., and Dulk, G.A. 1982. Electron-cyclotron masers as the source of certain solar and stellar radio bursts. *Astrophys. J.* **259**, 844–858.

Meredith, N.P., Horne, R.H., Iles, R.M., *et al.*, 2002. Outer zone relativistic electron acceleration associated with substorm-enhanced whistler-mode chorus. *J. Geophys. Res.* **107**, 1144, doi: 10.1029/2001JA900146.

Meredith, N.P., Cain, M., Horne, R.B. *et al.*, 2003. Evidence for chorus-driven electron acceleration to relativistic energies from a survey of geomagnetically periods. *J. Geophys. Res.* **108**, 1248, doi:10.1029/2202JA009764.

Mikhailovskii, A.B., and Pokhotelov, O.A. 1975a. A new generation mechanism of geomagnetic pulsations by fast particles. *Fizika plasmi* **1**, 786–792.

Mikhailovskii, A.B., and Pokhotelov, O.A. 1975b. Influence of whistlers and ion-cyclotron waves on excitation of Alfvén waves in the magnetospheric plasma. *Fizika plasmi* **1**, 1004–1012.

Mikhailovskii, A.B. 1992. *Electromagnetic Instabilities in an Inhomogeneous Plasma*. New York: Adam Hilger.

Mikhailovskii, A.B. 1998. *Instabilities in a Confined Plasma*. Bristol: Institute of Physics Publishing.

Molchanov, O.A. 1985. *Low Frequency Waves and Induced Emissions in Near-Space Plasma* (In Russian). Moscow: Nauka.

Molvig, K., Hilfer, G., Miller, R.H., and Myczkowski, J. 1988. Self-consistent theory of triggered whistler emissions. *J. Geophys. Res.* **93**, 5665–5683.

Mursula, K., Rasinkangas, S., Bösinger, T., *et al.*, 1997. Nonbouncing Pc-1 wave bursts. *J. Geophys. Res.* **102**, 17611–17624.

Nagano, I., Yagitani, S., Kojima, H, and Matsumoto, H. 1996. Analysis of wave normal and Poynting vectors of the chorus emissions. *J. Geomagn. Geoelectr.* **48**, 299–307.

Nakamura, R., Yamamoto, S., Kokubun, S., *et al.*, 1990. Pulsating auroral activity and energetic electron injections. *EOS Trans. AGU* **71**, 913.

Northrop, T., and Teller, E. 1960. Stability of adiabatic motion of charged particles in the Earth's magnetic field. *Phys. Rev.* **117**, 112–225.

Northrop, T. 1963. *The Adiabatic Motion of Charged Particles*. New York, John Wiley and Sons.

Nunn, D. 1971. A theory of VLF emissions. *Planet. Space Sci.* **19**, 1141–1167.

Nunn, D. 1973. The sideband stability of electrostatic waves in an inhomogeneous medium. *Planet. Space Sci.* **21**, 67–88.

Nunn, D. 1974. A self-consistent theory of triggered VLF emissions. *Planet. Space Sci.* **22**, 349–378.

Nunn, D. 1990. The numerical simulation of VLF nonlinear wave-particle interactions in collision-free plasmas using the Vlasov hybrid simulation technique. *Computer Physics Com.* **60**, 1–20.

Nunn, D., and Smith, A.J. 1996. Numerical simulation of whistler-triggered VLF emissions observed in Antarctica. *J. Geophys. Res.* **101**, 5261–6277.

Nunn, D., Omura, Y., Matsumoto, H., Nagano I., and Yagitani, S. 1997. The numerical simulation of VLF chorus and discrete emissions observed on the Geotail satellite using a Vlasov code. *J. Geophys. Res.* **102**, 27083–27097.

Nunn, D., Demekhov, A.G., Trakhtengerts, V.Y., and Rycroft, M.J. 2003. VLF emission triggering by a highly anisotropic energetic electron plasma. *Ann. Geophys.* **21**, 481–492.

Nunn, D, Rycroft, M., Trakhtengerts, V. 2005. A parametric study of the numerical simulations of triggered VLF emissions. *Ann. Geophys.* **23**, 3655–3666.

O'Neil, T.M. 1965. Collisionless damping of nonlinear plasma oscillations. *Phys. Fluids* **8**, 2255–2262.

Oguti, T. 1981. TV observations of auroral arcs. In *Physics of Auroral Arc Formation*, Geophys. Monogr. Ser., 25, S.-I. Akasofu and J.R. Kan (Eds) Washington, D.C.: AGU, 31–41.

Ohta, K., Kitagawa, T., Hayakawa, M., Dowden, R.L. 1997. A new type of mid-latitude multi-path whistler trains including non-ducted whistlers. *Geophys. Res. Lett.* **24**, 2937–2940.

Omura, Y., and Matsumoto, H. 1989. Particle simulation of nonlinear plasma wave instabilities: amplitude modulation, decay, soliton and inverse cascading. In *Plasma waves and instabilities at comets and in magnetospheres*, Geophys. Monogr, 53, Tsurutani, B.T. and Oya H. (Eds). Washington, D.C.: AGU.

Omura, Y., Nunn, D., Matsumoto, H., and Rycroft, M.J. 1991. A review of observational, theoretical and numerical studies of VLF triggered emissions. *J. Atmos. Terr. Phys.* **53**, 351–368.

Ondoh, T. 1976. Magnetospheric whistler ducts observed by ISIS satellites. *J. Radio Res. Lab. Japan* **23**, 139–147.

Ostapenko, A.A., Polyakov, S.V. 1990. The dynamics of the reflection coefficient of Alfvén waves in Pc 1 range from ionosphere under variations of electron concentration of the lower ionosphere. *Geomagn. Aeron.* **30**, 50–56.

Palmadesso, P., and Schmidt, G. 1971. Collisionless damping of a large amplitude whistler wave. *Phys. Fluids* **14**, 1411–1418.

Parrot, M., Santolik, O., Cornilleau-Wehrlin, N., *et al.*, 2003. Source location of chorus emissions observed by CLUSTER. *Ann. Geophys.* **21**, 473–480.

Pasmanik, D.L., Trakhtengerts, V.Y., Demekhov, A.G., Lyubchich, A.A., Titova, E.E., Yahnina, T.A., Rycroft, M.J., Manninen, J., Turunen, T. 1998. A quantitative model for cyclotron wave–particle interactions at the plasmapause. *Ann. Geophys.* **16**, 322–330.

Pasmanik, D.L., and Trakhtengerts, V.Y. 2001. Cyclotron wave–particle interactions in the whistler-mode waveguide. *Radiophys. Quantum Electron.* **43**, 117–128.

Pasmanik, D.L., Demekhov, A.G., Trakhtengerts, V.Y., Nunn, D., Rycroft, M.J. 2002. Cyclotron amplification of whistler waves: a parametric study relevant for the nature of discrete VLF emissions. *J. Geophys. Res.* **107**, 1162, doi:10.1029/2001JA000256.

Pasmanik, D.L., Demekhov, A.G., Trakhtengerts, V.Y., and Parrot, M. 2004a. Modeling whistler wave generation regimes in magnetospheric cyclotron maser. *Ann: Geophys*, **22**, 3561–3570.

Pasmanik, D.L., Titova, E.E., Demekhov, A.G. *et al.*, 2004b. Quasi-periodic ELF/VLF emissions in the Earth's magnetosphere: comparison of satellite observations and modeling. *Ann. Geophys.* **22**, 4351–4361.

Pease, R.S. 1993. Survey of fusion plasma physics. In *Plasma Physics: An Introductory Course*, Dendy R.O. (Ed.) Cambridge: Cambridge University Press. 475–507.

Pereira, N., and Stenflo, L., Nonlinear Schrödinger equation including growth and damping. *Phys. Fluids*, **20** (10), 1733–1734.

Perkins, W.A., Barr, W.L. 1966. Observation of a velocity distribution instability. *Proc. Int. Conf. Plasma Phys. and Controlled Nucl. Fusion Res.* Vienna: IAEA **2**, 115–134.

Petelin, M.I. 1961. Propagation of electromagnetic waves in a non-equilibrium magnetoactive plasma. *Izv. Vyssh. Uchebn. Zaved. Radiofiz.* **4**, 455–464.

Pilipenko, V.A., and Pokhotelov, O.A. 1975. Effect of high-frequency turbulence on the generation of magneto-acoustic waves in the magnetosphere. *Geomagn. Aeronom.* **15**, 1117–1119.

Polyakov, S.V., and Rapoport, V.O. 1981. Ionospheric Alfvén resonator. *Geomagn. Aeron.* **21**, 610–614.

Polyakov, S.V., Rapoport, V.O., and Trakhtengerts, V.Y. 1983. Alfvén sweep maser. *Fizika Plazmi* **9**, 371–377.

Post, R.F., Fowler, T.K., Killeen, J., and Mirin, A.A. 1973. Concept for a high power density mirror fusion reactor. *Phys. Rev. Letters* **31**, 280–282.

Priest, E.R. 1982. *Solar Magnetohydrodynamics*. Dordrecht: Kluwer.

Reeve, C.D., and Rycroft, M.J. 1971. Expedition to Iceland 1969. *Phys. Bull.* **22**, 145–147.

Roberts, C.S., and Bushsbaum, S.J. 1964. Motion of a charged particle in a constant magnetic field and a transverse electromagnetic wave propagating along the field. *Phys. Rev. A* **135**, 381–389.

Rosenbluth, M.N., and Sagdeev R.Z. (General Editors). 1984. *Handbook of Plasma Physics*, Vol. 1 and Vol. 2. Amsterdam: North-Holland.

Roux A., Solomon, J. 1971. Self-consistent solution of quasi-linear theory: Application to the spectral shape and intensity of VLF waves in a magnetosphere. *J. Atmos. Terr. Phys.* **33**, 1457–1471.

Roux, A., and Pellat, R. 1978. A theory of triggered emissions. *J. Geophys. Res.* **83**, 1433–1441.

Roux, A., Cornilleau-Wehrlin, N., and Rauch, J.L. 1984. Acceleration of the thermal electrons by ICW's propagating in a multicomponent plasma. *J. Geophys. Res.* **89**, 2267–2273.

Rudakov, L.I., Sagdeev, R.Z. 1958. *Physics of Plasmas and the Problem of Controlled Thermo-nuclear Fusion*. Moscow: AN SSSR, **3**, 268–277.

Sa, L.A.D., and Helliwell, R.A. 1988. Structure of VLF whistler mode sideband waves in the magnetosphere. *J. Geophys. Res.* **93**, 1987–1992.

Sa, L.A.D. 1990. A wave-particle-wave interaction mechanism as a cause of VLF triggered emissions. *J. Geophys. Res.* **95**, 12277–12286.

Sagdeev, R.Z., Shafranov, V.D. 1960. On the instability of plasma with an anisotropic velocity distribution in a magnetic field. *Sov. Phys. JETP* **12**, 130.

Sagdeev, R.Z., and Galeev, A.A. 1969. *Nonlinear Plasma Theory*. New York: Benjamin.

Sagdeev, R.Z., Shapiro, V.D., and Shevchenko, V.I. 1985. A mechanism of triggered emissions in the magnetosphere. *Zh. Eksp. Theor. Fiz.* **89**, 22–34.

Sagdeev, R.Z., Usikov, D.A., and Zaslavsky, C.M. 1989. *Nonlinear Physics*. New York: Harwood Academic Publishers.

Sandahl, J. 1984. Pitch-angle scattering and particle precipitation in a pulsation aurora an experimental study: KGI Report No. 185. *Kiruna Geophys. Inst.* 1056–1062.

Santolik, O., Gurnett, A., Pickett, J.S., *et al.*, 2003. Spatio-temporal structure of storm-time chorus. *J. Geophys. Res.* **108**, 1278, doi:10.1029/2002JA009791.

Santolik, O., Gurnett, D.A., and Pickett, J.S. 2004. Multipoint investigation of the source region of storm-time chorus. *Ann. Geophys.* **22**, 2555–2563.

Sato, N., Hayashi, K., Oguti, T., and Fukunishi, H. 1974. Relationship between quasi-periodic VLF emission and geomagnetic pulsation. *J. Atmos. Terr. Phys.* **36**, 1515.

Sato, N., and Fukunishi, H. 1981. Interaction between ELF-VLF emissions and magnetic pulsations: classification of quasi-periodic ELF-VLF emissions based on frequency-time spectra. *J. Geophys. Res.* **86**, 19–29.

Sazhin, S. 1993. *Whistler-mode Waves in a Hot Plasma*. Cambridge: Cambridge University Press.

Sazhin, S.S., and Hayakawa, M. 1992. Magnetospheric chorus emissions: a review. *Planet. Space Sci.* **40**, 681–697.

Sazhin, S.S., and Hayakawa, M. 1994. Periodic and quasi-periodic VLF emissions. *J. Atmos. Terr. Phys.* **56**, 735–753.

Schneider, J. 1959. Stimulated emission of radiation by relativistic electrons in a magnetic field. *Phys. Rev. Lett.* **2**, 504–505.

Schulz, M., Lanzerotti, L.J. 1974. *Particle Diffusion in Radiation Belts*. New York: Springer.

Schulz, M. 1991. The magnetosphere. In *Geomagnetism* **4**, Jacobs, J.A (Ed.), San Diego, Calif: Academic 87–293.

Semenova, V.I., Trakhtengerts, V.Y. 1980. Some peculiarities of wave guide propagation of low frequency waves in the magnetosphere. *Geomagn. Aeron.* **20**, 1021–1027.

Shafranov, V.D. 1963. Electromagnetic waves in plasma. *Voprosi Teorii Plasmi* **3**, 3–140.

Shapiro, V.D., and Shevchenko, V.I. 1968. Nonlinear theory of cyclotron damping of electromagnetic waves in plasma. *Ukr. Fiz. Zh.* **13**, 1989–1996.

Shapiro, V.D., and Shevchenko, V.I. 1969. Stability of a monochromatic wave in a plasma. *Sov. Phys. JETP* **57**, 2066–2078.

Shevchik, V.N., and Trubetskov, D.I. (Eds). 1975. *Electronics of Backward-Wave Oscillators* (in Russian). Saratov State University.

Simpson, J.A., Hamilton, D.C., Lentz, G.A., *et al.*, 1975. Jupiter revisited: First results from the University of Chicago Charged Particle experiment on Pioneer 11. *Science* **188**, 455–459.

Simpson, J.A., Anglin, J.D., Balogh, A., *et al.*, 1992. Energetic charged particle phenomena in the Jovian magnetosphere: First results from the Ulysses COSPIN collaboration. *Science* **257**, 1547–1550.

Smith, R.L., Helliwell, R.A., and Yabroff, I.W. 1960. Theory of trapping of whistlers in field-aligned columns of enhanced ionization. *J. Geophys. Res.* **65**, 815–823.

Smith, R.L. 1961. Propagation characteristics of whistlers trapped in field-aligned columns of enhanced ionization. *J. Geophys. Res.* **66**, 3699–3707.

Smith, A.J., and Nunn, D. 1998. Numerical simulation of VLF risers, fallers and hooks observed in Antarctica. *J. Geophys. Res.* **103**, 6771–6784.

Sonnerup, B.U.O., and Su, S.Y. 1967. Large amplitude whistler waves in a hot collision-free plasma. *Phys. Fluids* **10**, 462–464.

Stiles, G.S., and Helliwell, R.A. 1975. Frequency-time behavior of artificially stimulated VLF emissions. *J. Geophys. Res.* **80**, 608–618.

Stix, T.H. 1992. *Waves in Plasmas*. New York: AIP.

Storey, L.R.O. 1953. An investigation of whistling atmospherics. *Phil. Trans. Roy. Soc. London* **A246**, 113–141.

Sturrock, P.A. 1994. *Plasma Physics*. Cambridge: Cambridge University Press.

Sudan, R.N., and Ott, E. 1971. Theory of triggered VLF emissions. *J. Geophys. Res.* **76**, 4463–4475.

Summers, D., Ma, C., Meredith, N.P., *et al.*, 2002. Model of the energization of outer-zone electrons by whistler-mode chorus during the October 9, 1990 geomagnetic storm. *Geophys. Res. Lett.* **29**, 2174, doi: 10.1029/2002GL016039.

Tagirov, V.R., Trakhtengetrs, V.Yu., and Chernouss, S.A. 1986. On the nature of pulsating auroral patches. *Geomagn. Aeron.* **26**, 600–605.

Titova, E.E., Maltseva, O.A., and Kleimenova, N.G. 1985. About conducting of VLF emissions of chorus type by a plasmapause. *Geomagn. Aeron.* **25**, 520–522.

Titova, E.E., Yahnina, T.A., Yahnin, A.G., Gvozdevsky, B.B., Lyubchich, A.A., Trakhtengerts, V.Y., Demekhov, A.G., Horwitz, J.L., Lefeuvre, F., Lagoutte, D., Manninen, J., Turunen, T. 1998. Strong localized variations of the low-altitude energetic electron fluxes in the evening sector near plasmapause. *Ann. Geophys.* **16**, 25–33.

Tixier, M., and Cornilleau-Wehrlin, N. 1986. How are the VLF quasi-periodic emissions controlled by harmonics of field line oscillations? The result of a comparison between ground and GEOS satellites measurements. *J. Geophys. Res.* **91**, 6899–6919.

Trakhtengerts, V.Y. 1963. On the mechanism of VLF radiation generation in the external radiation belt of the Earth. *Geomagn. Aeron.* **3**, 442–451.

Trakhtengerts, V.Y. 1966. Stationary states of the Earth's outer radiation zone. *Geomagn. Aeron.* **6**, 827–836.

Trakhtengerts, V.Y. 1967. On the nonlinear theory of the cyclotron instability of the Earth radiation belts. *Geomagn. Aeron.* **7**, 339–341.

Trakhtengerts, V.Y. 1984. Relaxation of plasma with anisotropic velocity distribution. In: *Handbook of Plasma Physics* (General Editors: M.N. Rosenbluth and R.Z. Sagdeev), 519–552.

Trakhtengerts, V.Y., Tagirov, V.R., Chernous, S.A. 1986. Flow cyclotron maser and impulsive VLF emissions. *Geomagn. Aeron.* **26**, 99–106.

Trakhtengerts, V.Y. 1995. Magnetosphere cyclotron maser: Backward wave oscillator generation regime. *J. Geophys. Res.* **100**, 17205–17210.

Trakhtengerts, V.Y. 1996. Cyclotron resonance maser as a solar flare trigger. *Radiophys. Quantum Electron.* **39**, 463–471.

Trakhtengerts, V.Y., Rycroft, M.J., and Demekhov, A.G. 1996. Interrelation of noise-like and discrete ELF/VLF emissions generated by cyclotron interactions. *J. Geophys. Res.* **101**, 13293–13303.

Trakhtengerts, V.Y., Demekhov, A.G., Grafe, A. 1997. Three-dimensional magnetospheric currents due to energetic particle precipitation. *Geomagn. Aeron.* **37**, 9–16.

Trakhtengerts, V.Y., and Shalashov, A.G. 1999. Acceleration of electrons during the magnetic compression of coronal plasmas. *Astron. Rep.* **43**, 540–548.

Trakhtengerts, V.Y., Hobara, Y., Demekhov, A.G., and Hayakawa, M. 1999. Beam-plasma instability in inhomogeneous magnetic field and second order cyclotron resonance effects. *Phys. Plasmas* **6**, 692–698.

Trakhtengerts, V.Y. 1999. A generation mechanism for chorus emission. *Ann. Geophys.* **17**, 95–100.

Trakhtengerts, V.Y., Demekhov, A.G., Polyakov, S.V., Belyaev, P.P., Rapoport, V.O. 2000. A mechanism of Pc 1 pearl formation based on Alfvén sweep maser. *J. Atmos. Sol.-Terr. Phys.* **62**, 231–238.

Trakhtengerts, V.Y., Rycroft, M.J. 2000. Whistler-electron interactions in the magnetosphere: new results and novel approaches. *J. Atmos. Sol.-Terr. Phys.* **62**, 1719–1733.

Trakhtengerts, V.Y., Hobara, Y., Demekhov, A.G., and Hayakawa, M. 2001. A role of the second-order resonance effect in a self-consistent approach to triggered VLF emissions. *J. Geophys. Res.* **106**, 3897–3904.

Trakhtengerts, V.Y., Rycroft, M.J., Nunn, D., and Demekhov, A.G. 2003a. Cyclotron acceleration of radiation belt electrons by whistlers. *J. Geophys. Res.* **108**, 1138, doi:10.1029/2002JA009559.

Trakhtengerts, V.Y., Demekhov, A.G., Hobara, Y., and Hayakawa, M. 2003b. Phase-bunching effects in triggered VLF emissions: Antenna effect. *J. Geophys. Res.* **108**, 1160, doi:10.1029/2002JA009415.

Trakhtengerts, V.Y., Demekhov, A.G., Titova, E.E., Kozelov, B.V., Santolik, O., Gurnett, D., Parrot, M. 2004. Interpretation of Cluster data on chorus emissions using the backward wave oscillator model. *Phys. Plasmas* **11**, 1345–1351.

Trakhtengerts, V.Y., and Demekhov, A.G. 2005. Discussion paper: Partial ring current and polarization jet. *Int. J. Geomagn. Aeron.* **5**, GI3007, doi:10.1029/2004GI000091.

Trakhtengerts, V.Y., Demekhov, A.G., Titova, E.E., Kozelov, B.V., Santolik, O., Macusova, E., Gurnett, D., Pickett, J.S., Rycroft, M.J., Nunn, D. 2007. Formation of VLF chorus frequency spectrum: Cluster data and comparison with the backward wave oscillator model. *Geophys. Res. Lett.* **34**, L02104, doi:10.1029/2006GL027953.

Trakhtengerts, V.Y., and Demekhov, A.G. 2007. Generation of Pc 1 pulsations in the regime of backward wave oscillator. *J. Atmos. Sol.-Terr. Phys.* **69**, 1651–1656.

Trefall, H., Ullaland, S., Stadsnes, J., *et al.*, 1975. Morphology and fine time structure of an early-morning precipitation event. *J. Atmos. Terr. Phys.* **37**, 83–105.

Trigilio, C., Leto, P., and Umana, G. 1998. Strong radio flaring period in UXArietis. *Astron. Astrophys.* **330**, 1060–1066.

Tsuruda, K. 1973. Penetration and reflection of VLF waves through the ionosphere: Full wave calinlations with ground effect. *J. Atmos. Terr. Phys.* **35**, 1377–1405.

Tsuruda, K., Machida, S., Oguti, T., *et al.*, 1981. Correlation between the VLF chorus and pulsating aurora observed by low level television at $L = 4.4$. *Can. J. Phys.* **59**, 1042–1048.

Tsurutani, B.T., and Lakhina, G.S. 1997. Some basic concepts of wave-particle interactions in collisionless plasmas. *Rev. Geophys.* **35**, 491–502.

Tverskoy, B.A. 1968. *Dynamics of the Earth's Radiation Belts* (in Russian). Moscow: Nauka.

Twiss, R.Q. 1958. Radiation transfer and the possibility of negative absorption in radio astronomy. *Austral. J. Phys.* **11**, 564–579.

Van Allen, J.A., Ludwig, G.H., Ray, E.C., McIlwain, C.E. 1958. Observations of high intensity radiation by satellites 1958 Alpha and Gamma. *Jet Propulsion* **28**, 588–592.

Varotsou, A., Boscher, D., Bourdarie, S., *et al.*, 2005. Simulation of the outer radiation belt electrons near geosynchronous orbit including both radial diffusion and resonant interaction with whistler-mode chorus waves. *Geophys. Res. Lett.* **32**, L19106, doi:10.1029/2005GL023282.

Vasyliunas, V. 1970. Mathematical models of magnetospheric convection and its coupling to the ionosphere, in *Particles and Fields in the Magnetosphere*, edited by B.M. McComack, pp. 60–71. Norwell, Mass: Reidel.

Vedenov, A.A., Velikhov, E.P., and Sagdeev, R.Z. 1962. Quasilinear theory of weakly nonlinear plasma processes. *Nucl. Fusion Suppl.* **2**, 465.

Vodopyanov, A.V., Golubev, S.V., Demekhov, A.G., Zorin, V.G., Mansfeld, D.A., Razin, S.V., and Trakhtengerts, V. Yu. 2005. Laboratory modeling of nonstationary processes in space cyclotron masers: First results and prospects. *Plasma Phys. Rep.* **31**, 927–937.

Vomvoridis, J.L., and Denavit, J. 1979. Test particle correlation by a whistler wave in a non-uniform magnetic field. *Phys. Fluids* **22**, 367–377.

Vomvoridis, J.L., and Denavit, J. 1980. Nonlinear evolution of a monochromatic whistler wave in a non-unifom magnetic field. *Phys. Fluids* **23**, 174–183.

Vomvoridis, J.L., Crystal, T.L., and Denavit, J. 1982. Theory and computer simulations of magnetospheric VLF emissions. *J. Geophys. Res.* **87**, 1473–1489.

Wachtel, J.M., and Wachtel, E.J. 1980. Backward wave oscillation in the gyrotron. *Appl. Phys. Lett.* **37**, 1059–1061.

Walker, A.D.M. 1971. The propagation of very low frequency radio waves in ducts, in the magnetosphere. *Proc. Roy. Soc. London* **321**, 69–93.

Walker, A.D.M. 1994. *Plasma Waves in the Magnetosphere*. Berlin: Springer.

Walt, M. 1994. *Introduction to Geomagnetically Trapped Radiation*. Cambridge: Cambridge Universtity Press.

Walt, M., Voss, H.D. 2004. Proton precipitation during magnetic storms in August through November 1998. *J. Geophys. Res.* **109**, A02201, doi:10.1029/2003JA010083.

Winckler, R.J., and Nemzek, R.J. 1993. Observations of the pulsation phase of auroras observed at Minneapolis during the peak of solar cycle 22. In *Auroral Plasma Dynamics, Geophys. Monogr. Ser.*, 80, R. L. Lysak (Ed.), Washington, D.C.: AGU 220–225.

Winglee, R.M. 1985. Enhanced growth of whistlers due to bunching of untrapped electrons. *J. Geophys. Res.* **90**, 5141–5151.

Wu, C.S., and Lee, L.C. 1979. A theory of the terrestrial kilometric radiation. *Astrophys. J.* **230**, 621–626.

Yamamoto, T. 1988. On the temporal fluctuations of pulsating auroral luminosity. *J. Geophys. Res.* **93**, 897–911.

Yulpatov, V.K. 1965. Nonlinear theory of interaction of a non-rectilinear periodic electron beam with electromagnetic field. Part I. *Voprosi Radioelectroniki* No. 12, 15–23.

Zarka, P. 1992. The auroral radio emissions from planetary magnetospheres – What do we know, what don't we know, what do we learn from them? *Adv. Space Res.* **12**, 99–115.

Zaslavsky, G.M. 1985. *Chaos in Dynamic Systems*. New York: Harwood Academic Publishers.

Zheleznyakov, V.V. 1960a,b. About instability of magnetoactive plasma relative high-frequency electromagnetic disturbances. I–II. *Izv. Vyssh. Uchebn. Zaved. Radiofiz.* **3**, 57–67, 180–192.

Zheleznjakov, V.V. 1996. *Radiation in Astrophysical Plasmas* (Astrophysics and Space Science Library. V. 204), Dordrecht; Kluwer Academic Publishers.

Index